DNA Methylation

METHODS IN MOLECULAR BIOLOGY™

John M. Walker, Series Editor

507. DNA Methylation: *Methods and Protocols, Second Edition*, edited by *Jörg Tost 2009*

502. Bacteriophages: *Methods and Protocols, Volume 2: Molecular and Applied Aspects*, edited by *Martha R. J. Clokie and Andrew M. Kropinski 2009*

501. Bacteriophages: *Methods and Protocols, Volume 1: Isolation*, Characterization, and Interactions, edited by *Martha R. J. Clokie and Andrew M. Kropinski 2009*

496. DNA and RNA Profiling in Human Blood: *Methods and Protocols*, edited by *Peter Bugert, 2009*

493. Auditory and Vestibular Research: *Methods and Protocols*, edited by *Bernd Sokolowski, 2009*

489. Dynamic Brain Imaging: *Methods and Protocols*, edited by *Fahmeed Hyder, 2009*

485. HIV Protocols: *Methods and Protocols*, edited by *Vinayaka R. Prasad and Ganjam V. Kalpana, 2009*

484. Functional Proteomics: *Methods and Protocols*, edited by *Julie D. Thompson, Christine Schaeffer-Reiss, and Marius Ueffing, 2008*

483. Recombinant Proteins From Plants: *Methods and Protocols*, edited by *Loïc Faye and Veronique Gomord, 2008*

482. Stem Cells in Regenerative Medicine: *Methods and Protocols*, edited by *Julie Audet and William L. Stanford, 2008*

481. Hepatocyte Transplantation: *Methods and Protocols*, edited by *Anil Dhawan and Robin D. Hughes, 2008*

480. Macromolecular Drug Delivery: *Methods and Protocols*, edited by *Mattias Belting, 2008*

479. Plant Signal Transduction: *Methods and Protocols*, edited by *Thomas Pfannschmidt, 2008*

478. Transgenic Wheat, Barley and Oats: *Production and Characterization Protocols*, edited by *Huw D. Jones and Peter R. Shewry, 2008*

477. Advanced Protocols in Oxidative Stress I, edited by *Donald Armstrong, 2008*

476. Redox-Mediated Signal Transduction: *Methods and Protocols*, edited by *John T. Hancock, 2008*

475. Cell Fusion: *Overviews and Methods*, edited by *Elizabeth H. Chen, 2008*

474. Nanostructure Design: *Methods and Protocols*, edited by *Ehud Gazit and Ruth Nussinov, 2008*

473. Clinical Epidemiology: *Practice and Methods*, edited by *Patrick Parfrey and Brendon Barrett, 2008*

472. Cancer Epidemiology, Volume 2: *Modifiable Factors*, edited by *Mukesh Verma, 2008*

471. Cancer Epidemiology, Volume 1: *Host Susceptibility Factors*, edited by *Mukesh Verma, 2008*

470. Host-Pathogen Interactions: *Methods and Protocols*, edited by *Steffen Rupp and Kai Sohn, 2008*

469. Wnt Signaling, Volume 2: *Pathway Models*, edited by *Elizabeth Vincan, 2008*

468. Wnt Signaling, Volume 1: *Pathway Methods and Mammalian Models*, edited by *Elizabeth Vincan, 2008*

467. Angiogenesis Protocols: *Second Edition*, edited by *Stewart Martin and Cliff Murray, 2008*

466. Kidney Research: *Experimental Protocols*, edited by *Tim D. Hewitson and Gavin J. Becker, 2008*

465. Mycobacteria, *Second Edition*, edited by *Tanya Parish and Amanda Claire Brown, 2008*

464. The Nucleus, Volume 2: *Physical Properties and Imaging Methods*, edited by *Ronald Hancock, 2008*

463. The Nucleus, Volume 1: *Nuclei and Subnuclear Components*, edited by *Ronald Hancock, 2008*

462. Lipid Signaling Protocols, edited by *Banafshe Larijani, Rudiger Woscholski, and Colin A. Rosser, 2008*

461. Molecular Embryology: *Methods and Protocols, Second Edition*, edited by *Paul Sharpe and Ivor Mason, 2008*

460. Essential Concepts in Toxicogenomics, edited by *Donna L. Mendrick and William B. Mattes, 2008*

459. Prion Protein Protocols, edited by *Andrew F. Hill, 2008*

458. Artificial Neural Networks: *Methods and Applications*, edited by *David S. Livingstone, 2008*

457. Membrane Trafficking, edited by *Ales Vancura, 2008*

456. Adipose Tissue Protocols, Second Edition, edited by *Kaiping Yang, 2008*

455. Osteoporosis, edited by *Jennifer J. Westendorf, 2008*

454. SARS- and Other Coronaviruses: *Laboratory Protocols*, edited by *Dave Cavanagh, 2008*

453. Bioinformatics, Volume 2: *Structure, Function, and Applications*, edited by *Jonathan M. Keith, 2008*

452. Bioinformatics, Volume 1: *Data, Sequence Analysis, and Evolution*, edited by *Jonathan M. Keith, 2008*

451. Plant Virology Protocols: *From Viral Sequence to Protein Function*, edited by *Gary Foster, Elisabeth Johansen, Yiguo Hong, and Peter Nagy, 2008*

450. Germline Stem Cells, edited by *Steven X. Hou and Shree Ram Singh, 2008*

449. Mesenchymal Stem Cells: *Methods and Protocols*, edited by *Darwin J. Prockop, Douglas G. Phinney, and Bruce A. Brunnell, 2008*

448. Pharmacogenomics in Drug Discovery and Development, edited by *Qing Yan, 2008*

447. Alcohol: *Methods and Protocols*, edited by *Laura E. Nagy, 2008*

446. Post-translational Modifications of Proteins: *Tools for Functional Proteomics, Second Edition*, edited by *Christoph Kannicht, 2008*

445. Autophagosome and Phagosome, edited by *Vojo Deretic, 2008*

METHODS IN MOLECULAR BIOLOGY™

DNA Methylation

Methods and Protocols

Second Edition

Edited by

Jörg Tost, PhD

CEA - Institut de Génomique, Evry, France

Humana Press

Editor
Jörg Tost
CEA - Institut de Génomique
Evry
France
tost@cng.fr

Series Editor
John M. Walker
University of Hertfordshire
Hatfield, Herts
UK

ISSN: 1064-3745 e-ISSN: 1940-6029
ISBN: 978-1-61737-938-3 e-ISBN: 978-1-59745-522-0
DOI 10.1007/978-1-59745-522-0

Printed on acid-free paper

springer.com

Preface

Epigenetics can be defined as the study of heritable changes in gene expression without alteration of the DNA sequence itself. This means that epigenetic variants are stable alterations that are heritable during somatic cell divisions (and possibly transmitted through germ line transmissions in some occasions) but do not involve mutations of the DNA itself. Epigenetic phenomena are mediated by various molecular mechanisms, including histone modifications and core histone variants; ATP-dependent chromatin-remodeling complexes; polycomb/trithorax protein complexes; small RNAs, including siRNA and miRNAs as well as other noncoding RNAs; and last but not least DNA methylation. This volume in the *Methods in Molecular Biology*™ series focuses entirely on protocols for the analysis of DNA methylation, which is the only genetically programmed DNA modification in mammals occurring almost exclusively at the carbon 5 position of cytosines followed by a guanine.

Realizing the importance of epigenetic changes in development and disease, a variety of techniques for the study of DNA methylation have been developed over the last few years. **Figure 1** gives an overview of many of the commonly used technologies, but many more methods and variants of the named assays do exist. No single method has emerged as the "gold" standard technique unifying quantitative accuracy and high sensitivity or possibilities for whole genome analysis and precise investigations of individual CpG positions. The choice of the method mainly depends on the desired application. Although by no means complete, this second edition of "DNA methylation" gives a comprehensive overview of available technologies together with detailed step-by-step protocols for all experimental procedures required to successfully perform DNA methylation analysis.

This is the second edition of the DNA methylation protocols; however, the field has dramatically changed within the 6 years that have passed since the first edition edited by K.I. Mills and B.H. Ramsahoye was published. As DNA methylation technologies and our knowledge of DNA methylation patterns have been advancing at a breathtaking pace over the past few years and most of the techniques described in the first edition have been further optimized and/or replaced by novel, easier, refined, and/or more quantitative technologies, I have entirely remodeled the contents of this book. The increase in available methods is also reflected in the great expansion of the number of chapters within this book. While the first edition contained 14 chapters, this second edition consists now of 27 chapters. Only three chapters have been retained from the first edition and these have been completely rewritten by the authors to accommodate the changes and improvements made in the last years. The analysis of gene-specific DNA methylation patterns has been complemented or superseded by genome-wide approaches and epigenomics has taken a central place in many laboratories.

The selection of different technologies enables the analysis of the global DNA methylation content as well as precise quantitative data on single CpG positions. Methods for the high-resolution analysis of CpG positions within a target region identified by one of the multiple available genome-wide technologies are presented, and emphasis has been placed on array-based approaches that permit a hypothesis-free-driven research to identify

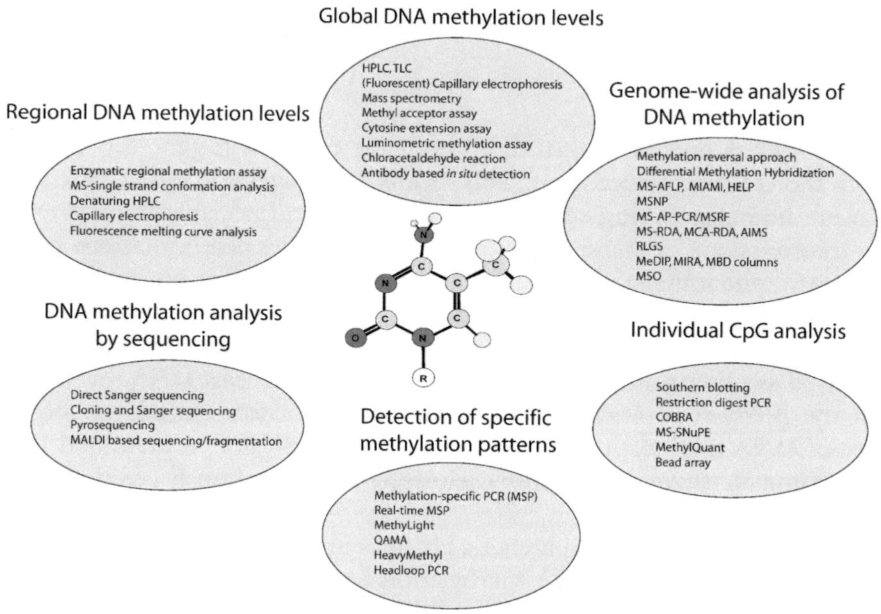

Fig. 1. An overview of the different technologies used for the analysis of DNA methylation. MS: Methylation sensitive; HPLC: High-performance Liquid Chromatography; TLC: Thin-layer Chromatography; MS-AFLP: Methylation-sensitive Amplified Fragment Length Polymorphism; MIAMI: Microarray-based Integrated Analysis of Methylation by Isochizomers; HELP: *Hpa*II tiny fragment Enrichment by Ligation-mediated PCR; MSNP: Methylation Single Nucleotide Polymorphism; MS-AP-PCR: Methylation-sensitive Arbitrarily-primed PCR; MSRF: Methylation-sensitive Restriction Fingerprinting; MS-RDA: Methylation-sensitive Representational Difference Analysis; MCA-RDA: Methylated CpG island Amplification—Representational Difference Analysis; AIMS: Amplification of intermethylated Sites; RLGS: Restriction Landmark Genomic Scanning; MeDIP: Methylated DNA ImmunoPrecipitation; MIRA: Methylated CpG Island Recovery Assay; MSO: Methylation-specific Oligonucleotide array; MALDI: Matrix-assisted Laser Desorption/Ionization mass spectrometry; COBRA: Combined Bisulfite Restriction Analysis, MS-SNuPE: Methylation-sensitive Single Nucleotide Primer extension; QAMA: Quantitative Analysis of Methylated Alleles. Reproduced with permission from Tost, J. (2008) Methods for the genome-wide and gene-specific analysis of DNA methylation levels and patterns. In: *Epigenetics* (Tost, J., ed.), Horizon Scientific Press, Norwich, UK, pp 63–103.

DNA methylation patterns of interest. In the final chapters of this book, more specialized applications like the sensitive detection of aberrant methylation patterns in body fluids, prevention of contamination, and whole genome amplification of bisulfite-treated DNA are described. Methods requiring special instruments are presented along technologies that can be performed with a simple thermocycler. This volume of the *Methods in Molecular Biology*^{TM} series contains widely used methods, such as cloning and sequencing and methylation-specific PCR as well as novel and promising techniques such as the immunodetection array that have only very recently passed the proof-of-principle stage.

This book is addressed to postdoctoral investigators and research scientists that are implicated in the different aspects of genetics and cellular and molecular biology as well as to clinicians involved in diagnostics or choice of treatment of diseases that have an epigenetic component. The presentation in this volume is equally suited for laboratories that already have a great deal of expertise in a certain technology to analyze DNA methylation, but might want to obtain other or complementary data using a second technique, and for genetics/genomics/biology groups that want to initiate research in this exciting area and want to identify the method best suited to answer their question. Notes and tips from the experts and/or pioneers of the different methods will enable a rapid implementation of the different protocols in the laboratory and avoid time-consuming and cost-intensive mistakes. With the tools and protocols available, our knowledge and understanding of DNA methylation will increase rapidly, and this book will contribute to spreading of the "savoir faire" to analyze DNA methylation.

I am indebted to all the authors for their hard work and outstanding contributions to this second edition of "DNA methylation". It was a pleasure to work with them on this project. I hope that the protocols described in detail in this volume will help to accelerate the analysis and description of the "methylome" of different species and will enhance our understanding of the molecular processes that determine the genomic DNA methylation landscape.

Evry, March 2008 ***Jörg Tost***

Contents

Part V. Special Applications

Contributors

CURT BALCH • *Department of Cellular and Integrative Physiology, Medical Sciences Program, Indiana University School of Medicine, Bloomington, IN, USA; Indiana University Cancer Center, Indianapolis, IN, USA*
MARÍA BERDASCO • *Epigenetics Group, Molecular Pathology Program, Spanish National Cancer Center (C.N.I.O.), Madrid, Spain*
MARINA BIBIKOVA • *Illumina, Inc., San Diego, CA, USA*
ROMULO MARTIN BRENA • *Division of Human Cancer Genetics, The Ohio State University, Columbus, OH, USA*
PAUL CAIRNS • *Department of Surgical Oncology, Fox Chase Cancer Center, Philadelphia, PA, USA*
MIHAELA CAMPAN • *Departments of Surgery and of Biochemistry and Molecular Biology, University of Southern California, Keck School of Medicine, Norris Comprehensive Cancer Center, Los Angeles, CA, USA*
JOSEPH F. COSTELLO • *Department of Neurological Surgery, University of California San Francisco Comprehensive Cancer Center, San Francisco, CA, USA*
DANIEL E. DEATHERAGE • *Human Cancer Genetics Program, The Ohio State University Comprehensive Cancer Center, Columbus, OH, USA*
EMELYNE DEJEUX • *Laboratory for Epigenetics, Centre National de Génotypage, CEA – Institut de Génomique, Evry, France*
JÜRGEN DISTLER • *Epigenomics AG, Berlin, Germany*
ALEXANDER DOBROVIC • *Molecular Pathology Research and Development Laboratory, Department of Pathology, Peter MacCallum Cancer Centre, Melbourne, Australia; Department of Pathology, University of Melbourne, Parkville, Australia*
CLAIRE DUGAST-DARZACQ • *Institut Jacques Monod du CNRS, Université Paris 7, Paris, France*
MATHIAS EHRICH • *SEQUENOM, Inc., San Diego, CA, USA*
HAFIDA EL ABDALAOUI • *Laboratory for Epigenetics, Centre National de Génotypage, CEA – Institut de Génomique, Evry, France*
MANEL ESTELLER • *Epigenetics Group, Molecular Pathology Program, Spanish National Cancer Center (C.N.I.O.), Madrid, Spain*
JIAN-BING FAN • *Illumina, Inc., San Diego, CA, USA*
MATHILDE FÖDERMAYR • *Elisabethinen Hospital, 1st Department of Internal Medicine, Linz, Austria*
MARIO F. FRAGA • *Epigenetics Group, Molecular Pathology Program, Spanish National Cancer Center (C.N.I.O.), Madrid, Spain*
JORDI FRIGOLA • *Institut de Medicina Predictiva i Personalitzada del Càncer (IMPPC), Barcelona, Spain*
THIERRY GRANGE • *Institut Jacques Monod du CNRS, Université Paris 7, Paris, France*
JOHN M. GREALLY • *Departments of Medicine (Hematology) and Molecular Genetics, Albert Einstein College of Medicine, Bronx, NY, USA*

IVO GLYNNE GUT • *Department of Technology Development, Centre National de Génotypage, CEA – Institut de Génomique, Evry, France*
JAMES G. HERMAN • *Cancer Biology Program, Sidney Kimmel Comprehensive Cancer Center, Johns Hopkins, Baltimore, MD, USA*
CHIBO HONG • *Department of Neurological Surgery, University of California, San Francisco Comprehensive Cancer Center, San Francisco, CA, USA*
DAVE S. B. HOON • *Department of Molecular Oncology, John Wayne Cancer Institute and Breast Center, Saint John's Health Center, Santa Monica, CA, USA*
TIM H.-M. HUANG • *Human Cancer Genetics Program, The Ohio State University Comprehensive Cancer Center, Columbus, OH, USA*
ALBERT JELTSCH • *School of Engineering and Science, Jacobs University Bremen, Bremen, Germany*
MIREIA JORDÀ • *Institut de Medicina Predictiva i Personalitzada del Càncer (IMPPC), Barcelona, Spain*
ZACHARY KAMINSKY • *The Krembil Family Epigenetics Laboratory, Centre for Addiction and Mental Health, Toronto, Ontario, Canada and University of Toronto, Toronto, Ontario, Canada*
PETER W. LAIRD • *Departments of Surgery and of Biochemistry and Molecular Biology, University of Southern California, Keck School of Medicine, Norris Comprehensive Cancer Center, Los Angeles, CA, USA*
SOPHIE LA SALLE • *Department of Pharmacology and Therapeutics and The Montreal Children's Hospital Research Institute, McGill University, Montreal, Quebec, Canada*
JULIEN D. F. LICCHESI • *Cancer Biology Program, Sidney Kimmel Comprehensive Cancer Center, Johns Hopkins, Baltimore, MD, USA*
SHILI LIN • *Department of Statistics, The Ohio State University, Columbus, OH, USA*
DIETER LUTZ • *Elisabethinen Hospital, 1st Department of Internal Medicine, Linz, Austria*
JONATHAN MILL • *The Krembil Family Epigenetics Laboratory, Centre for Addiction and Mental Health, Toronto, Ontario, Canada and Institute of Psychiatry, London, UK*
FABIO MOHN • *Friedrich Miescher Institute for Biomedical Research, Basel, Switzerland*
KENNETH P. NEPHEW • *Department of Cellular and Integrative Physiology, Medical Sciences Program, Indiana University School of Medicine, Bloomington, IN, USA; Indiana University Cancer Center, Indianapolis, IN, USA*
SANDY L. NGUYEN • *Department of Molecular Oncology, John Wayne Cancer Institute and Breast Center, Saint John's Health Center, Santa Monica, CA, USA*
CHRISTOPHER C. OAKES • *Department of Pharmacology and Therapeutics and The Montreal Children's Hospital Research Institute, McGill University, Montreal, Quebec, Canada*
MAYUMI ODA • *Departments of Medicine (Hematology) and Molecular Genetics, Albert Einstein College of Medicine, Bronx, NY, USA*
MIGUEL A. PEINADO • *Institut de Medicina Predictiva i Personalitzada del Càncer (IMPPC), Barcelona, Spain*
ARTURAS PETRONIS • *The Krembil Family Epigenetics Laboratory, Centre for Addiction and Mental Health, Toronto, Ontario, Canada and University of Toronto, Toronto, Ontario, Canada*
GERD P. PFEIFER • *Division of Biology, Beckman Research Institute of the City of Hope, Duarte, CA, USA*

CHRISTOPH PLASS • *Toxicology and Cancer Risk Factors, Division C010, German Cancer Research Center (DKFZ), Heidelberg, Germany*
DUSTIN POTTER • *Human Cancer Genetics Program, The Ohio State University Comprehensive Cancer Center, Columbus, OH, USA*
JOHANNES PRÖLL • *Red Cross Transfusion Service of Upper Austria, Linz and Elisabethinen Hospital, 1st Department of Internal Medicine, Linz, Austria*
TIBOR A. RAUCH • *Division of Biology, Beckman Research Institute of the City of Hope, Duarte, CA, USA*
RICHARD REINHARDT • *Max Planck Institute for Molecular Genetics, Berlin, Germany*
BERNARD ROBAIRE • *Department of Pharmacology and Therapeutics and Department of Obstetrics and Gynecology, McGill University, Montreal, Quebec, Canada*
JAIRO RODRIGUEZ • *Institut de Medicina Predictiva i Personalitzada del Càncer (IMPPC), Barcelona, Spain*
CHRISTIAN ROHDE • *School of Engineering and Science, Jacobs University Bremen, Bremen, Germany*
TIM-CHRISTOPH ROLOFF • *Friedrich Miescher Institute for Biomedical Research, Basel, Switzerland*
DIRK SCHÜBELER • *Friedrich Miescher Institute for Biomedical Research, Basel, Switzerland*
DAVID G. SKALNIK • *Indiana University Cancer Center, Indianapolis, IN, USA and Department of Biochemistry and Molecular Biology, Herman B. Wells Center for Pediatric Research, Indiana University School of Medicine, Indianapolis, IN, USA*
DOMINIC J. SMIRAGLIA • *Roswell Park Cancer Institute, Buffalo, NY, USA*
HEINRICH STAMERJOHANNS • *School of Engineering and Science, Jacobs University Bremen, Bremen, Germany*
EIJI SUNAMI • *Department of Molecular Oncology, John Wayne Cancer Institute and Breast Center, Saint John's Health Center, Santa Monica, CA, USA*
REIMO TETZNER • *Epigenomics AG, Berlin, Germany*
SASCHA TIERLING • *Institut für Genetik, FB Biowissenschaften, Universität des Saarlandes, Saarbrücken, Germany*
JÖRG TOST • *Laboratory for Epigenetics, Centre National de Génotypage, CEA – Institut de Génomique, Evry, France*
JACQUETTA M. TRASLER • *Department of Pharmacology and Therapeutics, Department of Pediatrics, Department of Human Genetics and The Montreal Children's Hospital Research Institute, McGill University, Montreal, Quebec, Canada*
BINH TRINH • *Departments of Surgery and of Biochemistry and Molecular Biology, University of Southern California, Keck School of Medicine, Norris Comprehensive Cancer Center, Los Angeles, CA, USA*
TOSHIKAZU USHIJIMA • *Carcinogenesis Division, National Cancer Center Research Institute, Tokyo, Japan*
DIRK VAN DEN BOOM • *SEQUENOM, Inc., San Diego, CA, USA*
ANH-THU VU • *Department of Molecular Oncology, John Wayne Cancer Institute and Breast Center, Saint John's Health Center, Santa Monica, CA, USA*
JÖRN WALTER • *Institut für Genetik, FB Biowissenschaften, Universität des Saarlandes, Saarbrücken, Germany*
MICHAEL WEBER • *Institut de Génétique Moléculaire, CNRS UMR 5535, 1919 Route de Mende, 34293 MONTPELLIER Cedex 5, France*

CHRISTIAN WECHSELBERGER • *Upper Austrian Research GmbH, Center for Biomedical Nanotechnology, Linz, Austria*
DANIEL J. WEISENBERGER • *Departments of Surgery and of Biochemistry and Molecular Biology, University of Southern California, Keck School of Medicine, Norris Comprehensive Cancer Center, Los Angeles, CA, USA*
TOMASZ K. WOJDACZ • *Institute of Human Genetics, University of Aarhus, Aarhus, Denmark*
SATOSHI YAMASHITA • *Carcinogenesis Division, National Cancer Center Research Institute, Tokyo, Japan*
PEARLLY S. YAN • *Human Cancer Genetics Program, The Ohio State University Comprehensive Cancer Center, Columbus, OH, USA*
OTTO ZACH • *Elisabethinen Hospital, 1st Department of Internal Medicine, Linz, Austria*
YINGYING ZHANG • *School of Engineering and Science, Jacobs University Bremen, Bremen, Germany*

Part I

Introduction

Chapter 1

DNA Methylation: An Introduction to the Biology and the Disease-Associated Changes of a Promising Biomarker

Jörg Tost

Abstract

DNA methylation occurring on the 5 position of the pyrimidine ring of cytosines in the context of the dinucleotide sequence CpG forms one of the multiple layers of epigenetic mechanisms controlling and modulating gene expression through chromatin structure. It closely interacts with histone modifications and chromatin-remodeling complexes to form the genomic chromatin landscape. DNA methylation is essential for proper mammalian development, crucial for imprinting, and plays a role in maintaining genomic stability as well as in dosage compensation. DNA methylation patterns are susceptible to change in response to environmental stimuli such as diet or toxins whereby the epigenome seems to be most vulnerable during early *in utero* development. Aberrant DNA methylation changes have been detected in several diseases, particularly cancer where genome-wide hypomethylation coincides with gene-specific hypermethylation. DNA methylation patterns can be used to detect cancer at very early stages, to classify tumors as well as predict and monitor the response to antineoplastic treatment. As a stable nucleic acid-based modification with limited dynamic range that is technically easy to handle, DNA methylation is a promising biomarker for many applications.

Key words: DNA methylation, nutritional epigenetics, environmental epigenetics, complex disease, epigenetics, imprinting, cancer.

1. Introduction

All cells of a multicellular organism carry the same genetic material coded in their DNA sequence, but cells obviously display a broad morphological and functional diversity. This heterogeneity is caused by differential expression of genes. Epigenetics can be defined as the study of heritable changes of a phenotype such as the gene expression patterns of a specific cell type that are not

Jörg Tost (ed.), *DNA Methylation: Methods and Protocols, Second Edition, vol. 507*

DOI 10.1007/978-1-59745-522-0_1 Springerprotocols.com

caused by changes in the nucleotide sequence of the genetic code itself.

These changes are mitotically and in some cases meiotically heritable. Epigenetic regulation mediates genomic adaption to an environment, thereby ultimately contributing toward the phenotype. They "bring the phenotype into being" as said by the developmental biologist Conrad H. Waddington in the 1940s *(1)*.

Epigenetic phenomena are mediated by a variety of molecular mechanisms including posttranslational histone modifications, histone variants, ATP-dependent chromatin remodeling complexes, polycomb/trithorax protein complexes, small and other noncoding RNAs, including siRNA and miRNAs, and DNA methylation, as described in detail in *(2)*. These diverse molecular mechanisms have all been found to be closely intertwined and stabilize each other to ensure the faithful propagation of an epigenetic state over time and especially through cell division. Nonetheless, epigenetic states are not definitive and changes occur with age in a stochastic manner as well as in response to environmental stimuli. Chromatin modulations play a central role to shape the epigenome and delineate a functional chromatin topology, which serves as the platform forming regulatory circuits in all cells. Open (euchromatin) and closed (heterochromatin) chromatin states are controlled by histone modifications and histone composition in close cross talk with the binding of a myriad of nonhistone proteins. The basic building block of chromatin is the nucleosome which is formed of an octamer of histone proteins containing an H3–H4 tetramer, flanked on either side with an H2A–H2B dimer around which 146 base pairs of DNA are spooled in a 1.65 left-handed superhelical turn. The protruding N-terminal tails of these histones are extensively modified by various modifications such as acetylation, methylation, phosphorylation, and ubiquitylation *(3)*. The combination of different N-terminal modifications and the incorporation of different histone variants, which have distinct roles in gene regulation, have led to the proposition of a regulatory histone code which determines at least partly the transcriptional potential for a specific gene or a genomic region *(4)*. DNA methylation is highly related to certain chromatin modifications; and enzymes that modify DNA and histones have been shown to directly interact and constitute links between local DNA methylation and regional chromatin structure.

This chapter briefly describes the DNA methylation landscape and the enzymes responsible for adding and potentially removing methyl groups to the DNA and touches upon the various biological processes in which DNA methylation plays a key role. Different pathologies for which changes in DNA methylation patterns have been investigated will be mentioned with a certain emphasis on cancer, as most of the disease-associated DNA methylation literature concerns changes in tumorigenesis. This chapter concludes with the reasons why DNA methylation is a very promising biomarker. Due to space restrictions and the large field of research described in this introduction, oversimplifications and omissions are inevitable. This chapter is addressed to scientists who have not been in close contact with this field before, and most of the references direct the reader to more exhaustive review articles. People with experience in DNA methylation research can directly jump to the protocol of their choice.

2. The Biology of DNA Methylation

DNA methylation is the only genetically programmed DNA modification in mammals. This postreplication modification is almost exclusively found on the 5 position of the pyrimidine ring of cytosines in the context of the dinucleotide sequence CpG *(5)*. 5-Methylcytosine accounts for ~1% of all bases, varying slightly in different tissue types and the majority (75%) of CpG dinucleotides throughout mammalian genomes are methylated. Other types of methylation such as methylation of cytosines in the context of CpNpG or CpA sequences have been detected in mouse embryonic stem cells and plants, but are generally rare in somatic mammalian/human tissues. CpGs are underrepresented in the genome, probably because they act as a mutation hotspot (deamination of methylated CpGs to TpGs). Mutation rates at CpG sites have been estimated to be about 10–50 × higher than other transitional mutations, as the mutation product is a naturally occurring DNA base which may not be appropriately repaired. The elevated mutation rate has led to depletion of the dinucleotide during evolution. Despite this general trend, relatively CpG-rich clusters of approximately 1–4 kb in length—so-called CpG islands—are found in the promoter region and first exons of many genes. They are mostly nonmethylated corresponding to the maintenance of an open chromatin structure and a potentially active state of transcription *(6)*. There are around 30,000 CpG islands in the human genome. As CpG islands are mainly unmethylated in the germline, they are less susceptible to deamination and have therefore retained the expected frequency of CpGs. It should be noted that a growing number of CpG islands have been identified that are methylated in nonpathological somatic tissues *(7)*. Depending on the employed set of parameters, a CpG island is defined as having a G+C content of more than 50% (55%), an observed versus expected ratio for the occurrence of CpGs of more than 0.6 (0.65) and a minimum size of 200 (500) bp *(8)*. About three-quarters of transcription start sites and 88% of active promoters are associated with CpG-rich sequences and might be regulated by DNA methylation. Promoter CpG islands of genes differ in their susceptibility to become methylated during normal development as well as during carcinogenesis, which might be due to intrinsic sequence properties *(9)*. This regulation is controlled in a tissue- and developmental-stage-specific manner and is maintained throughout the life of an individual.

2.1. DNA Methyltransferases and Methyl-Binding Proteins

The composition of the genome is reflected in and dictates the epigenetic machinery to establish particular local and global epigenetic patterns using both CpG spacing as well as sequence motifs and DNA structure *(10, 11)*. Mammalian one-carbon metabolism provides the methyl group for all biologic methylation reactions. These are dependent on methyl donors (methionine and choline) and cofactors (folic acid, vitamin B12, and pyridoxal phosphate) to synthesize the universal methyl donor S-adenosyl-L-methionine (SAM) *(12)*. During the methylation reaction, a methyl group is transferred from SAM to the DNA leaving S-adenosylhomocysteine which at high concentrations inhibits the action of DNA methyltransferases.

So far, four DNA methyltransferases have been identified (DNMT1, DNMT2, DNMT3A, and DNMT3B) as well as a DNMT-related protein (DNMT3L) *(13)*. They catalyze the transfer of a methyl group from SAM to the cytosine base. With the exception of DNMT2 which acts probably as RNA methyltransferase *in vivo*, all Dnmts are essential for embryonic viability as homozygous mutant mice die early. Simplified DNMT1 acts as maintenance methyltransferase as it prefers hemimethylated templates. It is located at the replication fork during the S phase of the cell cycle and methylates the newly synthesized DNA strand using the parent strand as a template. Consequently, it passes the epigenetic information through cell generations. *De novo* methylation is carried out by the methyltransferases DNMT3A and DNMT3B. These enzymes not only have certain preferences for specific targets (e.g., Dnmt3a together with Dnmt3L methylates maternal imprinted genes and Dnmt3b localizes at minor satellite repeats), but also work cooperatively to methylate the genome. Possible trigger mechanisms to initiate *de novo* methylation include preferred target DNA sequences, RNA interference, certain chromatin structures induced by histone modifications, and other protein–protein interactions *(13)*.

DNA methyl-binding domain (MBD1-4) proteins or methyl CpG-binding proteins (MeCP2) recognize and bind to methylated DNA. They recruit transcriptional corepressors such as histone-deacetylating complexes, polycomb proteins, and chromatin-remodeling complexes, and attract chromodomain-binding proteins. Besides the structurally related MBD proteins, methylated DNA can also be bound by some zinc finger proteins such as Kaiso and the more recently discovered ZBTB4 and ZBTB38 proteins that are also able to repress transcription in a methylation-dependent manner *(14)*.

Although active demethylation of DNA undoubtedly occurs during development, the exact mechanisms for global as well as for gene-specific demethylation events are still unclear. Demethylation might be caused by the replacement of methylated cytosines through an enzymatic process in which a glycosylase plays a major role *(15)* or by a deamination-induced repair process as the activation induced cytidine deaminase (AID) has been shown to deaminate cytidine in RNA as well as 5-methylcytosine in DNA *(16)*.

2.2. Development

Cytosine methylation is essential for mammalian embryogenesis, during which methylation levels change dynamically *(17)*. During development and differentiation the mammalian organism creates a number of cell-type-specific, differentially marked epigenomes, whose identity is *inter alia* defined by their respective DNA methylation patterns. Thus, the human body with one genome contains approximately 180 different epigenomes. Mammalian development is characterized by two waves of genome-wide epigenetic reprogramming, in the zygote and in the primordial germ cells. In mice (and probably other mammals) the genome becomes demethylated during preimplantation, probably to initiate cellular differentiation. Most of the paternal genome is actively and rapidly demethylated leading to the erasure of most paternal germline methylation marks, while the maternal genome remains methylated or may undergo further *de novo* methylation. After completion of the first cell cycle, loss of methylation on the maternal allele occurs passively through cell divisions until blastocyst formation. This demethylation removes most of the preexisting patterns of methylation inherited from the parental DNA. Around implantation, where cell lines start to commit to different developmental lineages, DNA methylation levels are then restored by *de novo* methylation. Disruption of any of the DNA methyltransferases results in embryonic lethality and hypomorphic alleles of *Dnmt1* result in genome-wide deregulation of gene expression. The second reprogramming event occurs also during embryogenesis but only in the primordial germ cells where DNA methylation patterns are erased at all single-copy genes (including imprinted genes) and some repetitive elements *(18)*. Depending on the sex of the newly formed germline, imprints at paternally methylated loci are restored shortly after birth, while maternally methylated loci occur only during the last stages of oogenesis.

Modifications to the environment during early development can also lead to permanent changes in the patterns of epigenetic modifications (see also the paragraph on environmental and nutritional epigenetics). For example, differences in the cell culture medium lead to differences in cleavage kinetics,blastocyst formation, and disturbed epigenetic profiles at imprinted gene loci *(19)*. This might also partly account for the increased incidence of imprinting disorders in children born from assisted reproduction technologies (ART) *(20)*. It should be noted that reduced fertility has also been linked to epigenetic changes. An alternative explanation is therefore that people with incorrect epigenetic information need to revert more often to reproductive technologies and the incorrectly programmed gametes rather than ART increase the risk for epigenetic disorders in their children. Truth lies probably in the combination of both explanations. An incomplete erasure and reprogramming of the epigenetic patterns might also be one of the reasons for the low success rate of cloning using somatic cell nuclear transfer, that is, the fusion of a somatic cell with an enucleated oocyte *(21)*. Global as well as gene-specific DNA methylation patterns – in particular at imprinted gene loci (see below) – are disturbed in cloned animals and lead to a multitude of pre- and perinatal developmental abnormalities.

Epigenetic changes are also an integral part of aging and cellular senescence, whereby the overall content of DNA methylation in the mammalian and human genome decreases with age. Simultaneously, distinct genes acquire methylation at specific sites such as their promoters, a situation that strikingly resembles the DNA methylation changes that are found in cancer *(22)*.

2.3. Transcription

Transcription does not occur on naked DNA but in the context of chromatin which critically influences the accessibility of the DNA to transcription factors and the DNA polymerase complexes. DNA methylation, histone modifications, and chromatin remodeling are closely interwoven and constitute multiple layers of epigenetic modifications to control and modulate gene expression through chromatin structure *(23)*. DNMTs and histone deacetylases (HDACs) are found in the same multiprotein complexes and MBDs interact with HDACs, histone methyltransferases, as well as with the chromatin-remodeling complexes. Furthermore, mutations or loss of members of the SNF2 helicase/ATPase family of chromatin-remodeling proteins such as *ATRX* or *LSH* lead to genome-wide perturbations of DNA methylation patterns and inappropriate gene expression programs.

Cytosine methylation of CpG dinucleotides is found in close proximity to critically important *cis*-elements within promoters and is often associated with a repressed chromatin state and inhibition of transcription. In many cases, methylated and silenced genes can be reactivated using DNA methylation inhibitors such as 5-azacytidine. However, it should be noted that an unmethylated state of a CpG island does not necessarily correlate with the transcriptional activity of the gene, but rather that the gene can be potentially activated. On the other hand, the simple presence of methylation does not necessarily induce silencing of nearby genes. Only when a specific core region of the promoter that is often – but not necessarily – spanning the transcription start site becomes hypermethylated, the expression of the associated gene is modified *(24)*. The methylation status at specific CpG dinucleotides in the core region might therefore better correlate with the expression of the gene than the overall methylation level of the entire CpG island. These islands are also found outside promoter regions and these appear generally to be more susceptible to methylation than the respective promoter sequences in various cancers as well as during cell culture. However, methylation of these CpG islands does not usually diminish transcription. It has been proposed that methylation begins in exonic regions and then progressively spreads to CpG islands in other locations, including promoter regions. The exact mechanism is still to be elucidated but it could be that the protection from methylation is lost through the absence of transcription facilitating the access of DNMTs to the DNA, as transcription factors and/or transcription/initiation complexes are absent. In some cases the methylation density in a promoter core region seems to be crucial to induce transcriptional silencing *(25)*, while in other cases the demethylation of specific CpG sites is sufficient for transcriptional reactivation *(26)*.

Methylation can interfere with transcription in several ways *(5)*. It can inhibit the binding of transcriptional activators with their cognate DNA recognition sequence such as Sp1 and Myc through sterical hindrance. The above-described MBD proteins and the DNMTs themselves bind to methylated DNA and prevent thereby the binding of potentially activating transcription factors. These two protein families also recruit additional proteins with repressive function such as histone deacetylases and chromatin-remodeling complexes to the methylated DNA to establish a repressive chromatin configuration.

In many cases, DNA methylation follows changes in the chromatin structure and is used as the molecular mechanism to permanently and thus heritably lock the gene in its inactive state *(5)*. Recent results have also shown that active histone marks such as H3K4Me3 at the transcription start site might permit transcription of a gene when stimulated even in the presence of a partly methylated CpG island immediately adjacent to the transcription start site *(27)*.

2.4. Genome Stability

5-Methylcytosine and other modified bases are also found in bacteria where they constitute an integral part of the modification-restriction enzyme that allows distinguishing between self and invading foreign DNA. DNA methylation plays an important role in the maintenance of genome integrity by transcriptional silencing of repetitive DNA sequences and endogenous transposons.

DNA methylation might prevent the potentially deleterious recombination events between nonallelic repeats caused by these mobile genetic elements. In addition, methylation increases the mutation rate leading to a faster divergence of identical sequences and disabling of many retrotransposons *(28)*.

2.5. Imprinting

In mammals, the maternal and paternal genomes are functionally not equal and both are required for normal development. A subset of genes is asymmetrically expressed from only the maternal or the paternal allele in a parent-of-origin-specific manner in all somatic cells of the offspring *(29)*. These imprinted genes are generally located in clusters and the alleles are differentially marked by DNA methylation, histone acetylation/deacetylation, and histone methylation, and often associated with antisense RNAs *(30, 31)*. About 50 imprinted genes are known in mouse and man, respectively, but up to 200 imprinted genes have been computationally predicted *(32)*. Imprints are established in the gametes by Dnmt3a and at least for maternally imprinted genes the regulatory cofactor Dnmt3L in a parent-of-origin-specific manner. These epigenetic marks in imprinting control regions are not erased in the zygote. Imprinted genes are probably the most important buffering factors for regulating the day-to-day flux between mother and fetus in placental mammals. The *H19/Igf2* locus is one paradigm for imprinting and has been extensively studied in mice demonstrating that the physical contacts between differentially methylated regions, containing insulators, silencers, and activators, create a higher-order chromatin structure, leading to transcriptional regulation of both *H19* and *Igf2 (33)*.

2.6. X Inactivation

Random silencing of one of the two X chromosomes in embryonic tissues of female mammals to achieve dosage compensation is another paradigm for a stable and heritable epigenetic state in somatic cells *(34)*. DNA methylation occurs quite late during the inactivation process. Only after expression of the large noncoding *Xist* RNA, its coating of the future inactive chromosome, changes in the patterns of histone modifications and variants, and gene silencing, DNA methylation patterns are established on the inactive X chromosome, where they are necessary to maintain the inactive X chromosome in its silent state.

2.7. Environmental and Nutritional Epigenetics

Epigenetics holds the promise to explain at least a part of the influences the environment has on a phenotype. Studies in monozygotic twins have demonstrated that epigenetic differences in genetically identical humans (monozygotic twins) accumulate with age and different environments create different patterns of epigenetic modifications *(35)*. Differences are therefore largest in twin pairs of old age that have been raised separately. Transient nutritional or chemical stimuli occurring at specific ontogenic stages may have long-lasting influences on gene expression by interacting with epigenetic mechanisms and altering chromatin compaction and transcription factor accessibility. Developmental stages in multicellular organisms proceed according to a tightly regulated temporal and spatial pattern of gene expression, accompanied by changes in DNA methylation patterns as described in the paragraph on development above. These changes occur in response to transient stimuli. Therefore, epigenetics provides a mechanism by which physiological homeostasis could be developmentally programmed and inherited.

DNA methylation is dependent on the diet-ingested methyl donor folate. DNA methylation levels correlate with the levels of available folate as well as the genotype-dependent activity of involved enzymes such as the 5,10-methylenetetrahydrofolate reductase gene *(36)*. In mice, an increase in folic acid intake leads to increased DNA methylation of an allele of the *agouti* locus, causing gene silencing and a modification of the phenotype *(37)*. Disorders like intrauterine growth retardation and neural tube defects as well as the adult onset of many complex diseases have been linked to aberrant methyl metabolism *in utero*. This modulation of epigenetic patterns *in utero* has given rise to the developmental origin of disease hypothesis, which postulates that the *in utero* environment can cause permanent changes to metabolic processes that directly affect postnatal phenotype, confers susceptibility to multifactorial disease at adult age, and may also be transmitted to subsequent generations *(38)*. Both chemical and environmental toxins have shown to induce changes to DNA methylation patterns without altering the genetic sequence and leading to epimutation-associated phenotypes *(39, 40)*. Environmental toxins such as benzpyrene and dioxin do appear to promote a transgenerational susceptibility to disease that remains unexplained by genetic means. Endocrine disruptors such as the antiandrogenic fungicide vinclozolin have been shown to alter the DNA methylation patterns in sperm and the effects persist for at least four generations *(41)*. Long after the stimulus is gone, "cellular memory" mechanisms enable cells to remember their chosen fate, thus perturbation at an early stage have long-lasting consequences.

2.8. Transgenerational Epigenetic Inheritance

Transgenerational epigenetic inheritance refers to the transfer of epigenetic information across generations, that is, through meiosis *(42)*. This mechanism would explain the inheritance of a phenotype in addition to the DNA sequence from the parents. The strongest evidence comes from a phenomenon called paramutation in plants where the epigenetic state at one locus is conferred to the homologous allele in a meiotically heritable manner, thus inducing a change in gene expression in the absence of a genetic mutation. Two models have been proposed either based on the pairing of homologous chromosomes or RNA-mediated silencing. A recent transgene model in mice lent support to the RNA model *(43)*. Two loci in mice have been shown to exhibit transgenerational epigenetic inheritance, the *agouti-viable yellow* and the*axin-fused* allele. In both cases the variable phenotype (coat color or presence or absence of a kinked tail, respectively) corresponds to the extent of DNA methylation of an IAP retrotransposon inserted at the respective locus. However, due to the clearing of DNA methylation patterns in primordial germ cells it is not the DNA methylation itself that is responsible for these metastable epialleles. So far, transgenerational epigenetic inheritance has not been clearly identified in humans despite some epidemiological evidence *(44)*.

3. DNA Methylation and Disease

DNA methylation and chromatin structure are strikingly altered in many pathological situations, particularly cancer and various mental retardation syndromes, and altered levels of folate and homocysteine have been repeatedly linked to disease. Although a number of genetic variations have recently been identified to confer susceptibility to a certain disease, in most cases even the worst combination of alleles of several disease susceptibility loci only explains a small percentage of disease occurrence. Consequently, environmental factors play undoubtedly a large role in the actual occurrence of disease. Epigenetic modifications constitute a memory of an organism to all the stimuli or insults it has ever been exposed to.

Disease-associated changes in epigenetic modifications can be classified into changes in genes that are epigenetically regulated and genes that are part of the molecular machinery responsible for correct establishment and propagation of the epigenetic modifications through development and cell division. Aberrant methylation patterns have been reported in various neurodevelopmental disorders, including ATRX (X-linked α-thalassemia and mental retardation), Fragile X, and ICF (Immune deficiency, centromeric instability, and facial abnormalities) *(45)*. The latter is caused by mutations in the DNA methyltransferase 3B. Mutations in the methyl-binding protein MeCP2 are found in Rett syndrome. Imprinting anomalies lead to disorders such as Prader–Willi, Angelman, and Beckwith–Wiedemann syndrome or transient neonatal diabetes *(45)*.

3.1. DNA Methylation Changes in Cancer

Cancer is probably the best-studied disease with a strong epigenetic component *(46, 47)*. In tumors, a global loss of DNA methylation (hypomethylation) of the genome is observed *(48)* and has been suggested to initiate and propagate oncogenesis by inducing chromosome instabilities and transcriptional activation of oncogenes and prometastatic genes such as *r-ras (49)*. The overall decrease in DNA methylation is accompanied by a region- and gene-specific increase of methylation (hypermethylation) of multiple CpG islands *(46, 47)*. Hypermethylation of CpG islands in the promoter region of a tumor suppressor or otherwise cancer-related gene is often associated with transcriptional silencing of the associated gene. The number of gene-associated promoters that are known to become hypermethylated during carcinogenesis is rapidly growing. Genes of numerous pathways involved in signal transduction (*APC*), DNA repair (*MGMT*, *MLH1*, and *BRCA1*), detoxification (*GSTP1*), cell cycle regulation (*p15*, *p16*, and *RB*), differentiation (*MYOD1*), angiogenesis (*THBS1* and *VHL*) and apoptosis (*Caspases*, *p14*, and*DAPK*) are often inappropriately inactivated by DNA methylation. It should be noted that so far no single gene has been identified that is always methylated in a certain type of cancer. Both hypo- and hypermethylation are found in the same tumor, but the underlying mechanisms for both phenomena have not yet been elucidated. A new dimension has recently been added to epigenetic cancer research with the demonstration of long-range gene silencing by epigenetic modifications *(50)*. Long-range epigenetic silencing seems to be a prevalent phenomenon during carcinogenesis, as a recent survey identified 28 regions of copy-number-independent transcriptional deregulation in bladder cancer that are potentially regulated through epigenetic mechanisms *(51)*. While the contribution of genetic factors to carcinogenesis such as the high-penetrance germline mutations in genes (e.g., *BRCA1* and *p53*) in familial cancers has long been recognized, it has become evident that epigenetic changes leading to transcriptional silencing of tumor suppressor genes constitute an at least equally contributing mechanism. For example, microarray expression profiles of breast tumors with *BRCA1* mutations are very similar to those of sporadic breast cancer cases with *BRCA1* promoter hypermethylation, demonstrating that disruption of *BRCA1* function by either genetic or epigenetic pathways leads to the same perturbations *(52)*. With the exception of haploinsufficient genes, "two hits" are necessary to inactivate the two alleles of a gatekeeper tumor suppressor gene to enable oncogenic progression, according to Knudson's two-hit hypothesis *(53)*. DNA methylation can act as one hit having the same functional effect as a genetic mutation or deletion as proven by numerous experiments, in which reestablishing expression of tumor suppressor genes could be achieved through drugs inducing demethylation. Epimutations can inactivate one of the two alleles, while the other is lost through genetic mechanisms or silence both alleles *(54)*. Epigenetic changes occur at higher frequency compared to genetic changes and maybe especially important in early-stage human neoplasia. They often precede malignancy as extensive CpG island hypermethylation can be detected in benign polyps of the colon, in low- as well as in high-grade tumors *(55, 56)*. It has therefore been suggested that epigenetic lesions in normal tissue set the stage for neoplasia. DNA hypermethylation could, for example, not only be detected in dysplastic epithelium of patients with ulcerative colitis, a condition associated with an increased risk for the development of colon cancer, but already in histological normal epithelium *(57)*. Aberrant DNA methylation patterns are, therefore, probably not a consequence or by-product of malignancy and contribute directly to the cellular transformation. It has been extrapolated that aberrant promoter methylation is initiated at ~1% of all CpG islands and as much as 10% become methylated during the multistep process of tumorigenesis *(55)*. Detection can be carried out in the tissue itself, but – more importantly – recent reports have demonstrated a high level of concordance of DNA methylation patterns in tumor biopsies and matched DNA samples extracted from body fluids such as serum, plasma, urine, and sputum. DNA methylation-based markers are therefore promising tools for noninvasive detection of different tumor types. The most effective way to detect the aberrant methylation is to analyze fluids that have been in physical contact with the site of the respective cancer. A large number of novel sources has been successfully tested, including nipple aspirate fluid, breast-fine needle washing, bronchial brush samples, buccal cells, needle biopsies, pancreatic juice, peritoneal fluid, prostate fluid or ejaculate, brochialveolar lavages, saliva, exfoliative cells from bladder or cervix, urine, peritoneal fluid, or stool samples *(58)*. Tumors release a substantial amount of genomic DNA into the systemic circulation and this freely circulating DNA contains the same genetic and epigenetic alterations that are specific to the primary tumor *(59)*. As the analyzed gene-specific methylation patterns are in most cases absent in control patients, methylation analysis of DNA recovered from plasma and serum can be used as a biomarker for molecular diagnosis and prognosis in various types of malignancies *(60)*. Besides early detection, the methylation status of CpG islands can be used to characterize and classify cancers. While, for example, head and neck, breast, or testicular tumors show overall low levels of methylation, some other tumor types such as colon tumors, acute myeloid leukemias, or gliomas are characterized by high levels of methylation, although some heterogeneity is observed in almost all tumor types. Methylation patterns can be shared by different types of tumors as well as being tumor type specific and methylation profiling can therefore identify distinct subtypes of human cancers *(55)*. Other important applications of DNA methylation analysis in cancer are the detection of tumor recurrence as well as the prediction and monitoring of patients response and effectiveness to a given anticancer therapy *(61)*. As DNA methylation is a nonmutational and therefore – at least in principle – a reversible modification, it can be used as point of departure for antineoplastic treatment by chemically or antisense oligonucleotide-induced demethylation *(62)*.

3.2. DNA Methylation and Complex Disease

While most of the interest has so far been focused on epigenetic changes in cancer, it is probable that epigenetic changes directly or indirectly contribute to the susceptibility and development of many complex or multifactorial diseases *(63)*. Epigenetic mechanisms are consistent with various nonmendelian features of multifactorial diseases such as the relatively high degree of discordance in monozygotic twins. Only few diseases have been studied in some detail. The promoter of the membrane-bound form of *COMT* is found to be hypomethylated in schizophrenia and bipolar disorder leading to hyperactivity of the gene, while the *RELN* promoter displays concomitant hypermethylation in schizophrenia *(64)*. The expression of the methyl-binding protein *MeCP2* is reduced in the frontal cortex of autistic patients and this correlates with increased methylation of the *MeCP2* promoter *(65)*. DNA methylation patterns are also globally disturbed in autoimmune diseases, such as the lupus erythematosus *(66)* or rheumatoid arthritis *(67)*. Epigenetic changes are probably also involved in the pathogenesis of diabetes, metabolic syndrome, and intermediate phenotypes, where disease susceptibility seems to be influenced by the maternal *in utero* environment and recent epidemiological evidence implicates also paternal behavior *(68)*. To further underline the scope of epigenetic alterations in disease, it is interesting to point out that recent research has shown that so-called monogenetic diseases such as α-thalassemia that have previously been attributed solely to genetic alterations can also be caused by epigenetic alterations at the same locus *(69)*. The field of epigenetics of complex diseases is still in its infancy, but epigenetics might provide the missing link between the genetic susceptibility and the phenotype by mediating and modulating environmental influences differentially depending on the epigenotype of a disease susceptibility locus.

4. DNA Methylation as a Biomarker

Research has so far mainly focused on the hypermethylation of promoter-associated CpG islands where hypermethylation is inversely correlated to their transcriptional activity. Correlation between DNA methylation and gene inactivation is a prerequisites for the identification and validation of novel functionally important genes, namely, tumor suppressor genes. However, a large number of promoters become hypermethylated during carcinogenesis where there is no evidence that the corresponding gene acts as a tumor suppressor gene. In this case, DNA methylation might still be a useful biomarker for tumor diagnosis or risk assessment if the methylation pattern is specific for a certain tumor type and/or correlates with clinically important parameters. A good example is the classic panel for the detection of the CpG island methylator phenotype defining a subtype of colorectal cancers with a distinct phenotype that comprises three MINT (Methylated IN Tumors) fragments *(70)*. These fragments have been identified through differential screening processes, but have only been later mapped to specific genomic loci. As described above, the analysis of DNA methylation patterns is complicated by the fact that some changes are due to exposure to environmental influences as well as accumulation of DNA methylation at some promoters during aging *(40)*. To be useful as biomarker, age-associated DNA methylation changes have therefore to be distinguished from cancer-predisposing alterations.

Biomarkers capable of distinguishing diseased or malignant cells from normal ones must be specific, sensitive, and detectable in specimens obtained through minimally invasive procedures to be clinically applicable. Many biomarkers on the protein, RNA, or DNA level fulfilling these criteria have been discovered. In routine clinical practice, most tumor diagnostics is carried out by biochemical assays determining the presence and/or quantity of enzymes, receptors, growth factors, or hormones. Despite the wide-spread use of (microarray-based) RNA detection techniques in research facilities, there are some potential pitfalls associated with its use in routine clinical diagnostics, such as the required preservation of mRNA from the tissue, tissue heterogeneity, and the need for normalization. Attention to numerous details of sample extraction, storage, and handling has to be paid to ensure intra- and interlaboratory reproducibility *(71)*. DNA-based molecular biomarkers can be more easily transferred from a research laboratory setting into routine diagnostics in a clinic due to the amplifiable and stable nature of DNA. Methylgroups on cytosines are part of the covalent structure of the DNA in contrast to other epigenetic marks such as chromatin. Once methylation is acquired, it is in most cases chemically and biologically stable over time, while expression of mRNA and/or proteins can be modified by non-disease-related environmental conditions and can vary over the cell cycle. As most methods determine the ratio between methylated and unmethylated CpGs, DNA methylation analysis is independent of the total amount of starting material. It provides a binary and positive signal that can be detected independent of expression levels. It is therefore easier to detect than negative signals like loss of heterozygosity. If the core region of a CpG island in a promoter that is controlling transcriptional activity is defined, the stable DNA-based analyte can be used as a proxy to monitor the (re)activation of gene expression during treatment. DNA methylation can be analyzed with a growing number of methods that are amenable to high throughput, and quantitative assays eliminate the need for normalization. They are applicable to formalin-fixed paraffin-embedded clinical specimens and other archived material. Epigenetic changes – similar to genetic alterations – lead to an altered phenotype of a certain cell conferring a selective advantage to those. However, in contrast to genetic DNA-based alterations such as point mutations that can theoretically occur at any position in the coding regions of a gene, DNA methylation changes are always confined to the same small regions of a gene (usually the promoter associated CpG island). Also, the assessment of the methylation status examines the lesion itself, which is the epigenetic inactivation of the promoter rather than the effect of this alteration, such as loss of protein expression or modified enzyme activity. A further advantage is the potential reversal of epigenetic changes by treatment with pharmacological agents, while genetic changes are irreversible *(62)*. One of the most important criteria for a clinically useful biomarker-enabling screening of individuals at potential risk and monitoring of therapy response or disease recurrence is the analysis of the reliable biomarker in surrogate tissues such as blood or body fluids that can be obtained through minimal invasive procedures. Similar approaches based on the detection of RNA have been complicated by the inherent instability of these molecules and difficulties in detecting changes in the level of tumor-derived RNA in the background of a large number of molecules derived from normal cells. The sensitive and specific detection of tumor-specific DNA methylation patterns at distal sites makes DNA methylation a biomarker of choice for the clinical management of cancer patients.

Although epigenetics in general and DNA methylation research in particular are advancing at a breathtaking speed, we are probably only at the tip of the iceberg and we will see a large increase in the number of identified epigenetic changes over the next few years together with first genome-wide mammalian epigenome maps. Elucidation of the epigenetic changes occurring during development, the investigation of their subtle but persistent alteration in response to exposure to environmental and chemical insults at doses far below those required for visible changes of the phenotype in toxicity tests, and analysis and mapping of the changes to the DNA methylation patterns occurring during tumorigenesis and in other pathologies will contribute to enhance our understanding of the importance of epigenetic changes in development and disease. The methods described within this book will help laboratories to set up the technology required to accelerate the identification of key genes that may allow the early detection and monitoring of disease-associated changes in the DNA methylation patterns. This knowledge might ultimately improve existing treatments and create new options to prevent, slow down the progress, or eventually cure some diseases.

References

1. Waddington, C. H. (1942) The epigenotype. *Endeavour* 1, 18–20.
2. Tost, J. (2008) *Epigenetics*, Horizon Scientific Press, Norwich, UK.
3. Kouzarides, T. (2007) Chromatin modifications and their function. *Cell* 128, 693–705.
4. Nightingale, K. P., O'Neill, L. P., Turner, B. M. (2006) Histone modifications: signalling receptors and potential elements of a heritable epigenetic code. *Curr Opin Genet Dev* 16, 125–136.
5. Bird, A. (2002) DNA methylation patterns and epigenetic memory. *Genes Dev* 16, 6–21.
6. Antequera, F. (2003) Structure, function and evolution of CpG island promoters. *Cell Mol Life Sci* 60, 1647–1658.
7. Shen, L., Kondo, Y., Guo, Y., et al. (2007) Genome-wide profiling of DNA methylation reveals a class of normally methylated CpG island promoters. *PLoS Genet* 3, 2023–2036.
8. Takai, D., Jones, P. A. (2002) Comprehensive analysis of CpG islands in human chromosomes 21 and 22. *Proc Natl Acad Sci USA* 99, 3740–3745.
9. Feltus, F. A., Lee, E. K., Costello, J. F., et al. (2006) DNA motifs associated with aberrant CpG island methylation. *Genomics* 87, 572–579.
10. Bock, C., Paulsen, M., Tierling, S., et al. (2006) CpG island methylation in human lymphocytes is highly correlated with DNA sequence, repeats, and predicted DNA structure. *PLoS Genet* 2, e26.
11. Jia, D., Jurkowska, R. Z., Zhang, X., et al. (2007) Structure of Dnmt3a bound to Dnmt3L suggests a model for de novo DNA methylation. *Nature* 449, 248–251.
12. Ulrey, C. L., Liu, L., Andrews, L. G., et al. (2005) The impact of metabolism on DNA methylation. *Hum Mol Genet* 14 Spec No 1, R139–147.
13. Klose, R. J., Bird, A. P. (2006) Genomic DNA methylation: the mark and its mediators. *Trends Biochem Sci* 31, 89–97.
14. Filion, G. J., Zhenilo, S., Salozhin, S., et al. (2006) A family of human zinc finger proteins that bind methylated DNA and repress transcription. *Mol Cell Biol* 26, 169–181.
15. Barreto, G., Schafer, A., Marhold, J., et al. (2007) Gadd45a promotes epigenetic gene activation by repair-mediated DNA demethylation. *Nature* 445, 671–675.
16. Morgan, H. D., Dean, W., Coker, H. A., et al. (2004) Activation-induced cytidine deaminase deaminates 5-methylcytosine in DNA and is expressed in pluripotent tissues: implications for epigenetic reprogramming. *J Biol Chem* 279, 52353–52360.
17. Reik, W., Dean, W., Walter, J. (2001) Epigenetic reprogramming in mammalian development. *Science* 293, 1089–1093.
18. Lees-Murdock, D. J., Walsh, C. P. (2008) DNA methylation reprogramming in the germ line. *Epigenetics* 3, 5–13.
19. Fauque, P., Jouannet, P., Lesaffre, C., et al. (2007) Assisted Reproductive Technology affects developmental kinetics, H19 Imprinting Control Region methylation and H19 gene expression in individual mouse embryos. *BMC Dev Biol* 7, 116.
20. Paoloni-Giacobino, A. (2007) Epigenetics in reproductive medicine. *Pediatr Res* 61, 51R–57R.
21. Niemann, H., Tian, X. C., King, W. A., et al. (2008) Epigenetic reprogramming in embryonic and foetal development upon somatic cell nuclear transfer cloning: focus on mammalian embryogenomics. *Reproduction* 135, 151–163.
22. Fraga, M. F., Esteller, M. (2007) Epigenetics and aging: the targets and the marks. *Trends Genet* 23, 413–418.
23. Geiman, T. M., Robertson, K. D. (2002) Chromatin remodeling, histone modifications, and DNA methylation-how does it all fit together? *J Cell Biochem* 87, 117–125.
24. Ushijima, T. (2005) Detection and interpretation of altered methylation patterns in cancer cells. *Nat Rev Cancer* 5, 223–231.
25. Cameron, E. E., Baylin, S. B., Herman, J. G. (1999) p15(INK4B) CpG island methylation in primary acute leukemia is heterogeneous and suggests density as a critical factor for transcriptional silencing. *Blood* 94, 2445–2451.

26. Pao, M. M., Tsutsumi, M., Liang, G., et al. (2001) The endothelin receptor B (EDNRB) promoter displays heterogeneous, site specific methylation patterns in normal and tumor cells. *Hum Mol Genet* 10, 903–910.
27. Brinkman, A. B., Pennings, S. W., Braliou, G. G., et al. (2007) DNA methylation immediately adjacent to active histone marking does not silence transcription. *Nucleic Acids Res* 35, 801–811.
28. Yoder, J. A., Walsh, C. P., Bestor, T. H. (1997) Cytosine methylation and the ecology of intragenomic parasites. *Trends Genet* 13, 335–340.
29. Reik, W., Walter, J. (2001) Genomic imprinting: parental influence on the genome. *Nat Rev Genet* 2, 21–32.
30. Holmes, R., Soloway, P. D. (2006) Regulation of imprinted DNA methylation. *Cytogenet Genome Res* 113, 122–129.
31. Lewis, A., Reik, W. (2006) How imprinting centres work. *Cytogenet Genome Res* 113, 81–89.
32. Luedi, P. P., Dietrich, F. S., Weidman, J. R., et al. (2007) Computational and experimental identification of novel human imprinted genes. *Genome Res* 17, 1723–1730.
33. Kurukuti, S., Tiwari, V. K., Tavoosidana, G., et al. (2006) CTCF binding at the H19 imprinting control region mediates maternally inherited higher-order chromatin conformation to restrict enhancer access to Igf2. *Proc Natl Acad Sci USA* 103, 10684–10689.
34. Heard, E., Disteche, C. M. (2006) Dosage compensation in mammals: fine-tuning the expression of the X chromosome. *Genes Dev* 20, 1848–1867.
35. Fraga, M. F., Ballestar, E., Paz, M. F., et al. (2005) Epigenetic differences arise during the lifetime of monozygotic twins. *Proc Natl Acad Sci USA* 102, 10604–10609.
36. Friso, S., Choi, S. W., Girelli, D., et al. (2002) A common mutation in the 5,10-methylenetetrahydrofolate reductase gene affects genomic DNA methylation through an interaction with folate status. *Proc Natl Acad Sci USA* 99, 5606–5611.
37. Waterland, R. A., Jirtle, R. L. (2003) Transposable elements: targets for early nutritional effects on epigenetic gene regulation. *Mol Cell Biol* 23, 5293–5300.
38. Tang, W. Y., Ho, S. M. (2007) Epigenetic reprogramming and imprinting in origins of disease. *Rev Endocr Metab Disord* 8, 173–182.
39. Bollati, V., Baccarelli, A., Hou, L., et al. (2007) Changes in DNA methylation patterns in subjects exposed to low-dose benzene. *Cancer Res* 67, 876–880.
40. Feil, R. (2006) Environmental and nutritional effects on the epigenetic regulation of genes. *Mutat Res* 600, 46–57.
41. Anway, M. D., Cupp, A. S., Uzumcu, M., et al. (2005) Epigenetic transgenerational actions of endocrine disruptors and male fertility. *Science* 308, 1466–1469.
42. Cuzin, F., Grandjean, V., Rassoulzadegan, M. (2008) Inherited variation at the epigenetic level: paramutation from the plant to the mouse. *Curr Opin Genet Dev*, 193–196.
43. Rassoulzadegan, M., Grandjean, V., Gounon, P., et al. (2006) RNA-mediated non-mendelian inheritance of an epigenetic change in the mouse. *Nature* 441, 469–474.
44. Chong, S., Whitelaw, E. (2004) Epigenetic germline inheritance. *Curr Opin Genet Dev* 14, 692–696.
45. Robertson, K. D. (2005) DNA methylation and human disease. *Nat Rev Genet* 6, 597–610.
46. Jones, P. A., Baylin, S. B. (2002) The fundamental role of epigenetic events in cancer. *Nat Rev Genet* 3, 415–428.
47. Laird, P. W. (2005) Cancer epigenetics. *Hum Mol Genet* 14 Spec No 1, R65–76.
48. Feinberg, A. P., Vogelstein, B. (1983) Hypomethylation distinguishes genes of some human cancers from their normal counterparts. *Nature* 301, 89–92.
49. Ehrlich, M. (2002) DNA methylation in cancer: too much, but also too little. *Oncogene* 21, 5400–5413.
50. Frigola, J., Song, J., Stirzaker, C., et al. (2006) Epigenetic remodeling in colorectal cancer results in coordinate gene suppression across an entire chromosome band. *Nat Genet* 38, 540–549.
51. Stransky, N., Vallot, C., Reyal, F., et al. (2006) Regional copy number-independent deregulation of transcription in cancer. *Nat Genet* 38, 1386–1396.
52. Hedenfalk, I., Duggan, D., Chen, Y., et al. (2001) Gene-expression profiles in hereditary breast cancer. *N Engl J Med* 344, 539–548.
53. Knudson, A. G. (2001) Two genetic hits (more or less) to cancer. *Nat Rev Cancer* 1, 157–162.
54. Balmain, A., Gray, J., Ponder, B. (2003) The genetics and genomics of cancer. *Nat Genet* 33 Suppl, 238–244.
55. Costello, J. F., Frühwald, M. C., Smiraglia, D. J., et al. (2000) Aberrant CpG-island methylation has non-random and tumour-type-specific patterns. *Nat Genet* 24, 132–138.

56. Goelz, S. E., Vogelstein, B., Hamilton, S. R., et al. (1985) Hypomethylation of DNA from benign and malignant human colon neoplasms. *Science* 228, 187–190.
57. Issa, J. P., Ahuja, N., Toyota, M., et al. (2001) Accelerated age-related CpG island methylation in ulcerative colitis. *Cancer Res* 61, 3573–3577.
58. Laird, P. W. (2003) Early detection: the power and the promise of DNA methylation markers. *Nat Rev Cancer* 3, 253–266.
59. Silva, J. M., Dominguez, G., Garcia, J. M., et al. (1999) Presence of tumor DNA in plasma of breast cancer patients: clinicopathological correlations. *Cancer Res* 59, 3251–3256.
60. Fleischhacker, M., Schmidt, B. (2007) Circulating nucleic acids (CNAs) and cancer – a survey. *Biochim Biophys Acta* 1775, 181–232.
61. Tost, J. (2007) Analysis of DNA methylation patterns for the early diagnosis, classification and therapy of human cancers, in (Kobayashi, T. B., ed.), *DNA Methylation Research Trends*, pp. 87–133, Nova Science Publishers, Hauppauge, NY.
62. Yoo, C. B., Jones, P. A. (2006) Epigenetic therapy of cancer: past, present and future. *Nat Rev Drug Discov* 5, 37–50.
63. van Vliet, J., Oates, N. A., Whitelaw, E. (2007) Epigenetic mechanisms in the context of complex diseases. *Cell Mol Life Sci* 64, 1531–1538.
64. Abdolmaleky, H. M., Cheng, K. H., Faraone, S. V., et al. (2006) Hypomethylation of MB-COMT promoter is a major risk factor for schizophrenia and bipolar disorder. *Hum Mol Genet* 15, 3132–3145.
65. Nagarajan, R. P., Hogart, A. R., Gwye, Y., et al. (2006) Reduced MeCP2 expression is frequent in autism frontal cortex and correlates with aberrant MECP2 promoter methylation. *Epigenetics* 1, 172–182.
66. Balada, E., Ordi-Ros, J., Vilardell-Tarres, M. (2007) DNA methylation and systemic lupus erythematosus. *Ann NY Acad Sci* 1108, 127–136.
67. Neidhart, M., Rethage, J., Kuchen, S., et al. (2000) Retro transposable L1 elements expressed in rheumatoid arthritis synovial tissue: association with genomic DNA hypomethylation and influence on gene expression. *Arthritis Rheum* 43, 2634–2647.
68. Junien, C., Nathanielsz, P. (2007) Report on the IASO Stock Conference 2006: early and lifelong environmental epigenomic programming of metabolic syndrome, obesity and type II diabetes. *Obes Rev* 8, 487–502.
69. Tufarelli, C., Stanley, J. A., Garrick, D., et al. (2003) Transcription of antisense RNA leading to gene silencing and methylation as a novel cause of human genetic disease. *Nat Genet* 34, 157–165.
70. Issa, J. P. (2004) CpG island methylator phenotype in cancer. *Nat Rev Cancer* 4, 988–993.
71. Simon, R. (2005) Roadmap for developing and validating therapeutically relevant genomic classifiers. *J Clin Oncol* 23, 7332–7341.

Part II

Global Methylation Levels

Chapter 2

Quantification of Global DNA Methylation by Capillary Electrophoresis and Mass Spectrometry

María Berdasco, Mario F. Fraga, and Manel Esteller

Abstract

Two approaches for the evaluation of the relative degree of global DNA methylation through the quantification of 2′ deoxynucleosides are described. Detection and quantification of 5-methyl 2′-deoxycytidine in genomic DNA is performed using both high-performance capillary electrophoresis (HPCE) with UV–Vis detection or liquid chromatography with electrospray ionization mass spectrometric detection (LC-ESI/MS). Treatment of genomic DNA with a ribonuclease and generation of nucleosides through enzymatic hydrolysis notably increases the specificity of both techniques. Both approaches have been demonstrated to be highly specific and sensitive, being useful for the rapid quantification of the degree of global DNA methylation and its exploitation for the analysis of poorly purified and/or concentrated DNA samples, such as tumor biopsies.

Key words: Capillary electrophoresis, mass spectrometry, global DNA methylation, 2′-deoxynucleosides, 5-methyl 2′ deoxycytidine.

1. Introduction

DNA methylation research can be approached from several standpoints since there are a wide range of techniques available for the study of the occurrence and localization of methylcytosine in the genome *(1)*. Each technique has its own peculiarities implying that there is a best-suited technique for each specific problem. The available methods for studying the degree of DNA methylation can be classified with respect to the type of information they produce: the degree of global genomic DNA methylation, the DNA methylation status of specific sequences, and the discovery of new methylation hot spots. With respect to the quantification of global levels of methylcytosine in the genomic

Jörg Tost (ed.), *DNA Methylation: Methods and Protocols, Second Edition, vol. 507*

DOI 10.1007/978-1-59745-522-0_2 Springerprotocols.com

DNA, measurements can be performed by high-performance separation techniques or by enzymatic/chemical means. The latter are never as sensitive as the former and sometimes their resolution is restricted to endonuclease cleavage sites *(2)*. Despite the drawbacks, enzymatic/chemical approaches are still commonly used since, unlike separation techniques, they do not require expensive and complex equipment, which is not always available. Although almost all actual efforts are focused on the characterization of the gene-specific methylation patterns or the construction of DNA methylation maps of the entire genome (*methylome*), global measurements of DNA methylation remain a valuable tool for understanding the molecular pathology of human cancer, for measuring the potential effect of tumor-preventive or -promoting compounds, and for monitoring therapeutic responses to hypomethylating agents undergoing evaluation in human clinical trials *(3)*.

Among high-performance separation techniques, capillary electrophoresis (HPCE) and liquid chromatography (HPLC) are used most frequently. The development of capillary electrophoretic (CE) techniques, based on the separation of molecules by the use of a narrow-bore fused-silica capillary, has given rise to a methodological approach that has several advantages over other current methodologies used for the separation of various DNA components, including a number of base adducts *(4)*. Molecules are separated on the basis of differences in size, charge, structure, and hydrophobicity under application of specific and strong voltages. CE has been shown to be extremely useful for the quantification of the extent of DNA methylation. Due to the sensitivity, specificity, and economy of these methods, HPCE had taken an advantage with regard to HPLC-based methods during the last years. However, the application of HPLC methods for the study of global DNA methylation has recently been enforced with the development of mass spectrometry (MS). LC/MS refers to the combination of liquid chromatographic (LC) separation with MS detection. The combination of these two powerful techniques enables the analysis of a great number of molecules, due to the resolution of each technique. In this way, it has been estimated that LC provides a consistent mechanism for the separation of molecules in over 80% of known organic species *(5)*. In addition, MS is a useful tool to provide information about structure, molecular weight, or the empirical formula about a specific analyte. The development of electrospray ionization enables LC/MS to be utilized for the quantitative determination and structural characterization of a great number of polar/ionic molecules, such as nucleic acids, in biological samples *(6)*.

2. Materials

All enzymes and reagents are available from Sigma–Aldrich if not otherwise stated.

2.1. Enzymes

1. Ribonuclease A (RNase A),
2. Nuclease P1: 200 U/mL in 30 mM sodium acetate, and
3. Alkaline phosphatase: 50 U/mL in 2.5 M ammonium sulphate.

2.2. Buffers and Other Reagents

1. 10 mM zinc sulphate,
2. 0.5 M Tris–HCl, pH 8.3,
3. Ethanol, and
4. 2-Isopropanol.

2.3. High-Performance Capillary Electrophoresis (HPCE)

1. 14 mM sodium bicarbonate (pH 9.6, equilibrated with 0.1 M sodium hydroxide) containing 20 mM sodium dodecyl sulphate (SDS),
2. 0.1 M sodium hydroxide,
3. 0.45-μm filters (Sartorius, Göttingen, Germany), and
4. Uncoated fused-silica capillary of 60.2 cm × 100 cm, with an effective length of 50 cm (Waters Chromatography S.A., Madrid, Spain).

2.4. High-Pressure Liquid Chromatography (HPLC)

1. 0.1% formic acid (HPLC grade) in water and
2. 0.1% formic acid in 50% water:50% methanol (HPLC grade).

2.5. Nucleotide Standards

All nucleosides standards are dissolved at 5 mM in Milli-Q grade water.

1. 2′-deoxyadenosine 5′monophosphate (dA),
2. 2′-deoxythymidine 5′monophosphate (dT),
3. 2′-deoxyguanosine 5′monophosphate (dG),
4. 2′-deoxycytidine 5′monophosphate (dC), and
5. 5-methyl 2′-deoxycytidine 5′monophosphate (5mdC).

2.6. Equipment

1. A HPCE P/ACE MDQ system (Beckman-Coulter, Fullerton, CA, USA) connected to a data-processing station (32 Karat™ Software);
2. An Agilent Serie 1100 HPLC system (Agilent Technologies, Palo Alto, CA, USA) equipped with an online vacuum-degassing system, a quaternary pumping system, an autosampler with internal refrigeration and ultraviolet and visible lamps for variable wavelength detection;
3. Reverse-phase column Atlantis dC18 column (2.1 × 150 mm; 5 μm particle size);

4. Guard column (2.1 × 20 mm; 5 μm particle size, Agilent); and
5. An Agilent LC/MSD VL MS equipped with an electrospray ionization source (Agilent) coupled to the HPLC system.

3. Methods

In this chapter, we describe two different approaches for the separation of nucleosides: a HPCE-based method and a HPLC-based method. As shown in **Fig. 2.1**, the first steps and the relative quantification of global DNA methylation signals are shared between both techniques.

3.1. Genomic DNA Extraction and RNase Treatment

DNA from animal tissues is extracted by standard methods *(7)*. It is important to obtain high-purity DNA to assure an effective

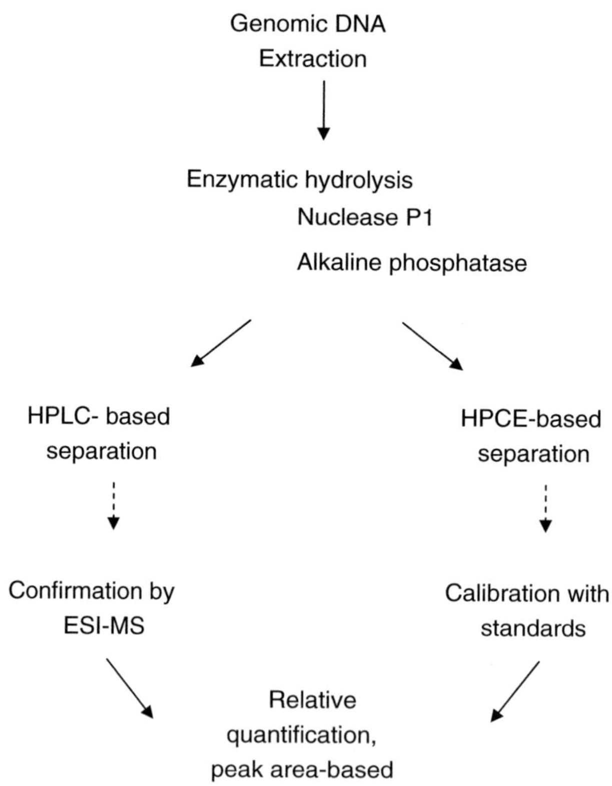

Fig. 2.1. Simplified representation of the two alternative procedures described in this chapter, which are used for separation of DNA nucleosides and quantification of global DNA methylation levels. After enzymatic hydrolysis of genomic DNA, nucleosides could be separated by high-performance capillary electrophoresis (HPCE) or liquid chromatography coupled to an electrospray ionization mass spectrometry (LC-ESI/MS). In both cases, relative quantification of 5-methyl-2′-deoxycytidine (5mdC) levels are extrapolated from HPCE or HPLC chromatograms.

action of the next steps of the protocol. A potential problem in the measurement of genomic DNA methylation is interference from RNA contamination (*see* **Note 1**); therefore, treatment with a ribonuclease is recommended before DNA hydrolysis.

1. Add RNase A to a final concentration of 20 μg/μL. Mix gently and incubate the mixture at 37°C for 30 min.
2. Following the incubation, add an equal volume of cold 2-isopropanol and mix thoroughly in order to enhance genomic DNA precipitation.
3. Centrifuge for 10 min at 11,000*g* and carefully decant the supernatant.
4. Wash the DNA pellet by adding cold 70% ethanol. Centrifuge for 5 min at 11,000*g* and resuspend the resulting pellet in Milli-Q grade water. Genomic DNA can be stored at 4°C till used.

3.2. DNA Hydrolysis

1. Prepare DNA samples (2–7 μg) in 10 μL of total volume. If necessary, dilute the samples in distilled water.
2. Denature the samples by heating for 2 min in a boiling water bath and cool rapidly in ice for 5 min.
3. Add nuclease P1 to a final concentration of 1.5 μg/μL and zinc sulphate to a final concentration of 1 mM (*see* **Note 2**). Incubate overnight at 37°C.
4. Add 0.75 μL of alkaline phosphatase and 1.25 μL of 0.5 M Tris–HCl, pH 8.3 (*see* **Note 2**). Incubate the mixtures for 2 h at 37°C.
5. In order to eliminate any solid residue, centrifuge samples at 10,000*g* for 3 min. Supernatant must be stored at 4°C till used.

3.3. Nucleoside Separation by High-Performance Capillary Electrophoresis (HPCE)

We have previously described the quantification of the relative methylcytosine content of the genomic DNA using a HPCE system to analyze hydrolyzed genomic DNA *(8, 9)*. In this context, separation and quantification of cytosine and methylcytosine is only possible by the use of a sodium dodecylsulphate (SDS) micelle system. This method is faster than HPLC (taking less than 10 min per sample) and is also reasonably inexpensive since it does not require continuous running buffers and displays a great potential for fractionation (up to 10^6 theoretical plates). Nevertheless, no or almost no preparative analyses are possible with HPCE systems because of the low injection volumes.

For the separation of nucleosides after genomic DNA hydrolysis, the following procedure must be applied:

1. Before each run, prepare all buffers and washing solutions with Milli-Q water and filter them through 0.45-μm filters (*see* **Note 3**).
2. Condition the capillary system just before each run by washing with 0.1 M NaOH for 3 min.

3. After washing, equilibrate the capillary system with the running buffer for 3 min. The optimal running buffer is 14 mM sodium bicarbonate, pH 9.6 containing 20 mM SDS, which allows for the micelle formation of the nucleosides.
4. Filtered hydrolyzed samples (*see* **Section 3.2**) through 0.45-μm pore filters.
5. Inject samples under pressure (0.3 psi) for 3 s. Running conditions, optimized in *(9)* consist of a temperature of 25°C and an operating voltage of 17 kV (*see* **Note 3**). Absorbance is monitored at 254 nm. **Figure 2.2** shows a representative electropherogram obtained for standard nucleosides and the DNA extracted from a human tumor cell line.

3.4. Nucleoside Separation by HPLC and Detection of Nucleosides Peaks by ESI/MS

The basic principles of both techniques are represented in **Fig. 2.3.** The separation mechanism in reverse phase (RP)-HPLC depends on the hydrophobic-binding interaction between the solute molecule of the sample in the mobile phase and the

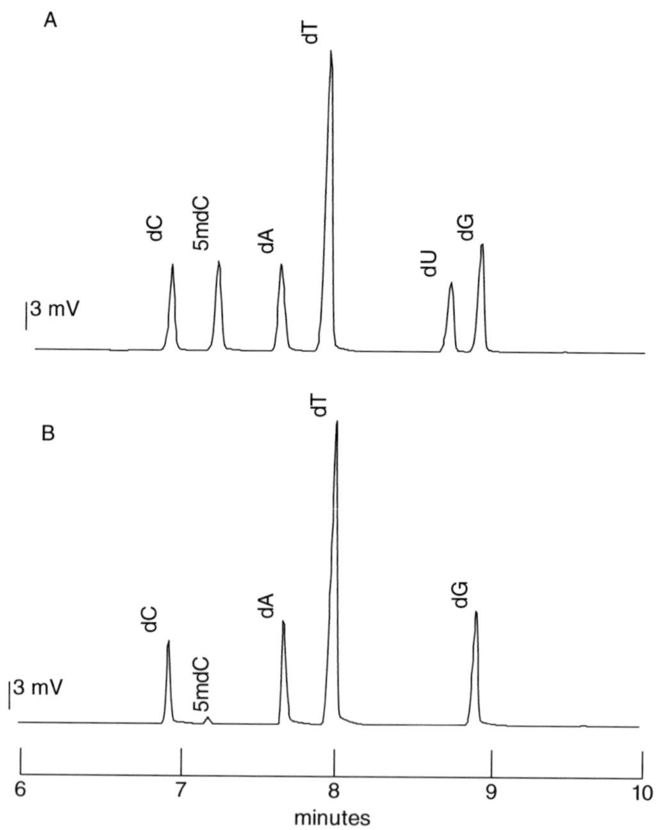

Fig. 2.2. Separation of nucleosides by HPCE. (**A**) Electropherogram for standard nucleosides (dC, 5mdC, dA, dT, and dG) dissolved in Mili-Q grade water at 5 mM. (**B**) Resolution of nucleosides obtained from enzymatic hydrolysis of genomic DNA from a human tumor cell line. Analytical conditions are described in **Section 3.3**.

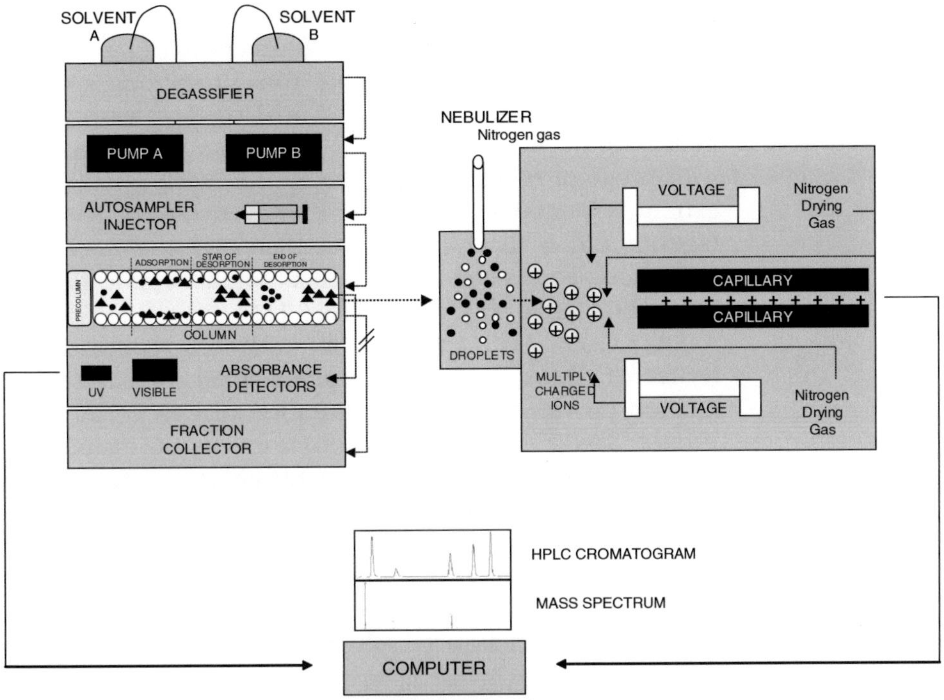

Fig. 2.3. Representative diagram of a LC-ESI/MS apparatus. First, samples are introduced into a HPLC system and analytes are separated in function of their individual hydrophobicity under specific conditions in a reverse-phase column. Then, the resulting mobile phase with the eluted molecules is introduced into the ESI/MS apparatus and subjected to fragmentation, ionization, and desorption processes under a constant nitrogen flow. HPLC and ESI/MS modules are connected to a computer, allowing the combined representation of HPLC chromatograms and mass spectra.

immobilized hydrophobic ligand (stationary phase) that constitutes the column. The capacity of solute molecule binding to the stationary phase can be controlled by manipulation of the hydrophobic properties of the mobile phase. The initial mobile phase-binding conditions used in RP-HPLC are primarily aqueous allowing the formation of a structured layer of water around both the matrix and the analyte. The sample must be applied to the column in such a flow rate that allows the optimal adsorption of the sample components. Transport and elution of analytes is achieved by increasing the concentration of the organic component in the mobile phase. Once the molecules are eluted from the column they get introduced into the electrospray system of the mass spectrometer. At this point, it is important to note that buffers must be free of salts, which could potentially damage the mass spectrometer. The electrospray ionization (ESI) and atmospheric pressure chemical ionization (APCI) methods are the major techniques based on the atmospheric pressure ionization (API). In the ESI, both solvent and sample are nebulized with the help of a gas stream and broken into small droplets. The mobile phase solvent evaporates from the droplets (desolvation). Droplets undergo Coulomb explosions when the charge density increases until the Raleigh limit (10^8 V/cm^3) and new smaller droplets are formed. Ions in solution are desorbed under the influence of high potential of the electrospray fields in the spray chamber. The ESI technique can be applied to a wide range of molecule sizes, except for small (<1000 mw) and extremely nonpolar molecules. However, one of the disadvantages of ESI is that the solution chemistry could influence the ionization process and some adducts could be generated, such as $[M + H]^+$, $[M + Na]^+$, and $[2M + H]^+$.

A LC-ESI/MS approach for analyzing enzymatic hydrolysates of DNA was previously described *(6)*. Although this method provided a good quantitative analysis of DNA methylation in less than 15 min, conditions for the LC included buffers with ammonium salts which are inconvenient for the maintenance of the LC-ESI/MS system and also favor the production of single ammonium adducts in the ESI/MS. Here we describe a protocol in which adequate separation of the DNA and RNA components is achieved within 25 min. Buffers without salts are employed, making the direct flow of solvents from LC to ESI/MS system feasible.

LC-ESI/MS conditions required for the analysis of the 2′-deoxyribonucleotide-5′-monophosphate levels are as follows:

1. Before each run, equilibrate the HPLC column with the running buffer. The mobile phase consists of two buffers: 0.1% formic acid in water (Solvent A) and 0.1% formic acid in 50% water:50% methanol (Solvent B) (*see* **Note 4**). Equilibration must be done by maintaining the initial conditions, 95% Solvent A–5% Solvent B in an isocratic mode during 5 min at constant flow of 0.220 mL/min. The employed Atlantis dC18 column permits to minimize the loss in retention in a 100%-aqueous mobile phase *(3)*. It is strongly recommended to protect the column by the use of a guard column (*see* **Note 3**).
2. Dilute the hydrolyzed DNA (*see* **Section 3.2**) in water to a final volume of 50 μL and filter it through a 0.45-μm pore filter just before injection (*see* **Note 3**).
3. HPLC separation must be performed with an initial gradient of 5% solvent B, then an increase of solvent B to 50% within 9 min and an isocratic gradient (50% of solvent B) during 25 min. The acquisition of HPLC signals is obtained by UV detection at 254 nm and 280 nm. It is important to point out that the HPLC separation under the previously described conditions is achieved in solvents without salt compounds. As a consequence, no desalting before the entry of the solvents into the ESI/MS is needed.
4. Source conditions for ESI/MS are as described in *(6)*, with minor modifications. A drying gas flow of 10.0 L/min was

employed, with auxiliary 35 psis gas to assist nebulization and a drying temperature of 350°C. The mass spectrophotometer was operated at a capillary voltage of 4,000 V, and spectra were collected in positive ion mode.

After 14 min, all the DNA and RNA compounds are completely separated as shown in the LC chromatogram (**Fig. 2.4**). The ESI/MS spectra are used to verify the identity of each HPLC peak used for the estimation of the DNA methylation levels. As expected, the ESI source with the mass spectrometer in positive ion detection mode shows protonated molecules as well as fragments ions and other known adducts derived from nucleosides. **Figure 2.4** shows the LC chromatogram and the product ion spectra of the five deoxyribonucleosides (5mdC, dC, dG, dA, and dT) and the five ribonucleosides (5mC, C, G, A, and U) after hydrolysis of a 4 μg of a tumor sample without RNase treatment during nucleic acid extraction. The transitions pairs of *m/z* 242.1/126.1, 228.1/112.1, 268.1/152.1, 252.1/136.0, and 243.1/127.0 corresponded to 5mC, 5mdC, dC, dG, dA, and T, respectively, while 258.1/126.0, 244.1/112.1, 284.1/152.2, 268.1/136.1, and 245.1/113.0 were acquired for 5mC, C, G, A, and U, respectively. The presence of T and U in the LC chromatogram is less prominent than the other nucleosides,

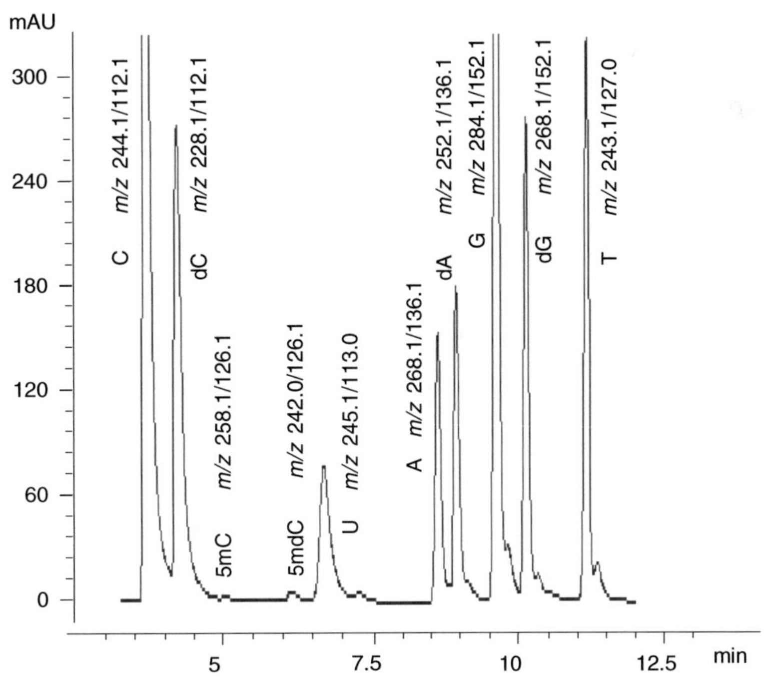

Fig. 2.4. LC-ESI/MS chromatogram and specific product ions of 10 nucleosides corresponding to the DNA and RNA compounds. DNA hydrolysis was carried out from 4 μg of DNA from a tumor cell line without RNase treatment. LC and ESI/MS conditions are described in **Section 3.4.**

which may be attributed to the weaker proton affinity of these nucleosides.

In the case of RNase-treated samples, the chromatogram shows only peaks corresponding to the five deoxyribonucleotides (**Fig. 2.5**). The HPLC peak eluting after 4.0±0.5 min corresponds to 2′-deoxycytidine (dC), and the HPLC peak eluting after 5.5±0.5 min correspond to 5-methyl-2′-deoxycytidine (5mdC). **Figure 2.5B** and **C** report the full-scan spectra (ESI/MS spectra) of dC and 5mdC, respectively. The $[M + H]^+$ adduct appears at *m/z* 228.1 and 242.1 for dC and 5mdC, respectively. Also present are the [2M] and $[2M + H]^+$ adducts at *m/z* 455.1 and 456.0 for dC and *m/z* 483.1 and 484.0 for 5mdC, respectively. In some samples, the $[M + 23]^+$ and the $[2M + 23]^+$adducts can also be found, which correspond to sodium adducts. It is important to point out that sodium adducts are frequently detected in ESI mass spectra of organic compounds, because they are normal compounds of glass vials used for HPLC

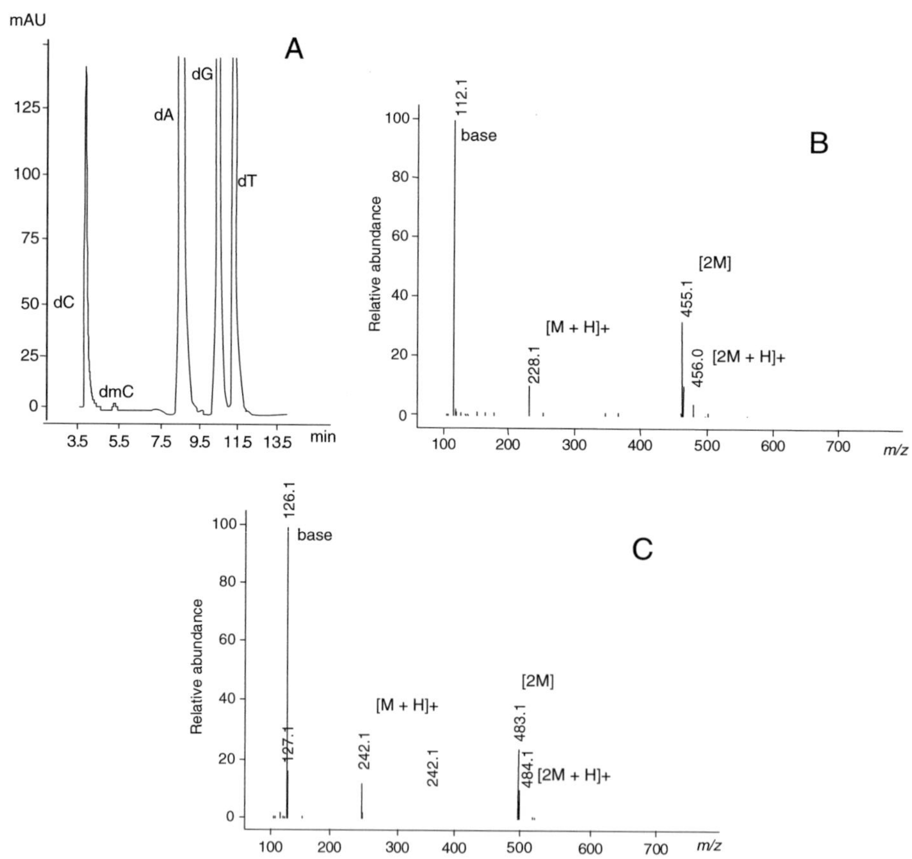

Fig. 2.5. LC-ESI/MS chromatogram of a human lymphocyte sample containing 3 μg of RNA-free genomic DNA. (**A**) Separation of the five deoxynucleosides in the HPLC chromatogram obtained by UV detection at an absorbance of 254 nm. (**B** and **C**) Full spectra obtained in ESI/MS for 2′-deoxycytidine (dC) and 5-methyl-2′-deoxycytidine (5mdC), respectively. LC and ESI/MS conditions are described in **Section 3.4**.

separation. After the separation of the DNA bases the fragmentation conditions established in the ESI/MS cause the separation of the pentose moiety from the pyrimidine ring of both dC and 5mdC resulting in the production of cytosine (*m/z* 112.1) and 5-methylcytosine (*m/z* 126.1). In this way, conditions for ESI/MS can be optimized to change the intensity of $[M + H]^+$adducts with respect to the formation of dimers, sodium adducts, and nitrogen bases *(3)*.

3.5. Quantification

To determine the 5mdC abundance, the percentages of global genomic DNA methylation are calculated by integration of the peak areas of 5mdC relative to global cytidine (methylated or not). Area peaks are obtained directly from HPCE or HPLC chromatograms, depending of the selected approach. The following equation was used in both cases: 5mdC peak area × 100/(dC peak area + mdC peak area).

4. Notes

The most common considerations for preventing failures in the separation of nucleosides by HPCE and HPLC techniques which could influence results are

1. One of the major problems of this technique is the incomplete digestion of RNA compounds. As the estimation of global DNA methylation is based on a relative index between methylated and unmethylated cytidines, this index could be underestimated in the presence of RNA compounds. Treatment with a ribonuclease assures the fidelity of the results as shown in **Fig. 2.5**.
2. Adjustment of the pH and molarity of the Tris and sulphate buffers is important to assure the complete and specific DNA hydrolysis. Unspecific hydrolysis could influence results, especially for the HPCE technique.
3. Temperature and voltage are the two main variables that determine the best separation of the nucleosides. Small particles can permanently block the capillaries. It is important to use filtered solvents always both for the HPCE and the HPLC method. Furthermore, in HPLC the employment of precolumns is strongly recommended. If not, the pressure of the system might not be constant and the resolution of the method might noticeably decrease. The temperature must be lower than 30°C for HPCE and column temperature should be controlled in the HPLC.
4. Solutions of organic acids, such as formic acid, in organic solvents act as corrosive factors of all steel components. Although the HPLC method uses a low concentration, a 0.1% solution of formic acid in methanol, the acid can attack steel.

Consequently, it is important to remove the running buffers by washing the system with methanol: water solutions before switching off the apparatus. However, a low concentration of acid is necessary for the positive ion mode detection in the mass spectrometry.

References

1. Esteller, M. (2007) Cancer epigenomics: DNA methylomes and histone-modification maps. *Nat Rev Genet* 8, 286–298.
2. Cañal, M. J., Fraga, M., Berdasco, M., et al. (2004) Epigenetics, the role of DNA methylation. *Curr Top Plant Biol* 4, 193–203.
3. Song, L., James, S.R., Kazim, L., et al. (2005) Specific method for the determination of genomic DNA methylation by liquid chromatography – electrospray ionization tandem mass spectrometry. *Anal Chem* 77, 504–510.
4. Norwood, C. B., Jackim, E., Cheer, S. (1993) DNA adduct research with capillary electrophoresis. *Anal Biochem* 213, 194–199.
5. Snyder, L. R, Kirkland, J. J. (1991) *Introduction to Modern Liquid Chromatography*. Wiley, New York.
6. Friso, S., Choi, S. W., Dolnikowski, G. G., et al. (2002) A method to assess genomic DNA methylation using high-performance liquid chromatography/electrospray ionization mass spectrometry. *Anal Chem* 74, 4526–4531
7. Maniatis, T., Fristsch, E. E., Sambrook, J. (1982) *Molecular Cloning: A Laboratory Manual*, Cold Spring Harbor Laboratory Press, Cold Spring Harbor, New York.
8. Fraga, M. F., Rodríguez, R., Cañal, M. J. (2000) Rapid quantification of DNA methylation by high performance capillary electrophoresis. *Electrophoresis* 21, 2990–2994.
9. Fraga, M. F., Uriol, E., Diego, L. B., et al. (2002) High-performance capillary electrophoretic method for the quantification of 5-methyl 2′-deoxycytidine in genomic DNA: application to plant, animal and human cancer tissues. *Electrophoresis* 23, 1677–1681.

Chapter 3

Methyl Group Acceptance Assay for the Determination of Global DNA Methylation Levels

Kenneth P. Nephew, Curt Balch, and David G. Skalnik

Abstract

DNA methylation levels are affected by numerous environmental influences, including diet and xenobiotic exposure, and neoplasia has been firmly associated with genomic hypomethylation and localized hypermethylation of tumor suppressor genes. To reverse methylation-induced gene repression, DNA hypomethylating agents are currently in clinical trials for various malignancies, with two of these now approved for the therapy of myelodysplastic syndrome, and the efficacy of these drugs can be assessed by the monitoring of global DNA methylation levels. Herein, we outline a simple, well-established method for the evaluation of genomic DNA methylation levels, based on the ability of isolated DNA to "accept" radiolabeled methyl groups from *S*-[^{3}H-methyl] adenosylmethionine, using the bacterial CpG methyltransferase SssI. As this enzyme methylates all unmethylated CpG dinucleotides in the genome, radiolabeled methyl group acceptance is inversely proportional to the level of preexisting methylation. This assay is applicable to a number of translational and basic research questions.

Key words: Genomic methylation, 5-methylcytosine, methyl acceptance, hypomethylation, hypermethylation, methyl transfer, methylation.

1. Introduction

Methylation of the pyrimidine deoxycytosine, within the dinucleotide CpG and resulting in 5-methylcytosine (5-mC), is the best-known epigenetic modification to DNA, and 5-mC is often referred to as a minor DNA base *(1)*. DNA methylation is strongly associated with transcriptional silencing, and normally acts to silence imprinted alleles and parasitic DNA elements, such as retrotransposons *(2)*. While the majority of CpG-restricted cytosines are methylated, distinct CG-rich regions known as CpG islands, are often found unmethylated. CpG islands are typically

Jörg Tost (ed.), *DNA Methylation: Methods and Protocols, Second Edition, vol. 507*

DOI 10.1007/978-1-59745-522-0_3 Springerprotocols.com

associated with the promoters and first exons of active genes *(3)*, although it remains unclear whether active transcription is directly responsible for the prevention of DNA methylation *(4–6)*. In tumors, however, this normal methylation state is found reversed, that is, with overall CpG hypomethylation and local hypermethylation, often within CpG islands associated with tumor suppressor genes *(7)*. Methylation levels have also been demonstrated to be altered by various xenobiotics and dietary factors *(8, 9)*, and altered metabolism of folic acid, the precursor to *S*-adenosylmethionine (the methyl donor for DNA methylation), has likewise been associated with a number of malignancies and developmental disorders *(10, 11)*.

To reverse CpG island methylation in cancer and various hematological disorders, a number of deoxycytosine analogs are currently being investigated for their ability to derepress epigenetically silenced genes. These analogs possess various structural modifications that preclude methylation at their 5-carbon positions, and two of these, 5-azacytidine (Vidaza) and 2′-deoxy-5-azacytidine (decitabine, Dacogen), are now approved for the therapy of the hematological disorder myelodysplastic syndrome *(12)*. To assess the bioactivity of these hypomethylating agents, the methylation status of peripheral blood mononuclear cells (PBMCs) is often monitored before and after treatment, and it has now been demonstrated that PBMC DNA demethylation is associated with clinical response in various hematologic malignancies *(13)*.

For assessing global DNA methylation levels, a number of methods have now been developed, including high-performance liquid chromatography (HPLC) or capillary electrophoresis (HPCE) *(14)*, **Chapter 2**). However, these two methods require expensive equipment and complete DNA hydrolysis to single mononucleotides. Other methods utilize an ELISA-based approach in concert with a highly specific monoclonal antibody *(15)* and chloroacetaldehyde labeling of bisulfite-treated DNA *(16)*. However, the latter method requires a depurination step, chloroacetaldehyde detoxification, and access to an expensive luminescence spectrometer *(16)*. One of the most inexpensive and straightforward methods was developed by Balaghi and Wagner, which employs the bacterial methyltransferase, CpG methylase, to transfer radiolabeled methyl groups to isolated DNA *(17)*. While this method requires a scintillation counter, the vast majority of institutions possess that instrument used for numerous applications involving radiolabel quantification. This "methyl acceptance assay" has been used for examining the relationship of folate status to rodent mammary tumorigenesis *(18)* and human colonic hypomethylation *(19)*, to investigate the role of a CpG-binding protein in mouse embryogenesis (**Fig. 3.1**) *(20)*, and to compare hypomethylating agents in ovarian cancer cells (**Fig. 3.2**) *(21)*. In other studies, methyl acceptance was used to

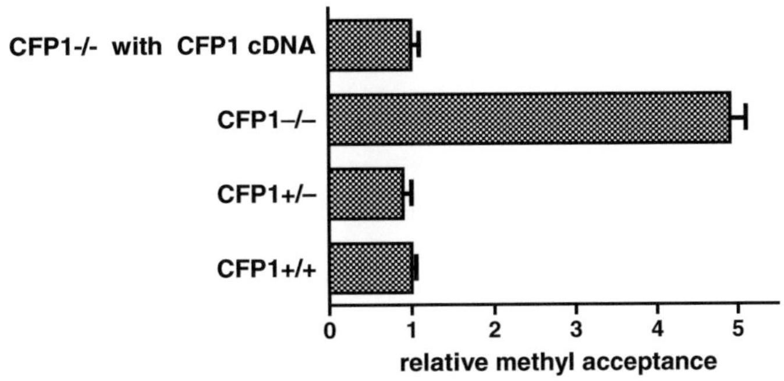

Fig. 3.1. Effects of CpG-binding protein (CFP1) knockout on genomic methylation levels in embryonic stem (ES) cells (adapted from *(20)*). Genomic ^{3}H-methyl acceptance was determined for DNA isolated from ES cell lines established from pregnant female CFP1$^{+/-}$ mice mated with CFP1$^{+/-}$ males. As shown, complete CFP1 knockout (CFP1$^{-/-}$) resulted in a significant ($p < 0.05$) increase in methyl acceptance (and thus, significant genomic demethylation). To rescue methyltransferase activity, CFP1$^{-/-}$ cells were also transfected with CFP1-encoding cDNA. This result demonstrates that CFP1 plays a vital role in mediating genomic DNA methylation during embryogenesis.

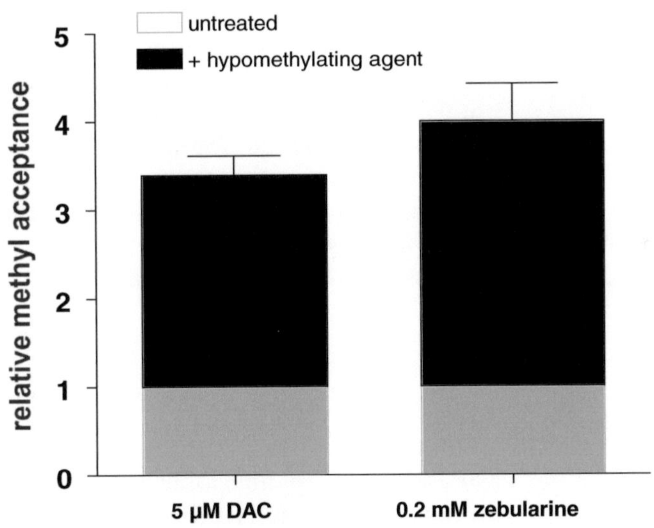

Fig. 3.2. Effects of hypomethylating agents on methyl acceptance in cancer cells (adapted from *(21)*). Bar graphs show ^{3}H-methyl acceptance by DNA isolated from ovarian carcinoma CP70 cells without treatment ("untreated", gray portions of bars) or treated with the methylcytosine analogs 2′-deoxy-5-azacytidine (DAC) or zebularine ("+ hypomethylating agent", black portions of bar). As shown, both agents significantly increased methyl acceptance in these cells; there was no statistical ($p > 0.05$) difference between the two drugs.

demonstrate arsenic-induced global demethylation in the livers of rats fed that specific toxin *(22)*, and to show global hypomethylation in human (potentially premalignant) uterine leiomyomas, as compared to the normal adjacent myometria *(23)*. This chapter describes in detail this laboratory genomic methylation assessment that is pertinent to a number of basic and translational research applications.

2. Materials

2.1. Nonstandard Laboratory Equipment

1. Vacuum filtration manifold (e.g., Millipore Model 1220) capable of holding 25-mm diameter filter disks.

2.2. Reagents

1. Desired source of DNA, for example, tumors, peripheral blood mononuclear cells, or tissue culture cells.
2. DNA isolation kit, such as DNeasy (Qiagen, Valencia, CA) or NucleoSpin (Clontech, Mountain View, CA). Alternatively, any standard DNA purification procedure may be used.
3. *S*-[^{3}H-methyl] adenosyl-L-methionine, 5–15 Ci/mmol (Perkin–Elmer, Waltham, MA). Store at −20°C in a non-frost-free freezer.
4. DE81 DEAE-cellulose ion-exchange filters (Whatman, Florham Park, NJ).
5. CpG methylase (M. SssI) (4 or 20 units/μL), supplied with its 10 × reaction buffer (0.1 M Tris–HCl, 0.5 M NaCl, 0.1 M $MgCl_2$, 10 mM dithiothreitol, pH 7.9) (New England Biolabs, Beverly, MA). Store both tubes at −20°C.
6. Wash buffer: 0.5 M phosphate buffer, pH 7.0.
7. Scintillation fluid, for example, CytoScint (MP Biochemicals, Solon, OH).
8. 6 or 8 mL scintillation vials (Perkin–Elmer).
9. 70% ethanol (optional).
10. Absolute ethanol (optional).

3. Method

The methyl acceptance assay exploits the genome-wide methylation of all CpG dinucleotides by the bacterial enzyme SssI (CpG methylase), using radiolabeled methyl groups donated by *S*-[^{3}H-methyl] adenosylmethionine (^{3}H-SAM). The labeled DNA is then immobilized on positively charged filter disks (DE81) for washing and quantification of ^{3}H incorporation. Consequently, the critical steps of the procedure are the incubation of the enzymatic reaction and adequate washing of filter-bound, labeled DNA.

1. Purify DNA to be assayed using a commercially available kit or standard purification procedure that includes incubation with proteinase K. DNA should be ≥0.5 mg/mL with an A_{260}/A_{280} ratio ≥1.8.
2. Methyl acceptance samples should be prepared in duplicate, in a total volume of 30 μL, as outlined below (*see* **Notes 1** and **2**):

DNA to be assessed (0.5 μg)
10 × SssI buffer, 3 μL
^{3}H-SAM 2.0 μCi
SssI (10 U/μL), 0.3 μL (3 U)
deionized H_2O to 30 μL

3. Incubate at 37°C for a minimum of 60 min.
4. Incubate at 65°C for 15 min to inactivate the enzyme.
5. Slowly spot the entire 30-μL reaction mixture onto the center of a 2.3-cm diameter DE81 filter disk placed on a paper towel. Allow to dry at room temperature.
6. Using forceps, place filters in the appropriate wells of a vacuum manifold. Wash three times with 10 mL wash buffer (0.5 M phosphate buffer, pH 7.0).
7. Remove filters from manifold and place them on a paper towel to air dry (*see* **Note 3**).
8. Place dried filters in 6 or 8 mL scintillation vials. Add 5 mL scintillant.
9. Assay in a scintillation counter using a counting program specific for the detection of ^{3}H.
10. Data are typically plotted as bar graphs, as shown in **Figs. 3.1** and **3.2** (*see* **Notes 5** and **6**). As SssI methylates all CpG cytosines throughout the genome, methyl group incorporation is inversely proportional to preexisting methylation (i.e., preexisting methylcytosines will be refractory to acceptance of labeled methyl groups).

4. Notes

1. To determine the background due to unincorporated radiolabel, prepare one filter lacking DNA and proceed with steps 2–9. The background should be <1% of the radioactivity of the disks containing DNA. In case of high background, incorporate **Note 3**.
2. To label more DNA, the reaction can be scaled up by correspondingly increasing the amounts of ^{3}H-SAM and SssI.
3. Following step 6, the filters may also be washed two times with 70% ethanol and two times with absolute ethanol, prior to drying.
4. For expediency, the filters may be dried at 95°C for 30 min (step 7).
5. In our experience (using triplicate samples), the technique is somewhat variable, with standard errors ranging from 5% to 30%. In case of high variability, the sample size may be increased.

6. To ensure the linearity of the assay, a standard curve should be generated using DNA methylated *in vitro* with SssI (100% methylated) and incrementally mixing this with whole genome-amplified DNA (0% methylated) to obtain 4–5 fractions between 0% and 100% methylated DNA .

Acknowledgements

The authors wish to thank Suzanne Young for technical assistance with the methyl acceptance assay. This work was supported by National Cancer Institute award CA085289 (to KN), a grant (to CB) from Ovar'coming Together (Indianapolis, IN), and National Science Foundation award MCB-0641851 (to DS).

References

1. Ehrlich, M., Gama-Sosa, M. A., Huang, L. H., et al. (1982) Amount and distribution of 5-methylcytosine in human DNA from different types of tissues of cells. *Nucleic Acids Res* 10, 2709–2721.
2. Wolffe, A. P., Matzke, M. A. (1999) Epigenetics: regulation through repression. *Science* 286, 481–486.
3. Jones, P. A., Takai, D. (2001) The role of DNA methylation in mammalian epigenetics. *Science* 293, 1068–1070.
4. Leu, Y. W., Yan, P. S., Fan, M., et al. (2004) Loss of estrogen receptor signaling triggers epigenetic silencing of downstream targets in breast cancer. *Cancer Res* 64, 8184–8192.
5. Mutskov, V., Felsenfeld, G. (2004) Silencing of transgene transcription precedes methylation of promoter DNA and histone H3 lysine 9. *EMBO J* 23, 138–149.
6. Weber, M., Hellmann, I., Stadler, M. B., et al. (2007) Distribution, silencing potential and evolutionary impact of promoter DNA methylation in the human genome. *Nat Genet* 39, 457–466.
7. Jones, P. A. (2002) DNA methylation and cancer. *Oncogene* 21, 5358–5360.
8. Feinberg, A. P. (2007) Phenotypic plasticity and the epigenetics of human disease. *Nature* 447, 433–440.
9. Jaenisch, R., Bird, A. (2003) Epigenetic regulation of gene expression: how the genome integrates intrinsic and environmental signals. *Nat Genet* 33 Suppl, 245–254.
10. Jang, H., Mason, J. B., Choi, S. W. (2005) Genetic and epigenetic interactions between folate and aging in carcinogenesis. *J Nutr* 135, 2967S–2971S.
11. Mattson, M. P., Shea, T. B. (2003) Folate and homocysteine metabolism in neural plasticity and neurodegenerative disorders. *Trends Neurosci* 26, 137–146.
12. Mack, G. S. (2006) Epigenetic cancer therapy makes headway. *J Natl Cancer Inst* 98, 1443–1444.
13. Momparler, R. L. (2005) Epigenetic therapy of cancer with 5-aza-2′-deoxycytidine (decitabine). *Semin Oncol* 32, 443–451.
14. Fraga, M. F., Esteller, M. (2002) DNA methylation: a profile of methods and applications. *Biotechniques* 33, 632–649.
15. Mizugaki, M., Itoh, K., Yamaguchi, T., et al. (1996) Preparation of a monoclonal antibody specific for 5-methyl-2′-deoxycytidine and its application for the detection of DNA methylation levels in human peripheral blood cells. *Biol Pharm Bull* 19, 1537–1540.
16. Oakeley, E. J., Schmitt, F., Jost, J. P. (1999) Quantification of 5-methylcytosine in DNA by the chloroacetaldehyde reaction. *Biotechniques* 27, 744–752.
17. Balaghi, M., Wagner, C. (1993) DNA methylation in folate deficiency: use of CpG methylase. *Biochem Biophys Res Commun* 193, 1184–1190.
18. Kotsopoulos, J., Sohn, K. J., Martin, R., et al. (2003) Dietary folate deficiency suppresses *N*-methyl-*N*-nitrosourea-induced mammary tumorigenesis in rats. *Carcinogenesis* 24, 937–944.
19. Pufulete, M., Al-Ghnaniem, R., Rennie, J. A., et al. (2005) Influence of folate status on genomic DNA methylation in colonic mucosa of subjects without colorectal adenoma or cancer. *Br J Cancer* 92, 838–842.

20. Carlone, D. L., Lee, J. H., Young, S. R., et al. (2005) Reduced genomic cytosine methylation and defective cellular differentiation in embryonic stem cells lacking CpG-binding protein. *Mol Cell Biol* 25, 4881–4891.
21. Balch, C., Yan, P., Craft, T., et al. (2005) Antimitogenic and chemosensitizing effects of the methylation inhibitor zebularine in ovarian cancer. *Mol Cancer Ther* 4, 1505–1514.
22. Chen, H., Li, S., Liu, J., et al. (2004) Chronic inorganic arsenic exposure induces hepatic global and individual gene hypomethylation: implications for arsenic hepatocarcinogenesis. *Carcinogenesis* 25, 1779–1786.
23. Li, S., Chiang, T. C., Richard-Davis, G., et al. (2003) DNA hypomethylation and imbalanced expression of DNA methyltransferases (DNMT1, 3A, and 3B) in human uterine leiomyoma. *Gynecol Oncol* 90, 123–130.

Part III

Methods for Whole Genome Analysis of DNA Methylation Patterns

Chapter 4

Immunodetection Array

Johannes Pröll, Christian Wechselberger, Mathilde Födermayr, Otto Zach, and Dieter Lutz

Abstract

A novel procedure for DNA methylation analysis is described that characterizes the extent of DNA methylation in CpG islands. The basic concept relies on direct immunodetection of 5′methylcytosines (5′mCs) without the need for bisulfite treatment utilizing a microarray format. This system is designed for the application of immunofluorescence using a monoclonal antibody that specifically recognizes 5′mC in single-stranded DNA hybridized to oligonucleotide microarrays. An ultrasensitive fluorescence scanner and 170-μm thin aldehyde-functionalized glass slides are used to optimize the signal-to-noise ratio and to minimize autofluorescence. These methodological improvements allow for the direct detection of 5′mC in genomic DNA hybridized to microarrays without prior PCR amplification with high analytical sensitivity.

Key words: Epigenetics, DNA methylation, oligonucleotide microarray, immunofluorescence.

1. Introduction

In situ immunodetection methods with antibodies directed against 5′methylcytosine (5′mC) allow for the measurement of the methylation content and its distribution on a cell-to-cell basis. The overall distribution of methylation creates a heterogeneous R-like banding pattern on metaphase chromosome spreads *(1)*. As only clustered methylated CpGs prevalent in, for example, repeat elements can be recognized at the chromosomal level, methylation patterns at relatively small loci such as CpG islands contribute little to the overall staining profile.

To address the technological challenges associated with the immunodetection of DNA methylation in specific sequences, we developed a novel approach for 5′mC analysis by applying

Jörg Tost (ed.), *DNA Methylation: Methods and Protocols, Second Edition, vol. 507*

DOI 10.1007/978-1-59745-522-0_4 Springerprotocols.com

a microarray format. 5′Methylcytosine is detected on genomic DNA without prior sequence conversion by sodium bisulfite treatment or target amplification by PCR-dependent methods (**Fig. 4.1**). The exploitation of recent improvements in the field of array manufacture (170-μm thin aldehyde glass slides) combined with ultrasensitive surface molecule detection at diffraction-limited resolution and the utilization of a fast scanning system with a maximal pixel size of 64-nm resolution made this approach feasible.

Proof-of-principle experiments provided information on the CpG methylation density of specific promoters *(2)*. Original microarray scans of genomic DNA hybridization after immunodetection for *p15/CDKN2b*, *p16/CDKN2a*, *E-cadherin*, and *β-actin* promoter methylation are shown in **Fig. 4.2**. Signal quantification of the spots revealed a clear discrimination between methylated and unmethylated promoters for all genes except *β-actin*. Mean fluorescence intensities (mfis) of promoter regions from *p15/CDKN2b* (mfi/SD 222/33) and *p16/CDKN2a* (mfi/SD 159/34) from HL-60 cells showed

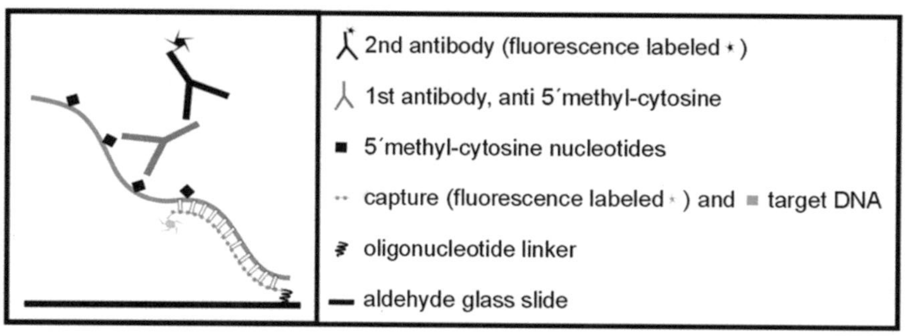

Fig. 4.1. Principle of 5′methylcytosine analysis with the immunodetection array.

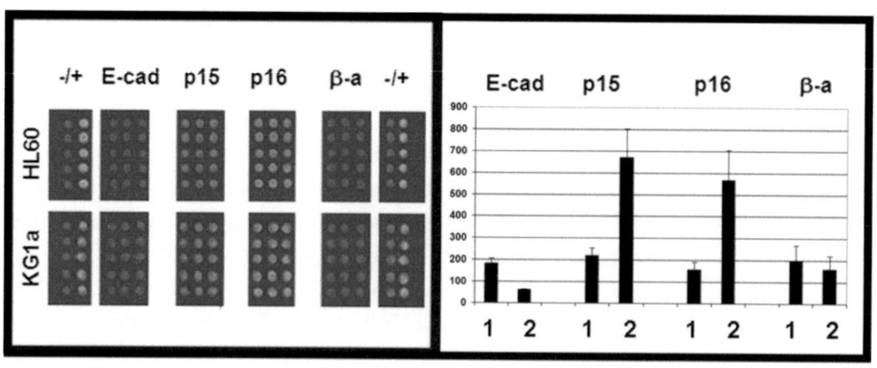

Fig. 4.2. Microarray 5′methylcytosine immunodetection in leukemia cell line-derived genomic DNA. Hybridization of restriction 1.5-μg restriction enzyme-digested genomic DNA of two AML tumor cell lines. Top scan of HL-60 (1) and bottom scan of KG1a (2) microarray. Scans display a significantly higher methylation signal for the KG1a DNA compared to HL-60 DNA. Methylation was measured by antibody staining (Cy3 labeled) of hybridized promoter DNA and overlaid to capture oligonucleotide scan (Cy5 labeled). Original magnification 100 ×; Spot distance 150 μm; Pixel resolution 0.512 μm.

significantly lower signals than from KG1a with *p15/CDKN2b* (mfi/SD 672/130) and *p16/CDKN2a* (mfi/SD 568/138) ($p < 0.0001$ in both cases), whereas *E-cadherin*-promoter region displayed inversed measures with significantly higher methylation signals in HL-60 (mfi/SD 183/23) than in KG1a (mfi/SD 63/4) ($p < 0.0001$) and reduced CpG methylation density compared to *p15/CDKN2b* and *p16/CDKN2a*.

This ultrasensitive assay for detecting 5′mCs allows the detection of DNA methylation patterns in the near-single molecular range. It contributes thereby to a better determination of epigenetic modifications in the dynamic structural organization of the genome. Using genomic DNA of the two well-characterized AML tumor cell lines HL-60 and KG1a, this novel microarray assay specifically distinguishes nonmethylated from methylated gene regulatory sequences. Therefore, the perfect exploitation of this technology allows envisioning the adaptation of this analytical system to a high-throughput methodology for direct detection of 5′mCs in genomic DNA.

2. Materials

2.1. Cell Lines and Cell Culture

1. The HL-60 cell line was obtained from the American Type Culture Collection (ATCC, Manassas, VA) and the KG1a cell line from the Deutsche Sammlung von Mikroorganismen und Zellkulturen (DSMZ, Braunschweig, Germany).
2. RPMI-1640 cell culture medium (Gibco Invitrogen, Paisley, UK) supplemented with 10% fetal calf serum (Gibco), L-glutamine (300 mg/L, Gibco) and antibiotics (Streptomycin (100 μg/mL, Gibco)/Penicillin (100 U/mL, Gibco)).

2.2. Oligonucleotide Sequences and Genomic DNA Preparation

1. Oligonucleotides (Table 4.1) are synthesized by VBC (Vienna, Austria). Positive control, nonsense, and all sense oligonucleotides carry a 3′-terminal amino group with a $(CH_2)_7$-linker for surface immobilization.
2. QIAmp Blood Kit (Qiagen, Hilden, Germany).
3. *MnlI* (New England Biolabs).

2.3. Microarray Surface

1. 50 × 24 × 0.17 mm borosilicate slides (Stölzle-Oberglas, Leoben, Austria). Borosilicate slides permit the reduction of background fluorescence signals (*see* **Note 1**).
2. Aldehyde functionalization of the slides was performed by Upper Austrian Research GmbH (Linz, Austria) permitting the coupling of amino-modified capture oligonucleotides to the slide surface by reductive amination involving the binding of a carbonyl group to an amine under the use of sodium cyanoborohydride or other reducing agents.

Table 4.1
Oligonucleotide sequences × in sequences marks 5′methylcytosine

Oligonucleotide	Sequence 5′-3′
Target oligonucleotide: (52mer)	ATxgatxgatxgatxgatxgcgctttttcccagaagcaatccaggagcgccc
Positive control: (33mer)	ATxgatxgatxgatxgatxgtccaggagcgccc
Nonsense: (24mer)	CCcagaagcaatccaggagcgccc
Sense *p15/CDKN2b*: (59mer; NT_008413.16)	gggcgcgcctggattgcttctgggaaaaagcgcctagcgcggacgcagccgagctcaaa
Sense *p16/CDKN2a* (58mer; NT_008413.16)	ccagccagtcagccgaaggctccatgctgctccccgccgccggctccatgctgctccc
Sense *E-cadherin*: (40mer; NT_010498.15)	CGccacggaggccccgagtggaccgacgtcggtgcgtggg
Sense β-actin: (60 mer; NT_007819.15)	tgcggcgaagccgcgtgagtgagcggcgcggggccaatcagcgtgcgccgttccgaaagt

2.4. The Spotting Process

1. Multigrid II microarray spotter (BioRobotics, Woburn, MA).
2. Tungsten Split Pins PT3000 (Point Technologies, Boulder, CO). Microscopic control of free-split width after incubation in the cleaning solution as recommended by the supplier and sonication in a water bath is helpful.
3. Amino-modified (3′-terminal amino group with $(CH_2)_7$-linker) and flurescently labeled (5′-terminal cyanine 5 modification) capture oligonucleotides were purchased from VBC (Vienna, Austria). Aliquots are stored at −80°C.
4. Spotting buffer: 1 × SSC (15 mM sodium citrate, 150 mM sodium chloride) with 2.5% glycerol was used to obtain a final oligonucleotide concentration of 5 μM (prepare fresh, *see* **Note 2**).
5. 384 microwell plates (Genetix, Munich, Germany).
6. Surface inactivation buffer: 0.2% BSA (Sigma–Aldrich) in Dulbecco's PBS (Gibco) (prepare fresh, *see* **Note 2**).
7. Surface wash buffer: 0.1 M carbonate–bicarbonate, pH 8.2 (Sigma-Aldrich); stable at room temperature for 1 month.

2.5. Microarray Hybridization

1. Corning® Microarray Hybridization Chambers (Corning Life Sciences).
2. Hybridization buffer: 4 × SSC, 1% SDS (Sigma–Aldrich). Make fresh as required (*see* **Note 2**).
3. Round cover slips with 5 mm in diameter (Menzel Gläser, Braunschweig, Germany) for single subgrid hybridizations or 22 × 22 mm cover slips (Assistent, Sondheim, Germany) for multiple subgrid hybridizations. Ensure that they are clean and free of dust (*see* **Note 1**).

4. Corning hybridization chambers (Fischer Scientific, Loughborough, UK).
5. Hybridization wash buffer 1: 2 × SSC, 0.1% SDS (Sigma–Aldrich). Prepare fresh (*see* **Note 2**).
6. Hybridization wash buffer 2: 2 × SSC (Sigma–Aldrich). Prepare fresh (*see* **Note 2**).
7. Hybridization wash buffer 3: 0.2 × SSC (Sigma–Aldrich). Prepare fresh (*see* **Note 2**).

2.6. Microarray Immunolabeling

1. Blocking buffer (make fresh): 2% (w/v) fraction V bovine serum albumin (BSA; 98% pure; Sigma–Aldrich) in Dulbecco's PBS (Gibco).
2. Antibody dilution buffer: PBS (Gibco) supplemented with 1% BSA and 0.05% Tween-20.
3. Monoclonal anti-5′mC antibody (clone: 33D3; mouse IgG1; Serotec, Düsseldorf, Germany) (*see* **Note 3**).
4. Control antibody, for example, anticollagen clone: COL-1; mouse IgG1.
5. Cyanine 3 (Cy3) dye conjugated secondary goat antimouse antibody (Amersham).

2.7. Ultrasensitive Microarray Readout

The device applies a novel readout technology that combines wide-field illumination as well as time delay and integration (TDI) mode for readout. The apparatus used for our measurements is based on an inverted microscope (Axiovert 200, Zeiss, Germany). This scanning device operates by synchronizing a high-precision scanning stage (Märzhäuser Wetzlar, Germany) with two CoolSNAP$_{HQ}$ CCD cameras with high-quantum efficiency and low readout noise running in TDI mode (CoolSNAP HQ, Photometrics, Roper Scientific, Tucson, AZ). A patented focus hold system allows for large-area scanning with single molecule sensitivity (Upper Austrian Research GmbH, Linz, Austria). A diode-pumped, continuous wave Nd:Yag laser (Neodymionen doted Yttrium-Aluminium-Granat) with a wavelength of ~532 nm (Millennia II, Spectra-Physics, Mountain View, CA) and a Kr-ion laser with a wavelength of ~647 nm (43 Series Ion Laser, Melles Griot, Carlsbad, CA) are used for illumination of cyanine 3 and cyanine 5, respectively. High-laser intensities, 100 × optical magnification and short (100 ms) exposure times allow ultrasensitive microarray readout within minutes. Further exploitation of this apparatus to biological questions is described by Hesse et al. *(3, 4)*.

2.8. Image Analysis

1. V^{++} (Digital Optics Limited, Auckland, New Zealand).
2. Matlab (The MathWorks Inc., Natick, MA).
3. Imagene 5.6 (BioDiscovery, CA).

3. Methods

3.1. Preparation of Genomic DNA

1. Isolate genomic DNA from cell lines with the QIAmp Blood Kit following the manufacturer's recommendations.
2. Digest 1 μg of genomic DNA with 1 U *MnlI* for 4 h at 37°C in NEB buffer 2 (*see* **Notes 4** and **5**).
3. Inactivate the enzyme by incubation at 65°C for 20 min and purify DNA by ethanol precipitation.

3.2. Microarray Production

Positive control oligonucleotides are spotted on each microarray and are used as internal standard. Nonsense and target oligonucleotides are used to test antibody and hybridization specificity. All sense oligonucleotides serve as capture oligonucleotides for the genomic DNA fragments. CpG island sequences of promoter regions are the basis for capture oligonucleotide sequence. The selection depends on target specificity and hybridization temperature and the CpG density as indicated in the NCBI's CpG map. Alternatively, methylation data can be used that has been determined by MS-PCR or other methods. Example sequences are given in Table 4.1.

1. Dilute oligonucleotides in spotting buffer to a final concentration of 5 μM.
2. Transfer 10 μL of each capture oligonucleotide to a 384 microwell plate.
3. Contact spot capture oligonucleotides with a Multigrid II microarray spotter using Tungsten Split Pins using a pitch distance of 150 μm and soft touch settings (1 s). These parameters result in spots of ~40 μm in diameter (*see* **Note 6**).
4. Perform oligonucleotide immobilization in the humid ambiance (60%) of the Multigrid II spotter for at least 2 h but no longer than overnight.
5. Inactivate the surface with blocking solution (BSA–PBS) by incubation for another 2 h at room temperature. Ensure full surface coverage by the blocking solution.
6. Wash microarrays in 0.1 M carbonate buffer (pH 8.2) for 15 min, dry them, and use them immediately for the hybridization experiments.

3.3. Microarray Hybridization

1. Add 100 pmol (or less) of oligonucleotides to 10 μL of hybridization buffer to obtain a final concentration of 10 μM. If genomic DNA is used for hybridization, dilute 1.5 μg of restriction digested DNA to 10 μL of hybridization buffer.
2. Use 2.5 μL of this hybridization solution for single subgrid hybridization; for multiple subgrid hybridization experiments use 10 μL hybridization solution.

3. Place the microarray in a hybridization chamber and submerge the chamber in a preheated waterbath.
4. Hybridize oligonucleotides at 55°C and genomic DNA at 65°C, respectively, without prehybridization or the addition of competitive DNA overnight.
5. After overnight hybridization, prepare fresh washing buffers 1, 2, and 3 and preheat them to 42°C.
6. Wash microarrays successively with washing buffer 1, 2, and 3 for 15 min each (*see* **Note 7**).

3.4. Microarray Immunolabeling

1. Block the slide surface by incubating the slide with 0.2% BSA in Dulbecco's PBS for 30 min at room temperature.
2. For 5′mC immunolabeling, incubate the microarrays with monoclonal anti-5′mC antibody or isotype-matched control antibody in antibody dilution buffer for 1 h at room temperature at a concentration of 1 μg/mL each. Use approximately 1 mL of antibody solution for full surface coverage.
3. Wash microarrays for 5 min in PBS.
4. Incubate with the secondary Cy3-labeled antimouse antibody diluted 1:2000 in antibody dilution buffer for 1 h at room temperature. Use approximately 1 mL of antibody solution for full surface coverage.
5. Wash slides three times for 5 min in PBS.
6. Dry microarrays (*see* **Note 8**) and scan them on the modified inverted microscope within the next 4 h without the use of coverslips and any embedding solution or antifading reagent resulting in better image quality when compared to wet microarray scanning.

3.5. Image Acquisition and Analysis

1. Scan the slides with the above described apparatus.
2. Perform spot recognition automatically using the Imagene 5.6 software.
3. Transfer mean fluorescence intensity data to Matlab.
4. Correct for background and perform statistical analysis of fluorescence signal intensities of the cell lines with a two-tailed Student's *t*-test.
5. An example of the spot quality obtained by the procedure is shown in **Fig. 4.3**. Conclusive assay results from oligonucleotide hybridization experiments are presented in **Fig. 4.4**, showing intra- and interassay variability of hybridization target (5′mC positive and 5′mC negative) and anti-5′mC antibody specificity. Examples for the direct detection of genomic DNA methylation after immunodetection are shown in **Fig. 4.2** for the *p15/CDKN2b*, *p16/CDKN2a*, *E-cadherin*, and *β-actin* promoters.

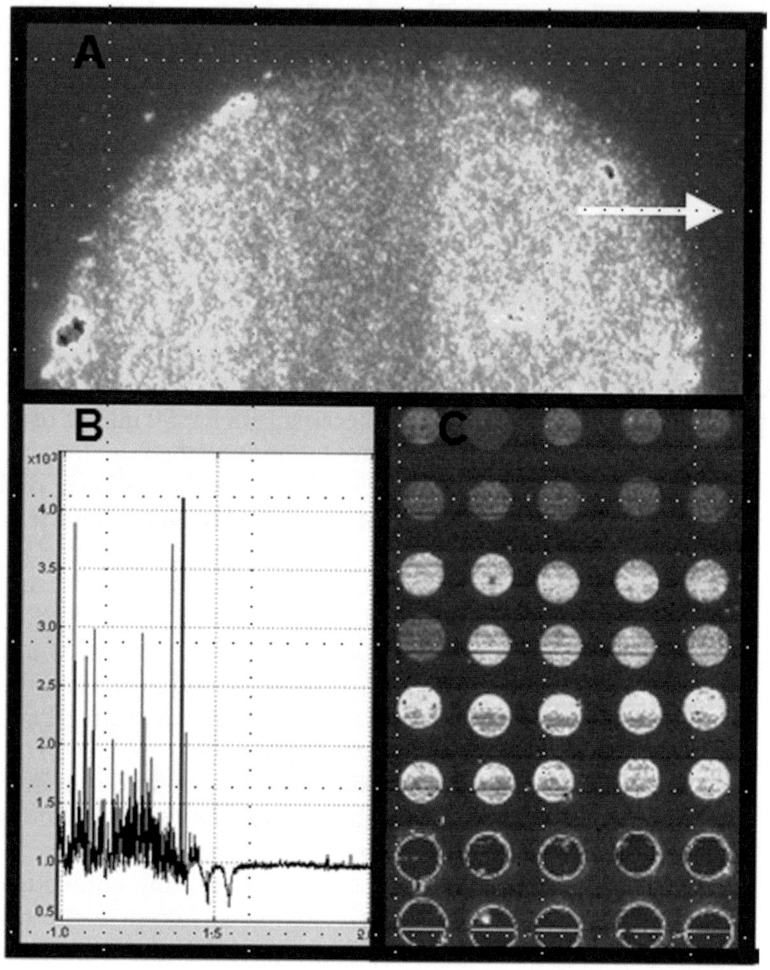

Fig. 4.3. Spot morphology at high scanning resolution. (**A**) Single molecule detection. Second antibody (cyanine 3 labeled) detection with high-sensitivity setting (binning 2; exposure 2 ms; laser 1 W; objective 100 ×; spot diameter ~ 100 μm). The image was acquired with Matlab (**B**). Fluorescence profile of single molecule spots and background regions as indicated in A (→). Visualization of fluorescence intensity in mean fluorescence counts (mfc) per pixel (pixel resolution, 128 nm). The image was acquired using the V^{++} software. (**C**) Capture oligonucleotide concentration (cyanine 3 labeled) used for spotting. 1.2, 2.5, 5, and 10 μM, respectively (two rows per [C] top down). Spot distance 200 μm. Image was acquired with Matlab.

4. Notes

1. If the surface of the glass slide is noticeably dirty or covered with precipitates, use pressurized N_2 carefully to support slide drying.
2. Avoid any microbial contamination by the use of fresh buffers only.

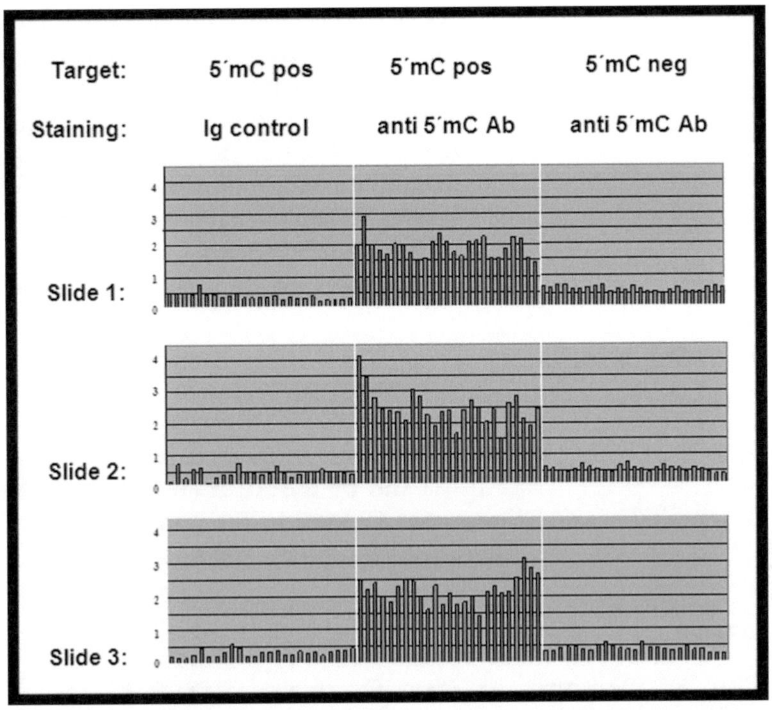

Fig. 4.4. Hybridization and antibody specificity of the immunodetection array. The 5′mC signal ratio compares 5′mC positive and 5′mC negative hybridization oligonucleotides on three individual aldehyde slides to an immunoglobulin control. Y axis = ratio of anti-5′methylcytosine antibody (Cy5 labeled) versus capture oligonucleotide (Cy3 labeled); X axis = each column represents the mean of 15 spots per subgrid (indicated without SD). Target oligonucleotides are used at a concentration of 1 μM for hybridization. The 5′mC positive target oligonucleotides contain five individual methylcytosines (Table 4.1). Quantification is performed with the ImaGene software.

3. Some anti-5′mC antibodies will give insufficient results with this type of surface (e.g., 5-mC antibody from Abcam, ab1884; sheep polyclonal antiserum). Test antibodies before use.
4. It is important to check that the selected restriction enzyme does not cut within the hybridization sequences of the regions you choose. Replace with another frequent cutting enzyme if required.
5. For ultra-high sensitivity measurements use repellent, low-bind tips and tubes (Eppendorf, Hamburg, Germany) for target preparation.
6. To determine the spotting efficiency we recommend controlling the spotting process for this slide type by differential interference contrast (DIC) microscopy to check wet-spot morphology as well as array symmetry immediately after the spotting process.

7. Surfaces should not fall dry when transferred from one washing solution to the other as salt precipitates diminish array quality significantly and are likely to occur during these steps. To avoid this problem, keep handling time and handling distance as short as possible, work with prearranged and preheated washing solutions in petri dishes, use a practically bent coverslip tweezers and quickly transfer the slides inverted (printed surface down).
8. Use pressurized N_2 carefully to support slide drying.

Acknowledgements

This study was supported by the federal government of Upper Austria and by the Austrian Federal Ministry for Education, Science and Culture in the context of the Gen-AU project "Ultrasensitive Proteomics and Genomics".

References

1. Pfarr, W., Webersinke, G., Paar, C., et al. (2005) Immunodetection of 5′-methylcytosine on Giemsa-stained chromosomes. *Biotechniques* 38, 527–528.
2. Pröll, J., Födermayr, M., Wechselberger, C., et al. (2006) Ultra-sensitive immunodetection of 5′methyl cytosine for DNA methylation analysis on oligonucleotide microarrays. *DNA Res* 28, 37–42.
3. Hesse, J., Sonnleitner, M., Sonnleitner, A., et al. (2004) Single-molecule reader for high-throughput bioanalysis. *Anal Chem* 76, 5960–5964.
4. Hesse, J., Jacak, J., Kasper, M., et al. (2006) RNA expression profiling at the single molecule level. *Genome Res* 16, 1041–1045.

Chapter 5

Methylated DNA Immunoprecipitation (MeDIP)

Fabio Mohn, Michael Weber, Dirk Schübeler, and Tim-Christoph Roloff

Abstract

Methylated DNA immunoprecipitation (MeDIP) is a versatile immunocapturing approach for unbiased detection of methylated DNA. In brief, genomic DNA is randomly sheared by sonication and immunoprecipitated with a monoclonal antibody that specifically recognizes 5-methylcytidine. The resulting enrichment of methylated DNA in the immunoprecipitated fraction can be determined by PCR to assess the methylation state of individual regions. Alternatively, MeDIP can be combined with large-scale analysis using microarrays as a genome-wide experimental readout. This protocol has been applied to generate comprehensive DNA methylation profiles on a genome-wide scale in mammals and plants, and further to identify abnormally methylated genes in cancer cells.

Key words: MeDIP, methylated DNA, DNA methylation, genome-wide analysis, epigenomics, 5mC antibody, CpG, immunoprecipitation.

1. Introduction

Techniques to map DNA methylation on a genome-wide scale were only developed recently. Several of these are based on classical methylation-sensitive (e.g., *Hpa*II and *Not*I) or methylation-specific (e.g., *Mcr*BC) restriction enzymes. However, these assays limit the analysis to certain sequence motifs, for example, only 3.9% of all nonrepeat CpGs in the human genome reside within recognition sites of the *Hpa*II enzyme *(1)*. To circumvent this motif bias, techniques have been developed that use affinity purification of methylated DNA either by monoclonal antibodies specific to 5-methylcytidine (5mC) or by methyl-binding protein (MBD) domains specific to methylated CpGs (for a more comprehensive list of methods see *(2,3)*). Importantly, such affinity-based approaches bear constraints that need to be taken into account

Jörg Tost (ed.), *DNA Methylation: Methods and Protocols, Second Edition, vol. 507*

DOI 10.1007/978-1-59745-522-0_5 Springerprotocols.com

when setting up experiments and choosing the most suitable technique. Namely, the fact that methylated CpG-rich sequences are more enriched by affinity approaches than methylated CpG-poor sequences which is simply because they contain more epitopes to be bound by the antibody or MBD domain.

Another caveat which applies to all techniques combined with genome-wide detection on microarrays is that allelic methylation and methylation of individual repetitive elements cannot be addressed comprehensively. Bisulfite genomic sequencing is less limited in that regard but requires extensive resources when applied genome-wide.

The methylated DNA immunoprecipitation (MeDIP) protocol (**Fig. 5.1**) is a fast and simple approach to determine DNA methylation on a genome-wide scale or for individual loci: genomic DNA is randomly sheared by sonication and methylated DNA is subsequently immunocaptured with a monoclonal

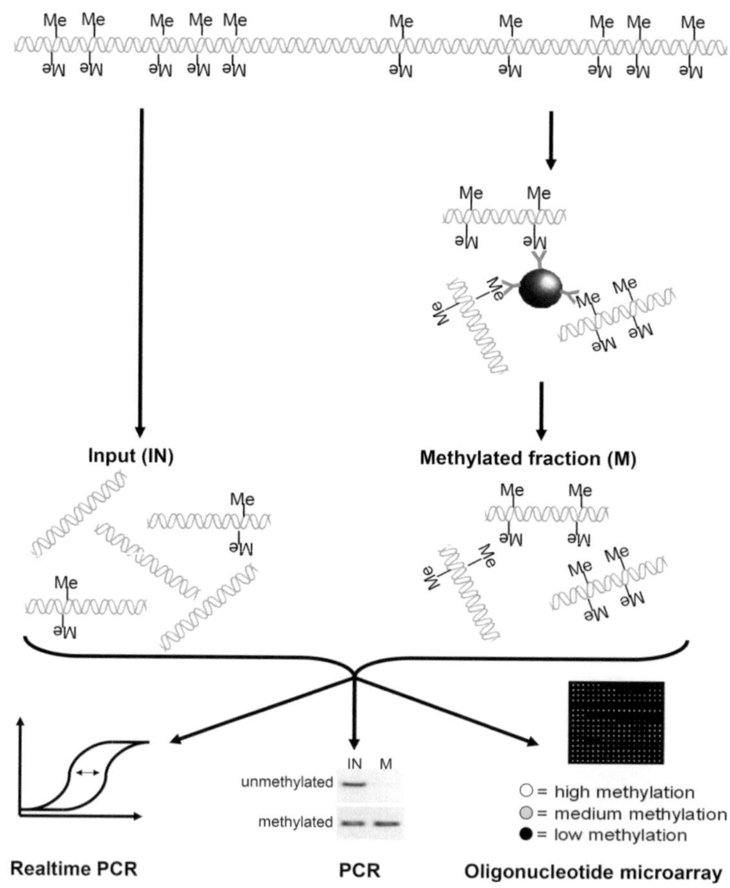

Fig. 5.1. Principle of MeDIP (methylated DNA immunoprecipitation). Total genomic DNA is sonicated and methylated DNA (Me) is immunoprecipitated with an antibody directed against 5-methylcytidine. Input DNA (IN) and methylated DNA (M) can be used for single-gene analysis by standard or real-time PCR or they can be differentially labeled with Cy5 and Cy3 and cohybridized as a two-color experiment on oligonucleotide microarrays.

antibody specific for 5mC *(4)*. For individual regions the enrichment of the methylated fraction as compared to the input can be assessed by PCR. In combination with DNA microarrays, MeDIP can be applied to generate large-scale maps of DNA methylation. The protocol has been successfully used with genomic DNA from various organisms that contain methylated cytosines in their genome (human, mouse, chimp, *Arabidopsis thaliana*, and *Neurospora crassa*). Comprehensive DNA methylation profiles on a genome scale have been published for mammals and plants *(5–7)*, and genes abnormally methylated in cancer cells could be identified *(5, 8)*.

2. Materials

Materials can be stored at room temperature (RT) for several weeks unless indicated otherwise.

2.1. Isolation of Genomic DNA

1. Lysis buffer: 20 mM Tris–HCl, pH 8.0, 4 mM EDTA, 20 mM NaCl, 1% SDS.
2. Proteinase K (Roche), store at −20°C.
3. RNase A (Sigma, Cat. No. 83833), store at −20°C.

2.2. Sonication of Genomic DNA

1. The protocol was established using a BRANSON digital Sonifier model 450 with a tapered Microtip. Other sonicators (including waterbath sonicators) work too, but conditions need to be adjusted.
2. Glycogen (Roche), store at −20°C.

2.3. Immunoprecipitation of Methylated DNA (MeDIP)

1. 1M Na-Phosphate buffer, pH 7.0: 39 mL 2 M monobasic sodium phosphate (NaH_2PO_4) (276 g/L), 61 mL 2 M dibasic sodium phosphate (Na_2HPO_4) (284 g/L), 100 mL H_2O.
2. 10 × IP buffer: 100 mM Na-Phosphate, pH 7.0, 1.4 M NaCl, 0.5% Triton X-100.
3. TE (Tris EDTA) buffer: 10 mM Tris–HCl, pH 8.0, 1 mM EDTA, pH 8.0.
4. 1 × IP buffer: one volume 10 × IP buffer and nine volumes TE buffer.
5. 5-Methylcytidine antibody: this protocol was established using the mouse monoclonal antibody against 5mC generated by Reynaud and colleagues *(4)* and supplied by EUROGENTEC (#BI-MECY-1000). A similar antibody is also available from other companies (CALBIOCHEM, DIAGENODE), which work with comparable efficiency. The antibody can be aliquoted and stored at −20°C to avoid freeze–thaw cycles.

6. PBS buffer: 2.67 mM KCl, 1.47 mM KH_2PO_4, 137.93 mM NaCl, 8.06 mM Na_2HPO_4, pH 7.3;
7. 0.1% BSA/PBS: 9 mL PBS, 1 mL BSA 10 mg/mL stock.
8. Magnetic beads: Dynabeads M-280 Sheep anti-mouse IgG (DYNAL BIOTECH #112.01).
9. A magnetic rack for microtubes is required for the washing steps following immunoprecipitation.
10. Proteinase K digestion buffer: 50 mM Tris–HCl, pH 8.0, 10 mM EDTA, 0.5% SDS.

2.4. PCR Verification of MeDIP

1. Taq polymerase (New England Biolabs, 5 U/μL).
2. 10 × ThermoPol buffer (New England Biolabs).
3. 2.5 mM dNTP mix

3. Methods

3.1. Isolation of Genomic DNA

1. Resuspend the cell pellet or the homogenized tissue in TE in a 2 mL microtube. We typically use 10^6–10^7 cells which are resuspended in 300 μL (the volume should be increased for big cell pellets or tissue samples).
2. Add 300 μL (one volume) lysis buffer containing 20 μL proteinase K (10 mg/mL stock) (*see* **Note 1**).
3. Incubate at 55°C for at least 5 h.
4. Extract with one volume phenol (600 μL). Transfer the upper, aqueous phase to a new tube.
5. Extract with one volume chloroform (600 μL). Transfer the upper, aqueous phase to a new tube.
6. Precipitate the DNA by slowly adding two volumes ethanol containing 75 mM Na Acetate pH 5.2 (1.2 mL) and let it sit at RT for a few minutes until the genomic DNA is fully precipitated.
7. Transfer the DNA with a pipette tip into a fresh tube containing 1 mL ethanol 70%, invert 5 ×, and centrifuge 5 min at top speed at RT.
8. Resuspend the DNA pellet in TE containing 20 μg/mL RNAse A.
9. Incubate at 37°C for 30 min (*see* **Note 2**).

3.2. Sonication of Genomic DNA

Genomic DNA is randomly sheared by sonication to generate fragments between 300 bp and 1000 bp (**Fig. 5.2**). Genomic DNA can also be fragmented with restriction enzymes like *Alu*I, but this is not recommended for unbiased microarray studies. The sonication efficiency varies with DNA concentration, sonicator model and settings, and size and quality of the sonication tip. Therefore, we recommend systematic checking of the size of the sheared DNA on an agarose gel to ensure equal sonication

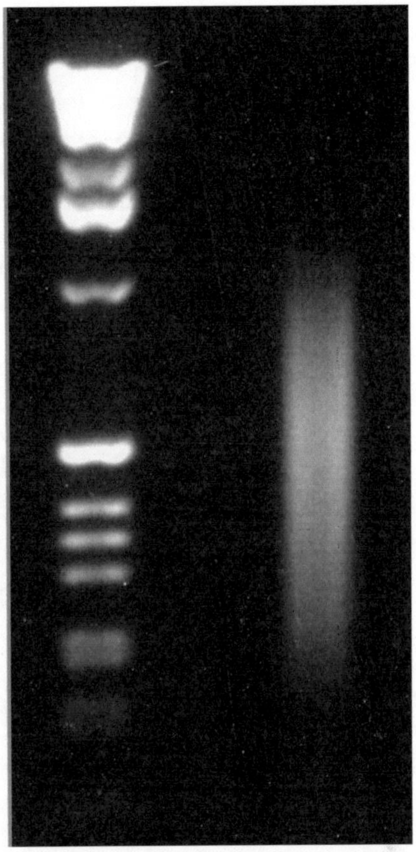

Fig. 5.2. The gel illustrates sonicated genomic DNA run on a 2% agarose gel stained with ethidium bromide. Most of the sheared fragments have a size between 300 bp and 1000 bp with an average fragment size around 500 bp.

between experiments. The following shearing settings are optimized for a Branson Sonifier model 450 with a tapered Microtip (*see* **Note 3**) and a Diagenode Bioruptor 200 waterbath sonicator, respectively.

3.2.1. Branson Sonifier

1. Dilute the genomic DNA in TE in a 1.5 mL microtube (10–20 μg DNA in 300 μL TE, 40–60 μg DNA in 700 μL TE).
2. Sonicate 5 × 10 s with amplitude 20% while the tube is cooled in ice water. Allow the sample to cool down after each pulse for 1 min (*see* **Note 3**).
3. Load 5–10 μL on an agarose gel to check the size of the DNA (mean size should be 300–1000 bp; **Fig. 5.2**). Alternatively, fragment size can be monitored by a labchip assay on a Bioanalyzer (Agilent). If necessary, sonicate for one or two additional pulses until the size of the DNA is 300–1000 bp (*see* **Note 4**).
4. Continue at 3.2.3

3.2.2. Bioruptor Waterbath Sonicator

1. Dilute 20 μg genomic DNA in 300 μL TE in a 1.5 mL microtube.
2. Sonicate for 12 cycles 30 s on/30 s off, the highest output level while cooling the tube to 1°C in the waterbath (*see* **Note 3**).
3. Load 5–10 μL on an agarose gel to check the size of the DNA (mean size should be 300–1000 bp; **Fig. 5.2**). Alternatively, fragment size can be monitored by labchip assay on a Bioanalyzer (Agilent). If necessary, sonicate for one or two additional pulses until the size of the DNA is 300–1000 bp (*see* **Note 4**).

3.2.3. Purification of Sonicated DNA

1. Precipitate the sonicated DNA with 400 mM NaCl (24 μL 5 M NaCl for 300 μL, 56 μL 5 M NaCl for 700 μL), two volumes 100% EtOH and 1 μL glycogen. Centrifuge for 60 min at 4°C at 16,100*g* in a table centrifuge. Add 300 μl of 70% EtOH to remove salts and centrifuge for 10 min at 16,100*g* (Note that the precipitation for 700 μL has to be done in 2-mL tubes or in two 1.5 mL microtubes).
2. Resuspend the DNA pellet in 50μL TE and measure DNA concentration.

3.3. Immunoprecipitation of Methylated DNA (MeDIP)

The sonicated DNA is immunoprecipitated with a monoclonal antibody against 5mC *(4)*. A portion of the sonicated DNA should be left untreated to serve as input control (for microarray experiments, typically 3–5 μg input material are required).

1. Dilute 4 μg of sonicated DNA in 450 μL TE (*see* **Note 5**).
2. Denature for 10 min in boiling water and immediately cool on ice for 10 min.
3. Add 51 μL of 10× IP buffer.
4. Add 10 μL of 5mC antibody.
5. Incubate 2–6 h at 4°C with overhead shaking.
6. Prewash 40 μL of Dynabeads with 800 μL of 0.1% BSA/PBS for 5 min at RT with overhead shaking to avoid sedimentation of the beads.
7. Trap the beads on the wall of the tube using a magnetic rack, discard supernatant, and repeat wash once with 800 μL 0.1% BSA/PBS.
8. Trap the beads on a magnetic rack, discard supernatant, and resuspend in 40 μL of 1× IP buffer.
9. Add Dynabeads to the sample.
10. Incubate 2 h at 4°C with overhead shaking (*see* **Note 6**).
11. Trap the beads on a magnetic rack, discard supernatant, and wash with 700 μL 1× IP buffer for 10 min at RT with shaking.
12. Repeat the washing with 700 μL 1× IP buffer twice.
13. Trap the beads on a magnetic rack, discard supernatant, and resuspend in 250 μL proteinase K digestion buffer.
14. Add 7 μL proteinase K (10 mg/mL stock).

15. Incubate 3 h at 50°C (use a shaking heat block at 800 rpm (Eppendorf Thermomixer) to prevent sedimentation of the beads).
16. Extract with one volume phenol (250 μL), transfer the upper, aqueous phase to a fresh microtube.
17. Extract supernatant with one volume chloroform (250 μL), transfer the upper, aqueous phase to a fresh microtube.
18. Precipitate the DNA with 400 mM NaCl (20 μL 5 M NaCl), glycogen (1 μL) and two volumes 100% ethanol (500 μL). Wash with 70% EtOH as described above.
19. Resuspend the DNA pellet in 60 μL TE and store at −20°C (*see* **Note 7**).

3.4. Analysis by PCR and Microarrays

Enrichments in the MeDIP fraction can be measured by PCR or by microarray analysis (**Fig. 5.1**). **Table 5.1** contains a set of control primers for human and mouse, which are suitable for testing the efficiency of MeDIP, and the PCR conditions as well as the PCR setup and cycling conditions.

Keep in mind that CpGs are unequally distributed in mammalian genomes and that the enrichment of any target sequence in the MeDIP fraction depends both on the methylation status of the target sequence and the number of CpGs it contains *(6)*. A low enrichment can thus reflect an unmethylated state or the absence of CpGs.

1. For PCR or real-time PCR, use 20 ng of total input DNA and 2 μL of MeDIP DNA.
2. Enrichments in the MeDIP fraction are calculated relative to an unmethylated control, which typically is a CpG island promoter of a housekeeping gene.
3. For genome-wide analyses, input and MeDIP fractions are differentially labeled with two dyes (e.g., Cy3 and Cy5) and cohybridized to oligonucleotide-tiling microarrays.

For the MeDIP fraction, DNA from parallel MeDIPs can be pooled to obtain the required quantity of DNA for labeling. Alternatively, if pooling of up to 8 MeDIPs is not feasible, amplification of a single MeDIP can be performed. The whole genome amplification (WGA) protocol (see NoE protocol PROT30 *(9)*) can be applied to single-stranded DNA samples and thus should be usable. However, we have not tested systematically if WGA introduces amplification biases when applied to MeDIP samples.

The array analysis depends largely on the type of array used and the questions to be answered. Thus, a general description of an analysis strategy could be misleading. An example for DNA methylation analysis on Nimblegen Promoter arrays can be found in *(6)*. For probe labeling, hybridization to NimbleGen 385 K chips and washing steps, the Nimblegen Hybridization Kit (KIT002-2) and the corresponding protocols were used. In brief, data analysis consists of the following steps: normalization of the

Table 5.1
Control PCR reaction scheme and primer to test MeDIP efficiency

PCR reaction setup

2 μL	Template (resuspended MeDIP/total input 10 ng/μL)			
0.5 μL	Taq polymerase (5 U/μL)			
5 μL	ThermoPol reaction buffer			
1.5 μL	Forward primer 5 μ*M*			
1.5 μL	Reverse primer 5 μ*M*			
2.5 μL	2.5 m*M* dNTP mix			
37 μL	H_2O			
PCR cycling conditions				
3 min.	95°C		1×	
30 sec.	95°C		|	
30 sec.	Annealing temperature (see below)		|× number of cycles (see below)	
30 sec.	72°C		|	
30 min.	72°C		1×	
Name	**Positive/negative**	**Sequence**	**Number of cycles**	**Annealing temp.**
Control primers for mouse				
Beta actin promoter	Neg.	AGCCAACTTTACGCCTAGCGT	32	60°C
		TCTCAAGATGGACCTAATACGGC		
Gapdh promoter	Neg.	CTCTGCTCCTCCCTGTTCC	32	60°C
		TCCCTAGACCCGTACAGTGC		
H19-ICR	Pos.	GCATGGTCCTCAAATTCTGCA	32	60°C
		GCATCTGAACGCCCCAATTA		
IAP (repeat)	Pos.	CTCCATGTGCTCTGCCTTCC	24	60°C
		CCCCGTCCCTTTTTTAGGAGA		
Control primers for human				
UBE2B	Neg.	CTCAGGGGTGGATTGTTGAC	36	60°C
		TGTGGATTCAAAGACCACGA		

(continued)

Table 5.1 (continued)

PCR reaction setup

HIST1H3B	Neg.	CCCACACTTCTTATGCGACA	34	60°C
		CTGTGCCTGGTTGCAGATTA		
H19 ICR	Pos.	GAGCCGCACCAGATCTTCAG	36	60°C
		TTGGTGGAACACACTGTGATCA		

data (e.g., loss normalization), averaging of signals in regions of interest (e.g., average over a region in a promoter), definition of a significant threshold for enrichment or depletion of DNA methylation signals and annotation of the identified regions to relate changes in DNA methylation to genomic information.

4. Notes

1. Due to the viscosity of the solution after cells lysis, it is recommended to add the proteinase K to the lysis buffer before mixing it with the sample.
2. It is important to completely remove any RNA, as the antibody also recognizes 5mC in the context of RNA molecules.
3. Keep the sample on ice during the sonication process to avoid denaturing and potential degradation of the DNA due to contaminations. Alternatively, waterbath sonicators can be used; however, conditions need to be adjusted for each apparatus.
4. Avoid shearing the DNA too small, that is, below an average size of 400 bp as this can affect the immunoprecipitation efficiency and also the detectability of methylated DNA by PCR and microarrays.
5. We recommend the use of 4 μg; however, smaller amounts down to 1 μg have been used successfully. For little starting material, the amount of antibody has to be adapted in order to avoid unspecific binding and increasing background in the IP.
6. Avoid incubation over 3 h as this might increase unspecific binding of DNA to the beads.
7. With the described conditions, the MeDIP procedure generally yields 5% of the original total DNA in mammalian cells (i.e., 200 ng of methylated DNA starting from 4 μg total DNA). The amount of recovered DNA can be increased by using amounts of Dynabeads to up to 100 μl. However, with increasing amounts of Dynabeads, the linear range

of enrichment might be changed. Thus, additional controls (e.g., bisulfite sequencing) need to be done to make sure that the enrichments are correctly reflecting the methylation status of the region of interest.

Acknowledgements

The authors would like to thank the members of the Schübeler lab for helpful comments on the protocol.

References

1. Fazzari, M. J., Greally, J. M. (2004) Epigenomics: beyond CpG islands. *Nat Rev Genet* 5, 446–455.
2. Weber, M. Schübeler, D. (2007) Genomic patterns of DNA methylation: targets and function of an epigenetic mark. *Curr Opin Cell Biol* 19, 273–280.
3. Zilberman, D., Henikoff, S. (2007) Genome-wide analysis of DNA methylation patterns. *Development* 134, 3959–3965.
4. Reynaud, C., Bruno, C., Boullanger, P., et al. (1992) Monitoring of urinary excretion of modified nucleosides in cancer patients using a set of six monoclonal antibodies. *Cancer Lett* 61, 255–262.
5. Weber, M., Davies, J. J., Wittig, D., et al. (2005) Chromosome-wide and promoter-specific analyses identify sites of differential DNA methylation in normal and transformed human cells. *Nat Genet* 37, 853–862.
6. Weber, M., Hellmann, I., Stadler, M. B., et al. (2007) Distribution, silencing potential and evolutionary impact of promoter DNA methylation in the human genome. *Nat Genet* 39, 457–466.
7. Zilberman, D., Gehring, M., Tran, R. K., et al. (2007) Genome-wide analysis of Arabidopsis thaliana DNA methylation uncovers an interdependence between methylation and transcription. *Nat Genet* 39, 61–69.
8. Keshet, I., Schlesinger, Y., Farkash, S., et al. (2006) Evidence for an instructive mechanism of de novo methylation in cancer cells. *Nat Genet* 38, 149–153.
9. The EPIGENOME Network of Excellence (2007) Whole genome amplification protocol for ChIP–chip (PROT30). (Accessed at http://www.epigenome-noe.net/ research-tools/pdfs/p30.pdf.)

Chapter 6

The MIRA Method for DNA Methylation Analysis

Tibor A. Rauch and Gerd P. Pfeifer

Abstract

DNA methylation patterns are often altered in human cancer and aberrant methylation is considered a hallmark of malignant transformation. Several methods have been developed for the characterization of gene-specific and genome-wide DNA methylation patterns. In this chapter, we describe the methylated-CpG island recovery assay (MIRA), which is based on the high affinity of the MBD2b/MBD3L1 complex for double-stranded CpG-methylated DNA. MIRA has been used in combination with microarray platforms to map DNA methylation patterns across the human genome.

Key words: DNA methylation analysis, methyl-CpG binding protein, microarrays, CpG islands, tiling arrays.

1. Introduction

Methylation of DNA at the 5-position of cytosines within CpG dinucleotides is an important component of epigenetic gene-silencing systems. Aberrations in DNA cytosine methylation may underlie human disease. In particular, the genome of cancer cells is known to undergo substantial changes in DNA methylation *(1)*. Most notable are genome-wide hypomethylation events that specifically target repetitive DNA elements, and gene-specific hypermethylation of CpG islands. CpG islands are sequences with greater than normal G+C DNA content *(2)*. They are usually between 0.5 kb and 2 kb long and contain a relatively high frequency of CpG dinucleotides. CpG sequences normally are underrepresented in mammalian genomes, owing to mutational pressure and/or lack of efficient DNA repair at methylated CpGs *(3)*. However, in most normal tissues and in the germline, CpG

Jörg Tost (ed.), *DNA Methylation: Methods and Protocols, Second Edition, vol. 507*

DOI 10.1007/978-1-59745-522-0_6 Springerprotocols.com

islands are unmethylated. Accordingly, they are not subject to erosion by mutational events and retain a close to expected frequency of CpG dinucleotides. In cancer tissue, CpG islands often are found methylated and each individual tumor may contain several hundred methylated CpG islands. The exact extent of DNA methylation changes and the mechanism that elicits these events are unknown and are subject to intense investigation. Therefore, it is of great importance to have technologies available that can interrogate the methylation status of normal and diseased tissues or cell types at a genome-wide level and at high resolution.

Several different techniques have been developed to analyze DNA methylation patterns on a genome-wide scale *(4)*. These methods include several restriction enzyme-based techniques, such as restriction landmark genomic scanning (*5*, **Chapter 11**), methylation-sensitive representational difference analysis (*6*, **Chapter 10**), and differential methylation hybridization (*7*, **Chapter 7**). Huang and colleagues were the first to apply microarrays for the analysis of DNA methylation, thus providing a significant advancement to researchers in this field *(7)*. The methods using methylation-sensitive restriction endonucleases are naturally limited by the occurrence of the respective sequences within a CpG island or any other target sequence. Another commonly used approach to identify methylated genes is based on mRNA expression arrays to identify genes reactivated by treatment with DNA methylation inhibitors, such as 5-aza-deoxycytidine (*8*, *9*, *10*, *11*, **Chapter 13**). This approach can only be used effectively with cell lines and some genes may be refractory to demethylation-coupled reactivation due to additionally imposed chromatin modifications that are independent of DNA methylation. Antibodies against 5mC have been used in immunoprecipitation experiments combined with microarrays in analogy with chromatin immunoprecipitation (ChIP on chip) assays (*12*, **Chapter 5**). Another variation of current methylation microarray approaches is the use of the methylation-dependent restriction enzyme *Mcr*BC to cleave methylated DNA *(13, 14)*. Finally, high-throughput direct sequencing of bisulfite-converted genomes can be used to derive high resolution and precise DNA methylation maps (*15*, **Chapter 14**). However, for complete genome-wide analysis, considerable bioinformatics challenges will need to be overcome since, with the exception of the rare 5mC bases, the genome consists of only three DNA bases (U or T, A, G) after sodium bisulfite conversion of cytosine.

Among the methods suitable for genome-wide mapping of DNA methylation, the methylated CpG island recovery assay (MIRA) represents an approach that is based neither on restriction endonucleases, antibodies, or sodium bisulfite treatment of the genomic DNA. MIRA depends on the facts that the methyl-CpG-binding protein MBD2b specifically recognizes methylated CpG dinucleotides *(16)* and that this interaction is strongly enhanced by the MBD3L1 protein *(17–19)*, a heterodimerization partner of MBD2 *(20)*. Among all methyl-CpG-binding proteins known, MBD2b has the highest affinity for methylated DNA and displays the greatest capacity to differentiate between methylated and unmethylated DNA. It recognizes a wide range of methylated CpG sequences with little sequence specificity *(21)*. In our lab, lack of a defined sequence specificity of the MBD2b/MBD3L1 complex was confirmed by cloning and random sequencing of MIRA-enriched DNA fragments. Pulldown of methylated fragments is most efficient when at least two methylated CpG sites are present *(17)*. In the

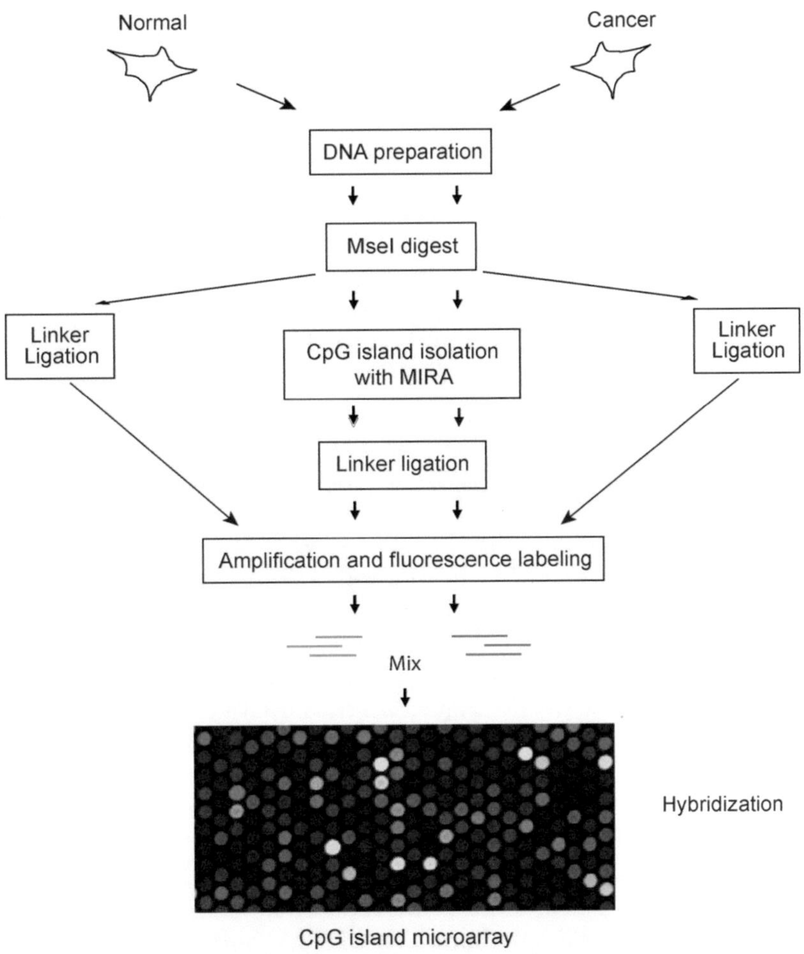

Fig. 6.1. Schematic diagram of the MIRA microarray method. Input and MIRA-enriched fractions are prepared, labeled with different dyes, mixed, and hybridized to the microarray slides. In a simplified version, MIRA-enriched DNA from normal and tumor cells can be mixed and hybridized directly. (*see* Color Plate 1)

MIRA procedure, fragmentized genomic DNA is incubated with the MBD2b/MBD3L1 high-affinity protein complex. Unlike the anti-5mC antibody precipitation technique, which requires single-stranded DNA for antibody recognition, MIRA works on normal double-stranded DNA; in fact the complex does not bind to single-stranded DNA. The CpG-methylated DNA is easily captured from the binding reaction via the GST-tagged MBD2b and glutathione beads. The isolated CpG-methylated fraction is linker ligated and then PCR amplified. The MIRA-enrichment method has been proven to be compatible with several types of microarray platforms. **Figure 6.1** outlines the procedure of the MIRA technology.

2. Materials

Buffer components and other reagents used in the MIRA procedure must be molecular biology-grade fine chemicals.

2.1. Expression and Purification of Recombinant Proteins

1. LB liquid media and agarose plates for bacterial work can be prepared according to standard bacterial protocols *(22)*.
2. BL21 (DE) Epicurian Coli Competent Cells (Agilent/Stratagene, Santa Clara, CA)
3. Lysozyme (Sigma; St. Louis, MO).
4. STE (Sodium–Tris-EDTA) buffer for GST-tagged protein purification (SGPP buffer): 10 mM Tris–HCl, pH 7.8, 150 mM NaCl, 1 mM EDTA.
5. PMSF (phenylmethylsulfonylfluoride) solution: 100 mM PMSF (Sigma) is dissolved in isobutanol and stored at −20°C. Add PMSF to buffers just before use.
6. Lysis solution: 10% (w/v) *N*-lauroylsarcosine in water.
7. Triton X-100 solution: 10% (v/v) Triton X-100 in water.
8. Glutathione Sepharose 4B beads (GE Healthcare; Uppsala, Sweden).
9. Glutathione bead wash buffer: Phosphate buffered saline (137 mM NaCl, 2.7 mM KCl, 10 mM Na_2HPO_4, 1.8 mM KH_2PO_4) and 0.1% (v/v) Triton X-100.
10. GST-tagged protein elution buffer: 50 mM Tris–HCl, pH 8.5, 150 mM NaCl, 20 mM reduced glutathion (Sigma) and 0.1% (v/v) Triton X-100.
11. Protein dialysis buffer: 50 mM Hepes, pH 7.4, 150 mM NaCl, 5mM β-mercaptoethanol and 50% (v/v) glycerol.
12. STE (Sodium-Tris-EDTA) buffer for His-tagged protein purification (SHPP buffer): 10 mM Tris–HCl, pH 7.8, 150 mM NaCl, 0.1 mM EDTA.

13. Ni-NTA agarose beads (Novagen/EMG; Darmstadt, Germany).
14. Ni-NTA agarose beads wash buffer: 50 mM NaH_2PO_4, pH to 8.0 with 1 M NaOH, 300 mM NaCl, 20 mM imidazole.
15. His-tagged protein elution buffer: (50 mM NaH_2PO_4, pH to 8.0 with 1 M NaOH, 300 mM NaCl, 250 mM imidazole).

2.2. MIRA Procedure and Amplicon Generation

1. MseI enzyme, NEB 2 buffer and 1 mg/mL BSA (New England Biolabs; Ipswich, MA).
2. QIAquick PCR purification kit (Qiagen; Valencia, CA).
3. Sonicated JM110 bacterial DNA (*see* **Note 1**). The JM110 bacterial strain can be purchased from Agilent/Stratagene (Santa Clara, CA). The JM110 strain can grow in antibiotic-free LB medium, and chromosomal DNA is prepared from bacteria according to a standard protocol *(23)*. Purified JM110 DNA is sonicated to ~500 bp-long fragments, ethanol precipitated, and redissolved in TE buffer.
4. TE buffer: 10 mM Tris–HCl, pH 8.0 and 1 mM EDTA.
5. 10 × MIRA binding buffer: 100 mM Tris–HCl (pH 7.9), 500 mM NaCl, 10 mM DTT, 100 mM $MgCl_2$, 1.0% (v/v) Triton X-100.
6. MagneGST beads and magnetic stand (Promega; Madison, WI). Magnetic beads should be washed before use to remove preservatives. Take 2.5 μL of MagneGST beads and wash 3 × with 1 mL of PBS containing 0.1% (v/v) Triton X-100. To reduce nonspecific binding to the surface of the beads, set up a blocking reaction similar to the MIRA reaction but containing only MIRA-binding buffer and sonicated JM110 DNA. Incubate it at 4°C for 20 min on a rotating platform. Capture the beads by using the magnetic stand and carefully remove the supernatant, letting ~10 μL buffer to remain on the beads.
7. MIRA wash buffer: 10 mM Tris–HCl (pH 7.5), 700 mM NaCl, 1 mM EDTA, 3 mM $MgCl_2$, 0.1% (v/v) Triton X-100.
8. Long-linker 23-mer: 5′-AGCAACTGTGCTATCCGAGG GAT-3′ and Mse-linker 12-mer: 5′-TAATCCCTCGGA-3′. Two unidirectional linkers are annealed by combining 50 μL of 100 μM Long- and 50 μL of 100 μM Mse-linkers. This mixture is boiled for 1–2 min in a water bath, then allowed to slowly cool to room temperature. The annealed double-stranded linker can be stored indefinitely at −20°C.
9. T4 DNA ligase and 10 × ligase buffer (New England Biolabs; Ipswich, MA).
10. Taq DNA polymerase, 10 × PCR buffer, and 5 × Q solution (Qiagen; Valencia, CA).

3. Methods

3.1. Preparation of GST-MBD2b and His-MBD3L1 Proteins for MIRA

3.1.1. Expression of GST-tagged MBD2b Protein

1. Recombinant plasmids for bacterial expression of GST-MBD2b and His-MBD3L1 proteins are available upon request.
2. Transform BL21 (DE3)-competent cells with GST-MBD2b protein-expressing plasmid and plate them on ampicillin-containing LB plates.
3. Inoculate 50 mL LB (amp) with 20 well-developed bacterial colonies and grow at 37°C until OD reaches 0.6 at fixed wavelength A 600.
4. Add 50 μL of 100 mM IPTG to induce expression of GST-tagged MBD2b protein.
5. Allow the cells to grow for an additional 4–6 h at 37°C.
6. Transfer the induced bacterial culture into a 50 mL tube and centrifuge at 3500*g* for 10 min at 4°C, pour off supernatant. Bacterial cells can be stored at −80°C for several months, or proceed with protein purification.

3.1.2. Purification of GST-Tagged MBD2b Protein

1. Resuspend bacterial pellet in 10 mL of SGPP buffer containing 100 μg/mL lysozyme.
2. Add 100 μL of PMSF solution and incubate on ice for 10 min.
3. Lyse bacterial cells by addition of 1 mL of lysis solution.
4. Sonicate bacterial lysate until it clears up and is not viscous anymore.
5. Add 1 mL Triton X-100 solution to the lysate and vortex it for 20 s.
6. Centrifuge the lysate at 3500*g* for 10 min.
7. Transfer supernatant into a new tube.
8. Add 0.1 mL of Glutathione Sepharose 4B beads (50% slurry – *see* **Note 2**) to 12 mL cleared lysate and mix gently by shaking at 4°C for 45 min.
9. Pellet the beads at 1000*g* for 1 min.
10. Add 10 mL of glutathione bead wash buffer and invert the tube several times.
11. Collect the beads by centrifugation at 1000*g* for 1 min.
12. Repeat the previous two steps two more times.
13. Elute the GST-tagged MBD2b protein from the beads with 1 mL of GST-elution buffer at 4°C for 4 h on a rotating platform.
14. The eluted GST-MBD2b protein should be dialyzed against 2 l of PBS in the cold-room for 5 h and then overnight against protein dialysis buffer. After dialysis, MBD2b can be kept at −20°C for 6 months.

15. Check the protein concentration on a 10% SDS-PAGE gel using BSA controls.

3.1.3. Expression of His-tagged MBD3L1 Protein

1. Transform BL21 (DE3)-competent cells with His-MBD3L1 protein-expressing plasmid and plate them on kanamycin-containing LB plates.
2. Inoculate 50 mL LB (Kan) with 20 well-developed bacterial colonies and grow at 37°C until OD reaches 0.6 at fixed wavelength A 600.
3. Add 50 μL of 100 mM IPTG to induce expression of His-tagged MBD3L1 protein.
4. Allow the cells to grow for an additional 4–6 h at 37°C.
5. Transfer the induced bacterial culture into a 50 mL tube and centrifuge at 3500*g* for 10 min at 4°C, pour off supernatant. Bacterial cells can be stored at −80°C for several months, or proceed with protein purification.

3.1.4. Purification of His-tagged MBD3L1 Protein

1. Resuspend bacterial pellet in 10 mL of SHPP buffer containing 100 μg/mL lysozyme.
2. Add 100 μL of PMSF solution and incubate on ice for 10 min.
3. Lyse bacterial cells by addition of 1 mL of lysis solution.
4. Sonicate bacterial lysate until it clears up and is not viscous anymore.
5. Add 1 mL of Triton X-100 solution to the lysate and vortex it for 20 s.
6. Centrifuge the lysate at 3500*g* for 10 min.
7. Save supernatant into a new tube.
8. Add 0.1 mL of Ni-NTA Agarose beads to 12 mL of cleared lysate and mix gently by shaking at 4°C for 30–45 min.
9. Pellet the beads at 1000*g* for 1 min.
10. Add 10 mL of Ni-NTA agarose bead wash buffer and invert tube several times (*see* **Note 3**).
11. Collect the beads by centrifugation at 1000*g* for 1 min.
12. Repeat the previous two steps two more times.
13. MBD3L1 can be eluted from the beads with His-elution buffer at 4°C for 30 min on a rotating platform.
14. The eluted MBD3L1 protein should be dialyzed against 2 l of PBS for 5 h in the cold room and then overnight against protein dialysis buffer. After dialysis, MBD3L1 can be kept at −20°C for 6 months.
15. Check the protein concentration on a 10% SDS-PAGE gel using BSA controls.

3.2. Genomic DNA Preparation for MIRA

3.2.1. Genomic DNA Purification

Genomic DNA can be isolated from cells or tissue samples by any standard proteinase K and phenol/chloroform-extraction protocol.

3.2.2. Genomic DNA Fragmentation with MseI Endonuclease

Purified high molecular weight genomic DNA must be fragmentized for MIRA. This can be achieved by restriction endonuclease digestion. We suggest using *Mse*I enzyme for fragmentation which cuts 5′-TTAA-3′ sequences and leaves most CpG islands intact (*see* **Note 4**).

1. Set up the following reaction: 2–3 μg of genomic DNA; 5.0 μL of 10 × NEBuffer 2 buffer; 5.0 μL of 1 μg/μL BSA; 2 μL of *Mse*I (10 U/μL); and add H_2O to 50 μL.
2. Incubate the reaction at 37°C overnight.
3. Check whether the digestion is complete by running the sample on a 1.5% agarose gel.
4. Purify digested genomic DNA with Qiagen PCR purification kits according to the company's recommendations.
5. Measure the concentration of MseI-cut genomic DNA by using a spectrophotometer.

3.3. MIRA-Binding Reaction

1. Set up the following binding reaction in a 1.5 mL Eppendorf tube: 40 μL of 10 × MIRA buffer; 10 μL of 50 ng/μL of JM110 DNA; 1 μg of purified GST-MBD2b; 1 μg of purified His-MBD3L1 and add H_2O to a final volume of 350 μL.
2. Mix by pipetting and preincubate at 4°C for 20 min on a rotating platform.
3. Add 250–500 ng of *Mse*I-cut genomic DNA in 50 μL. (The final volume is now 400 μL).
4. Incubate the binding reaction at 4°C at least for 4 h (or overnight) on a rotating platform.
5. Add 10.0 μL of preblocked MagneGST beads.
6. Incubate it at 4°C for 45 min on a rotating platform.
7. Retrieve MagneGST beads carrying the enriched methylated DNA fraction. Use the magnetic stand to capture the beads, and carefully remove the supernatant with a pipette.
8. Add 800 μL of MIRA wash buffer into the tube and invert 4–5 times.
9. Retrieve beads (methyl-CpG-rich fraction) by using the magnetic stand and carefully decant supernatant.
10. Repeat steps 8 and 9 two more times.
11. Elute and purify the mCpG-enriched fraction from the MagneGST beads with QIAquick PCR purification kit according to the company's protocol.
12. Elute methyl-CpG-rich fraction from the column with 50 μL of H_2O.
13. Reduce the volume of the eluted fraction to 5 μL in a Speed Vac concentrator.

3.4. Linker-Ligation and Amplification

1. Set up the following reaction: 5.0 μL of MIRA-enriched fraction; 1.0 μL of 10 × ligase buffer; 3.0 μL of 50 μM double-stranded linker; and 1.0 μL of T4 DNA ligase. For the ligation

of "input" samples, add 10 ng of the *Mse*I-digested genomic DNA into the reaction.

2. Incubate at 4°C overnight.
3. Set up the following PCR by adding to the previous ligation mix: 10.0 μL of 10 × PCR buffer; 20.0 μL of 5 × Q solution; 4.8 μL of 25 mM $MgCl_2$; 14.0 μL of 2.5 mM dNTPs; 2.0 μL of Taq polymerase (2.5 U/μL); and 39.2 μL of H_2O.
4. Before the amplification, let Taq polymerase work at 72°C for 7 min to fill in the 3′ ends of the ligated double-stranded linkers. The two strands of the unligated double-stranded linkers are separated at this temperature and can serve as primers in the subsequent PCR.
5. Cycling parameters: denaturation: 94°C for 20 s; annealing: 68°C for 30 s; and elongation: 72°C for 2 min. Repeat the previous cycle 10–12 times (*see* **Note 5**).
6. Purify the amplicon by using a Qiagen PCR purification kit according to the company's recommendations.
7. Measure the DNA content of the amplicon by using either a Nanodrop or another spectrophotometer. If the yield is lower than expected (less than 0.5–1 μg), take 10 ng of the first amplicon and repeat the amplification once more (*see* **Note 6**).
8. MIRA-enriched and input amplicons, for example, from control and tumor tissue, can be labeled with Cy3 and Cy5 dyes, respectively, and hybridized to commercially available CpG island or promoter arrays according to the manufacturer's instructions (*see* **Note 7**).

4. Notes

1. JM110 is a Dam and Dcm methylation minus bacterial strain. Sonicated JM110 DNA is used to reduce the background by blocking nonspecific binding of the mammalian DNA to the beads and proteins.
2. A 50% slurry of Glutathione Sepharose 4B beads is prepared according to the company's protocol.
3. Ni-NTA Agarose beads can strongly bind to some plastic tubes; 0.1% (v/v) Triton X-100 may be added into the Ni-NTA agarose beads wash buffer to minimize the loss.
4. As an alternative to restriction enzyme digestion, careful sonication to produce fragments 300–500 bp in length can be used. In this case, T4 DNA polymerase treatment is essential and blunt-ended, double-stranded linkers must be used for creation of amplicons. MIRA can also be used with unamplified DNA but much more DNA is needed as starting material.

Approximately 25 μg of input DNA have been used successfully *(19)*.

5. PCR should be performed in such a way that cycling is stopped right after the linear phase of amplification. The easiest way of monitoring the amplification is to perform the PCR in a real-time thermocycler. Adding SYBR green dye into the PCR does not interfere with any of the subsequent procedures.
6. It is necessary to add 0.5 μM long-linker oligonucleotide into the second PCR.
7. We have successfully used microarrays from the UHN Microarray Centre, University of Toronto, Canada *(17)*, Agilent *(19)*, Affymetrix (unpublished data), and NimbleGen *(19)*. We followed the companies' protocols for array hybridization. For NimbleGen arrays, the labeling of amplicons, microarray hybridization, and scanning were performed by the NimbleGen Service Laboratory as previously described *(24)*. Data were extracted from scanned images using NimbleScan 2.3 extraction software (NimbleGen Systems Inc.; Madison, WI) *(19)*. For Agilent CpG island microarrays, which contain 237,000 oligonucleotide probes covering 27,800 CpG islands, two micrograms each of the amplicons from MIRA-enriched DNA and control samples were labeled with the BioPrime Array CGH Genomic Labeling kit (Invitrogen; Carlsbad, CA) with either Cy5-dCTP (e.g., tumor) or Cy3-dCTP (e.g., control) following the manufacturer's instructions. The purified labeled samples were then mixed and microarray hybridization was performed according to the Agilent ChIP-on-chip protocol (v.9.0). The hybridized arrays were scanned on an Axon 4000B microarray scanner and the images were analyzed with Axon GenePix software v.5.1. Image and data analysis were done as described *(17)*. When screening for methylated CpG islands in cancer tissue, it is important to define a reliable cut-off value for methylation-positive CpG islands. We found good concordance of the array data with bisulfite-based methylation assays by considering individual CpG islands as methylation positive when at least two adjacent probes within the CpG island scored a fold-difference factor of >3.0 when comparing tumor and normal tissue DNA.

Acknowledgements

This work was supported by NIH grants CA104967 and CA128495 to GPP.

References

1. Jones, P. A., Baylin, S. B. (2002) The fundamental role of epigenetic events in cancer. *Nat Rev Genet* 3, 415–428.
2. Bird, A. P. (1986) CpG-rich islands and the function of DNA methylation. *Nature* 321, 209–213.
3. Pfeifer, G. P. (2006) Mutagenesis at methylated CpG sequences. *Curr Top Microbiol Immunol* 301, 259–281.
4. Esteller, M. (2007) Cancer epigenomics: DNA methylomes and histone-modification maps. *Nat Rev Genet* 8, 286–298.
5. Costello, J. F., Fruhwald, M. C., Smiraglia, D. J., et al. (2000) Aberrant CpG-island methylation has non-random and tumour-type-specific patterns. *Nat Genet* 24,132–138.
6. Ushijima, T., Morimura, K., Hosoya, Y., et al. (1997) Establishment of methylation-sensitive-representational difference analysis and isolation of hypo- and hypermethylated genomic fragments in mouse liver tumors. *Proc Natl Acad Sci USA* 94, 2284–2289.
7. Yan, P. S., Chen, C. M., Shi, H., et al. (2001) Dissecting complex epigenetic alterations in breast cancer using CpG island microarrays. Cancer Res 61, 8375–8380.
8. Sato, N., Fukushima, N., Maitra, A., et al. (2003) Discovery of novel targets for aberrant methylation in pancreatic carcinoma using high-throughput microarrays. *Cancer Res* 63, 3735–3742.
9. Shi, H., Wei, S. H., Leu, Y. W., et al. (2003) Triple analysis of the cancer epigenome: an integrated microarray system for assessing gene expression, DNA methylation, and histone acetylation. *Cancer Res* 63, 2164–2171.
10. Suzuki, H., Gabrielson, E., Chen, W., et al. (2002) A genomic screen for genes upregulated by demethylation and histone deacetylase inhibition in human colorectal cancer. *Nat Genet* 31, 141–149.
11. Yamashita, K., Upadhyay, S., Osada, M., et al. (2002) Pharmacologic unmasking of epigenetically silenced tumor suppressor genes in esophageal squamous cell carcinoma. *Cancer Cell* 2, 485–495.
12. Weber, M., Davies, J. J., Wittig, D., et al. (2005) Chromosome-wide and promoter-specific analyses identify sites of differential DNA methylation in normal and transformed human cells. *Nat Genet* 37, 853–862.
13. Lippman, Z., Gendrel, A. V., Colot, V., et al. (2005) Profiling DNA methylation patterns using genomic tiling microarrays. *Nat Methods* 2, 219–224.
14. Nouzova, M., Holtan, N., Oshiro, M. M., et al. (2004) Epigenomic changes during leukemia cell differentiation: analysis of histone acetylation and cytosine methylation using CpG island microarrays. *J Pharmacol Exp Ther* 311, 968–981.
15. Taylor, K. H., Kramer, R. S., Davis, J. W., et al. (2007) Ultradeep bisulfite sequencing analysis of DNA methylation patterns in multiple gene promoters by 454 sequencing. *Cancer Res* 67, 8511–8618.
16. Hendrich, B., Bird, A. (1998) Identification and characterization of a family of mammalian methyl-CpG binding proteins. *Mol Cell Biol* 18, 6538–6547.
17. Rauch, T., Li, H., Wu, X., et al. (2006) MIRA-assisted microarray analysis, a new technology for the determination of genome-wide DNA methylation patterns, identifies frequent methylation of homeodomain containing genes in lung cancer cells. *Cancer Res* 66, 7939–7947.
18. Rauch, T., Pfeifer, G. P. (2005) Methylated-CpG island recovery assay: a new technique for the rapid detection of methylated-CpG islands in cancer. *Lab Invest* 85, 172–180.
19. Rauch, T., Wang, Z., Zhang, X., et al. (2007) Homeobox gene methylation in lung cancer studied by genome-wide analysis with a microarray-based methylated CpG island recovery assay. *Proc Natl Acad Sci USA* 104, 5527–5532.
20. Jiang, C. L., Jin, S. G., Pfeifer, G. P. (2004) MBD3L1 is a transcriptional repressor that interacts with MBD2 and components of the NuRD complex. *J Biol Chem* 279, 52456–52464.
21. Fraga, M. F., Ballestar, E., Montoya, G., et al. (2003) The affinity of different MBD proteins for a specific methylated locus depends on their intrinsic binding properties. *Nucleic Acids Res* 31, 1765–1774.
22. Swords, W. E. (2003) Chemical transformation of*E. coli*. *Methods Mol Biol* 235, 49–55.
23. Bickley, J., Owen, R. J. (1995) Preparation of bacterial genomic DNA. *Methods Mol Biol* 46, 141–148.
24. Selzer, R. R., Richmond, T, A, Pofahl, N. J., et al. (2005) Analysis of chromosome breakpoints in neuroblastoma at sub-kilobase resolution using fine-tiling oligonucleotide array CGH. *Genes Chromosomes Cancer* 44, 305–319.

Chapter 7

The HELP Assay

Mayumi Oda and John M. Greally

Abstract

Genomic representations using ligation-mediated PCR have been used successfully as the foundation for a number of high-throughput assays. *Hpa*II tiny fragment enrichment by ligation-mediated PCR (HELP) is an example of the use of such representations to study cytosine methylation in the genome. The HELP assay differs from most other assays testing cytosine methylation because of its positive representation of hypomethylated DNA in the genome, whereas other assays infer the presence of hypomethylated sequences by the absence of signal, for which there can be confounding technical reasons. Hypomethylated sequences represent the minority of the genome and tend to be located at unique sequences with functionally interesting properties such as transcription start sites. By performing a comparative genomic hybridization using an *Msp*I representation from the same DNA sample, we represent all potential loci that could be generated by *Hpa*II in the situation of global hypomethylation; in practice, *Hpa*II generates a subset of loci from this population, allowing us to discriminate hypomethylated loci (represented by both *Hpa*II and *Msp*I) from methylated loci (represented by *Msp*I only).

Key words: Cytosine methylation, CG dinucleotide, microarray, epigenetic, epigenome.

1. Introduction

In general, assays testing cytosine methylation fall into three categories, those based on (1) selective sensitivity of certain restriction enzymes to methylation, (2) selective affinity of certain proteins for methylated DNA, and (3) selective conversion of unmethylated cytosines to uracil following bisulphite exposure. All three categories are described in this volume; the HELP assay *(1)* is an example of the use of methylation-sensitive restriction enzymes. Genomic DNA is digested with the

Jörg Tost (ed.), *DNA Methylation: Methods and Protocols, Second Edition, vol. 507*

DOI 10.1007/978-1-59745-522-0_7 Springerprotocols.com

methylation-sensitive restriction enzyme *Hpa*II, which only digests when the restriction site (CCGG) is unmethylated. Adapters are ligated to the fragments created by digestion that are subsequently used for ligation-mediated PCR amplification. A second aliquot of the sample is digested in parallel with the methylation-insensitive isoschizomer (recognizing the same restriction site) *Msp*I. The two digestion products are differentially labeled with two fluorophores and DNA methylation can be analyzed by cohybridization of the two fractions on various microarray products. The advantages offered by HELP are extreme technical ease of use and the ability to incorporate a number of internal controls, principal among which is the use of an *Msp*I representation for comparison. The signal at a given locus from a *Hpa*II representation can be influenced not only by the methylation status of that locus but also the size of the fragment, its base composition (both variables influencing PCR amplification), and whether the locus is mutated in any way (copy number, mutations of the CG-containing, and therefore highly mutable, restriction enzyme target site). As the *Msp*I representation will be influenced to the same extent by these variables, expressing the *Hpa*II signal as a function of the *Msp*I representation at each locus allows more robust comparison of different loci from the same DNA sample, an 'intragenomic' comparison. In common with other assays, HELP also allows two different samples to be compared ('intergenomic'), looking for differences in methylation between cell types.

Base composition affects all cytosine methylation assays in different ways. Affinity-based assays are influenced by the density of CG dinucleotides in a given sequence, while restriction enzyme-based assays will interrogate more or fewer loci depending on the number of restriction sites locally. The validation requirements for the microarray data also differ between affinity-based assays that require all CG dinucleotides over extended regions to be tested for their cumulative effect on binding, whereas restriction enzyme-based assays allow targeting of the restriction sites themselves.

Finally, it should be stressed that the degree of difficulty associated with performing these molecular assays is generally outweighed significantly by the challenges associated with their analysis. The relative simplicity of the readout of the HELP assay in terms of a bimodal distribution of *Hpa*II/*Msp*I ratios lends itself to a categorical data output (methylated and hypomethylated) that is of practical use in understanding the physiology of the epigenome and in terms of understanding the variability of methylation in different cell states.

2. Materials

2.1. Genomic DNA Extraction

1. 1 × PBS: 137 mM NaCl, 2.7 mM KCl, 4.3 mM Na_2HPO_4, and 1.4 mM KH_sPO_4.
2. 1.0 M Tris–HCl, pH 8.0: Dissolve 121.14 g Tris into approximately 800 mL of distilled water and adjust pH to 8.0 with HCl. Fill the solution up to 1.0 L and autoclave.
3. 0.5 M EDTA, pH 8.0: Dissolve 93.06 g EDTA into distilled water; adjust pH with HCl, make up to 1.0 L, and autoclave.
4. 20% SDS: Dissolve 20 g of SDS in autoclaved water. Bring the solution to a final volume of 100 mL.
5. RNAse A (Sigma, St. Louis, MO): Resuspend in water to a final concentration of 10 mg/mL. Aliquot and store at −20°C.
6. Extraction buffer: 10 mM Tris–HCl, pH 8.0, 0.1 M EDTA, pH 8.0, 0.5 % SDS, 20 μg/mL RNAse A. Prepare fresh each time.
7. Proteinase K 20 mg/mL (Invitrogen, Carlsbad, CA).
8. TE: 10 mM Tris–HCl, pH 8.0, 1.0 mM EDTA pH 8.0.
9. Tris-EDTA (TE)-saturated phenol.
10. Chloroform.
11. Isoamyl alcohol.
12. Phenol–Chloroform–Isoamyl alcohol (PCI): Phenol: Chloroform: Isoamyl Alcohol 25:24:1 by volumes. Prepare fresh each time.
13. Spectra/Por dialysis tubing (Spectrum Laboratories, Rancho Dominguez CA) MWCO 12K–14K kDa, pretreated according to manufacturer's recommendations and washed in double-distilled water prior to use.
14. 20 × SSC: 3.0 M NaCl, 0.3 M sodium citrate, pH 7.0. Dissolve 175.3 g NaCl and 88.2 g sodium citrate in 800 mL of water. Adjust pH to 7.0 with NaOH or HCl. Make up to 1.0 L with water and autoclave.
15. Polyethylene glycol (PEG) molecular weight 20,000 (Sigma).

2.2. Genomic Representations

1. *Hpa*II (New England Biolabs, Ipswich, MA).
2. *Msp*I (New England Biolabs).
3. Phenol–chloroform mix: TE-saturated phenol: chloroform 1:1 by volumes.
4. T4 DNA Ligase (Invitrogen, Carlsbad, CA).
5. Primer JHpaII 12 (HPLC purified): 5′-CGGCTGTTCATG-3′
   — Resuspended to a concentration of 6 OD/mL
6. Primer JHpaII 24 (HPLC purified): 5-CGACGTCGACT-ATCCATGAACAGC-3′

— Resuspended to a concentration of 12 OD/mL
7. Primer NHpaII 12 (HPLC purified): 5′- CGGCTTCC-CTCG -3′
   — Resuspended to a concentration of 6 OD/mL
8. NHpaII 24 (HPLC purified): 5′- GCAACTGTGCTATCC-GAGGGAAGC -3′
   — Resuspended to a concentration of 12 OD/mL
9. Pre-annealing of linkers: mix equal volumes of the pairs of 12mer and 24mer linkers (6 OD/ml and 12 OD/ml, respectively) in a screw-top Eppendorf. Boil for 5 min and then allow the reaction to cool down to room temperature. The annealed linkers can then be stored at −20°C (*see* **Note 1**).
10. 1.0 M Tris–HCl, pH 8.9: dissolve 121.14 g Tris into about 800 mL of distilled water and adjust pH to 8.9 with HCl. Fill up the solution up to 1.0 L and autoclave.
11. 1.0 M Ammonium Sulfate: dissolve 13.21 g of ammonium sulfate into distilled water, make up to 100 mL and autoclave.
12. 1.0 M Magnesium chloride.
13. Bovine serum albumin (BSA, Sigma).
14. β-mercaptoethanol.
15. 4 mM dNTP mix: dilute from 10 mM stock (Fisher, Pittsburgh, PA). Make 200 μL aliquots and store at −20°C.
16. Native Taq Polymerase and supplied buffer (Invitrogen).
17. QIAquick PCR purification kit (Qiagen, Valencia, CA).

3. Methods

Identification of hypomethylated regions by the HELP assay depends on the differential digestion and amplification of genomic DNA based on its 5mC content. For this difference to become evident, the assay relies on the selective digestion of the DNA at the 5′-CCGG-3′ restriction site recognized by *Hpa*II and *Msp*I. For this purpose, intact high molecular weight DNA must be used and the restriction digestion reaction must be carried out in such a fashion that digestion to completion is ensured. While any preparation technique that generates high-quality DNA is suitable, we recommend the following approach as a reliable, gentle means of generating material of excellent quality.

3.1. Genomic DNA Extraction

The recommended DNA preparation technique involves standard cell lysis, proteinase K digestion, and extraction in organic solvents, but proceeds to a dialysis step to purify and concentrate the material prior to use (*see* **Note 2**). If the sample is limited in amount, it may be more appropriate to use ethanol precipitation following the extraction step, which we have also found to be reliable.

1. Pellet down 2–3 million cells at room temperature for 5 min at 1500 rpm. Remove supernatant, resuspend the cell pellet in 1.0 mL of 1 × PBS, and wash once by spinning 5 min at 1500 rpm. Discard supernatant.
2. Resuspend the cell pellet in 50 μL of 1 × PBS. Pipette gently up and down until no cell clumps are visible. Add 500 μL of extraction buffer and incubate at 37°C for 1 h in a water bath.
3. Add 2.75 μL of proteinase K (20 mg/mL) to a final concentration of 100 μg/mL and incubate overnight at 50°C in a water bath.
4. Add one volume (~550 μL) of TE-saturated phenol and mix completely but gently by inversion (10 min on a rocking platform is best); centrifuge for 5 min at room temperature at top speed in microcentrifuge (16,000*g*). Do not mix by vortexing since this may shear the DNA.
5. Transfer the supernatant (aqueous phase) into a new tube, leaving behind any impurities, and being careful not to disturb the interface. Add an equal volume of PCI (25:24:1) and mix well by rocking for 10 min at room temperature, then centrifuge for 5 min at room temperature at 16,000*g*.
6. Transfer the supernatant into a new tube and measure the volume. If the supernatant is not clear, repeat step 5 until it becomes completely clear.
7. Transfer to cleared supernatant to pretreated dialysis tubing, clamp open end, and dialyze against 2.0 L of 2 × SSC at 4°C overnight, stirring gently.
8. When dialysis is complete, remove from 2 × SSC, dry outside of tubing gently, and dredge with PEG 20,000 to cause water to exit by osmosis. Do not overextract water; reduce volume to generate an expected DNA concentration in excess of 200 ng/μL, then wipe off PEG gently, unclamp the end of the tubing, and gently transfer DNA solution to an Eppendorf tube. Store at –20°C (*see* **Note 3**).
9. Quantify the DNA using a spectrophotometer and run 1 μL on a 1% agarose gel.
10. Do not proceed with digestion or amplification if the DNA does not appear to be intact (*see* **Note 4**).

3.2. Ligation-Mediated PCR

1. Set up a restriction digestion of 1 μg of genomic DNA with 2 μL (40 U) of either *Hpa*II or *Msp*I in separate 200 μL reactions, using NEB buffer #1 for *Hpa*II and buffer #2 for *Msp*I as recommended by the manufacturer. Incubate overnight at 37°C.
2. Run 10 μL of the digested DNA on a 1% agarose gel. The two digests should appear different: for the *Hpa*II digest, most of the DNA will remain high molecular weight, whereas with *Msp*I there should appear an almost even smear with

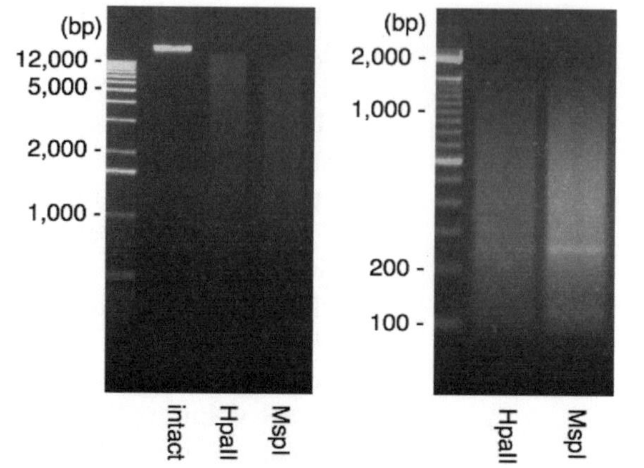

Fig. 7.1. Digestion of genomic DNA using *Hpa*II (*left panel*, third lane) and *Msp*I (fourth lane from left) generates different distributions of sizes despite cutting at the same 5′-CCGG-3′ target site. It is apparent that the *Hpa*II digestion leaves more DNA in the higher molecular-weight range, reflecting the methylation of the cytosine of the central CG dinucleotide within the target site. Lower molecular weight material in the *Hpa*II digestion is, by definition, enriched for unmethylated DNA. This fraction of the genome can be isolated using ligation-mediated PCR, as described in the protocol, generating the size distribution shown in the right panel. The bands observed within the *Msp*I representation are derived from Alu SINE sequences in the genome (Oda et al., unpublished). That they are not observed in the *Hpa*II representation reflects their frequent methylation in the genome.

no remnant of high molecular weight DNA (**Fig. 7.1**, *see* **Note 5**).

3. Add 200 μL of TE pH 8.0 to the digested DNA and 400 μL of saturated phenol:chloroform mix (1:1) and vortex briefly. Centrifuge at 16,000*g* for 10 min at room temperature.
4. Remove upper (aqueous) phase (about 400 μL) from last step and transfer into a clean tube. Add 1 μL of glycogen (20 μg/μL) and 40 μL of 3 M NaOAc pH 5.2 and mix well. Then add 1000 μL of 100% ethanol, vortex briefly, and spin at 16,000*g* for 45 min at 4°C (*see* **Note 6**).
5. Remove the supernatant and wash the pellet with 70% ethanol. Once you have carefully removed all of the ethanol, resuspend the pellet in 15.5 μL of 10 mM Tris pH 8.0. Set up the linker ligation on the same day, since the digested DNA will have single-stranded overhangs that may degrade.
6. Set up each ligation reaction in a PCR tube as follows:

5× T4 DNA ligase buffer	6 μL
DNA from last step	11 μL
Pre-annealed JHpaII linkers	7.5 μL
Pre-annealed NHpaII linkers	7.5 μL

Place the reaction mix in a thermocycler at 55°C for 5 min and then ramp down to 4°C over 1 h. Once the temperature in the thermocycler reaches 4°C, add 1.0 μL of T4 ligase into the reaction, mix, and incubate overnight at 16°C.

7. On the following day, remove the reactions from the thermocycler and transfer to a new Eppendorf tube. Dilute each reaction with 970 μL of 10 mM Tris–HCl pH 8.0. The linker-ligated DNA can be stored indefinitely at –20°C.
8. Set up the PCR reaction as follows in a 1.5 mL tube (*see* **Note 7**):

	MspI	***HpaII***
Diluted ligated DNA from last step	40 μL	80 μL
Pre-annealed JHpaII adapters	8 μL	8 μL
Pre-annealed NHpaII adapters	8 μL	8 μL
10 × Invitrogen Taq buffer	40 μL	40 μL
50 mM $MgCl_2$	16 μL	16 μL
4 mM dNTP mix	32 μL	32 μL
5 M betaine	80 μL	80 μL
Water	172 μL	132 μL

Divide the reaction mix into four PCR tubes (99 μL each) and incubate in a thermocycler as follows:

Step 1: 72°C for 10 min
Step 2: Add 1.0 μL of Invitrogen Taq polymerase (5 U/μL) to each tube, mix well, and return to thermocycler.
Step 3: 20 cycles of 30 s at 95°C, 3 min at 72°C
Step 4: Final extension of 10 min at 72°C
Step 5: Hold at 4°C.

9. Run 10 μL of the PCR product on a 1.5% agarose gel. A smear of PCR product from 100 to 2000 bp should be clearly visible (**Fig. 7.1**). Note that as the primers constitute ~50 bp of the product, the insert size range starts at ≥50 bp.
10. Clean the product using a QIAquick PCR purification kit, eluting in 50 μL of elution buffer. Quantify the DNA using a spectrophotometer and run 1 μL on a 1.5 % agarose gel to verify that the fragment size range was not altered during the clean-up process (*see* **Note 8**).

3.3. DNA Labeling and Microarray Hybridization

In order to determine the proportion of *Hpa*II (unmethylated) to *Msp*I (methylated) representations, the HELP PCR products must be labeled and both representations cohybridized onto a microarray. The array must be specifically designed to cover genomic regions contained between two consecutive *Hpa*II restriction sites. These sites occur at different intervals across the genome, but since PCR conditions have been optimized to amplify fragments between 50 bp and 2000 bp, only genomic regions contained between *Hpa*II sites located within this distance should be included in the design of the array. These fragments are referred to as "*Hpa*II amplifiable fragments" and there are >1.5 million in the March 2006 assembly of the human genome (hg18) at the UCSC Genome Browser, of which the majority is expected to be unique (*see* **Note 9**). The choice of which *Hpa*II amplifiable fragments should be included in the array design is highly dependent on the biological question of interest to the researcher. Both whole genome and focused arrays have been successfully used for DNA methylation studies using HELP by our group and others.

3.4. Data Analysis and Interpretation

After hybridization and image acquisition, two raw data files are generated for each sample: one for *Hpa*II and one for *Msp*I. The actual steps involved in the analysis will be highly dependent on the microarray platform and assay design chosen by the researcher. However, some general concepts will be common to all designs. As discussed above, proportional methylation at the different *Hpa*II sites can be determined by the relative abundance of the fragment contained between them in the two fractions. Three steps must be followed in order to arrive to that point:

(a) If more than one oligo was used to represent each *Hpa*II-amplifiable fragment, then the information (i.e., intensity) from all of these oligos must be summarized as one value per fragment.

(b) Since *Msp*I represents the total population of possible fragments and therefore the internal control of the method, any fragments that are not generating signal in the *Msp*I representation must be excluded from the analysis. A fragment may be absent in the *Msp*I fraction both for technical and biological reasons. Technical reasons include lack of amplification of the fragment during the PCR step, as well as failure during the labeling or hybridization reactions. Biologically, a given fragment can be absent due to the presence of a genomic deletion, a frequent event in cancer, or the presence of a mutation at the restriction site, which prevents it from being recognized by the restriction enzymes.

(c) Finally, the relative proportion of *Hpa*II to *Msp*I signal must be determined for each fragment. When the distribution of the population of these proportions is studied, a bimodal distribution becomes readily apparent, showing the presence of a methylated fraction of the genome (lacking a *Hpa*II signal)

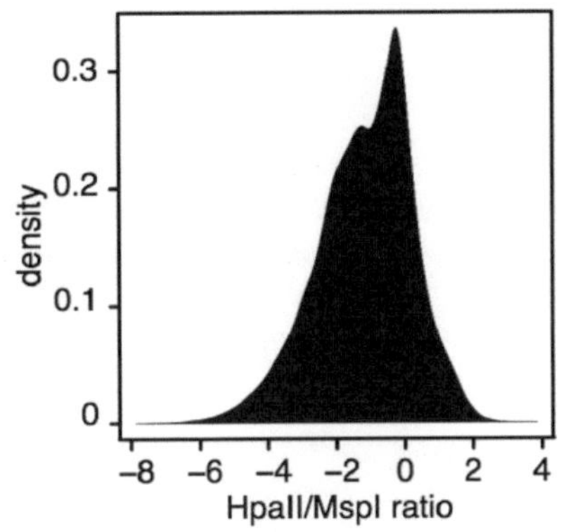

Fig. 7.2. The $\log_2$ ratio distribution of values obtained from a HELP assay manifests a bimodal distribution, as shown. The GM06990 cell line is unusual for having a substantial proportion of loci in the right peak, the higher ratio values of which indicate the hypomethylated subpopulation of loci.

and a hypomethylated fraction, with increasing amounts of *Hpa*II signal (**Fig. 7.2**, *see* **Note 10**).

When the protocol is followed as described above, we find that technical reproducibility is very high, with Pearson correlation coefficients between replicates greater than 0.96 (**Fig. 7.3**). By including random probes on the microarray, we can define background signal intensities, as shown by dashed (median) and solid (+2.5 median absolute deviations) lines in **Fig. 7.3**.

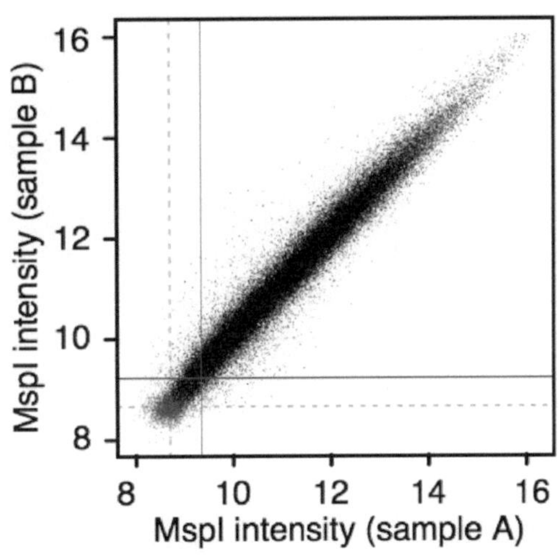

Fig. 7.3. The technical reproducibility of the HELP assay is extremely robust. The scatter plot shows a tight correlation between experimental replicates of *Msp*I ($r = 0.981$). The presence of random oligonucleotides on the microarray allows us to define background fluorescence, shown by the dashed lines (median) and solid lines (+2.5 median absolute deviations) for these random probes on the array.

4. Notes

1. The primers are pre-annealed in advance for convenience. If preferred, they can be pre-annealed immediately prior to use.
2. Dialysis is not absolutely required. We have had numerous instances of successful HELP assays with ethanol-precipitated DNA. We recommend dialysis because of the more consistent quality obtained.
3. Osmosis will cause increased salt concentration in the DNA solution, which would be a problem if TE were used at this step, and may be a problem even with SSC if excessive water removal and salt concentration occurs. If the volume needs to be reduced substantially (e.g., 10-fold), it is probably worth adding a further dialysis step against a more dilute buffer prior to osmotic water removal.
4. A small amount of DNA shearing is probably okay, as the later adapter ligation step involves compatible cohesive ends, so nonspecifically broken DNA molecules will not be amplified.
5. Some DNA samples (e.g., from DNA methyltransferase-deficient or 5-aza-2-deoxycytidine-treated cells) may have so little DNA methylation that the *Hpa*II digestions resemble those for *Msp*I.
6. The glycogen is not a requirement; it should help to precipitate limited amounts of DNA.
7. The increased amount of *Hpa*II ligation product reflects the lower amount of material it generates in the target size range compared with *Msp*I. If the amount of digestion by *Hpa*II looks unusually low or high, the amount of ligation product used at this stage may need to be adjusted accordingly.
8. Following clean-up, loss of any primer dimers is a desirable outcome; loss of product within the amplified size range is not.
9. Currently we use a design testing >1.32 million loci in the human genome.
10. When comparing two samples for methylation differences, we find the most significant changes to be those that involve moving from one category in one cell type to the other category in the other cell type, rather than absolute differences in ratio values.

Acknowledgements

This work was originally supported by a grant from the National Institutes of Health (NCI) R03 CA111577 and subsequently by NICHD grant R01 HD044078 to JMG. The author also acknowledges the contributions of the Bioinformatics Shared Resource and the Genomics Core Facility at the Albert Einstein College of Medicine, as well as resources from the Albert Einstein Cancer Center.

References

1. Khulan, B., Thompson, R. F., Ye, K., et al. (2006) Comparative isoschizomer profiling of cytosine methylation: the HELP assay. *Genome Res* 16, 1046–1055.

Chapter 8

Differential Methylation Hybridization: Profiling DNA Methylation with a High-Density CpG Island Microarray

Pearlly S. Yan, Dustin Potter, Daniel E. Deatherage, Shili Lin, and Tim H.-M. Huang

Abstract

Differential methylation hybridization (DMH) is a high-throughput DNA methylation screening tool that utilizes methylation-sensitive restriction enzymes to profile methylated fragments by hybridizing them to a CpG island microarray. This array contains probes spanning all the 27,800 islands annotated in the UCSC Genome Browser. Herein we describe a revised DMH protocol with clearly identified quality control points. In this manner, samples that are unlikely to provide good readouts for differential methylation profiles between the test and the control samples will be identified and repeated with appropriate modifications. In addition to the step-by-step laboratory DMH protocol, we also provide a detailed description regarding DMH data analysis. The suggested microarray platform contains 244,000 probes and it can be a daunting barrier for researchers with no prior experience in analyzing DNA methylation data. We have created a data analysis pipeline available in a user friendly, publicly available interface, the Broad Institute's GenePattern software, which can be accessed at http://bisr.osumc.edu:8080/gp. This permits scientists to use our existing data analysis modules on their own data. As we continue to update our analysis algorithm and approaches to integrate high-throughput methylation data with other large-scale data types, we will make these new computation protocols available through the GenePattern platform.

Key words: DNA methylation, differential methylation hybridization (DMH), CpG islands (CGI), microarray, M-score, GenePattern modules.

Pearlly S. Yan and Dustin Potter contribute equally to the work described in this chapter.

Jörg Tost (ed.), *DNA Methylation: Methods and Protocols, Second Edition, vol. 507*

DOI 10.1007/978-1-59745-522-0_8 Springerprotocols.com

1. Introduction

The current version of the Differential Methylation Hybridization (DMH) protocol is designed to survey densely methylated fragments in human CpG islands. It is reasonable to suppose that extensive methylation marks in a promoter CpG island will result in reduced gene transcription activity. This protocol is therefore well suited for the genome-wide identification of differential methylation between two samples with functional importance. The DMH protocol was first described by our group in 1997 *(1)*. Since then, we have applied this protocol as a screening tool for genome-wide DNA methylation analysis in a multitude of cancer samples from different organ sites *(2–12)*. The protocol is detailed in **Fig. 8.1** indicating also the quality-control check points described in this chapter. The genome is interrogated with two methylation-sensitive restriction enzymes: *Hpa*II and *Hin*P1I. Using two restriction enzymes in a sequential manner will reduce the chance of retaining fragments due to incomplete enzymatic digestion. As such, the DMH protocol will enrich fragments with fully methylated recognition sites between the ligated linkers. As these recognition sites are often found in the CpG Islands (CGIs), this renders the linker-mediated PCR fragments highly methylated.

When we originally converted the DMH protocol to a high-throughput method, we hybridized the DMH probes with microarray targets derived from PCR products prepared from a human CGI clone library. Data analysis using such a system is less complicated as both the array targets and the sample probes are created from fragments with the same external restriction site (*Mse*I). A major drawback from this scheme is the inability to ensure array quality between batches as the printing targets needs to be periodically regenerated (more or less of each target spot due to inconsistent PCR efficiencies) and the continual risk of cross-contamination due to carry-over by array pins (medium- to high-copy number targets may mask out low-copy targets). These concerns prompted us to search for a more dependable and robust platform for the DMH protocol. We take advantage of the availability of the long oligonucleotide-based CGI array, initially devised for chromatin immunoprecipitation (ChIP)-on-chip experiments, for DMH hybridization. The quality of these arrays is excellent and dependable. However, data analysis becomes more challenging as the total array targets increase from 21,000 to 244,000 and the target size decreases from an average length of 870 bp to targets ranging from 45 bp to 60 bp.

The revamping of the DMH protocol requires a parallel investment in data-analysis effort. The data visualization and

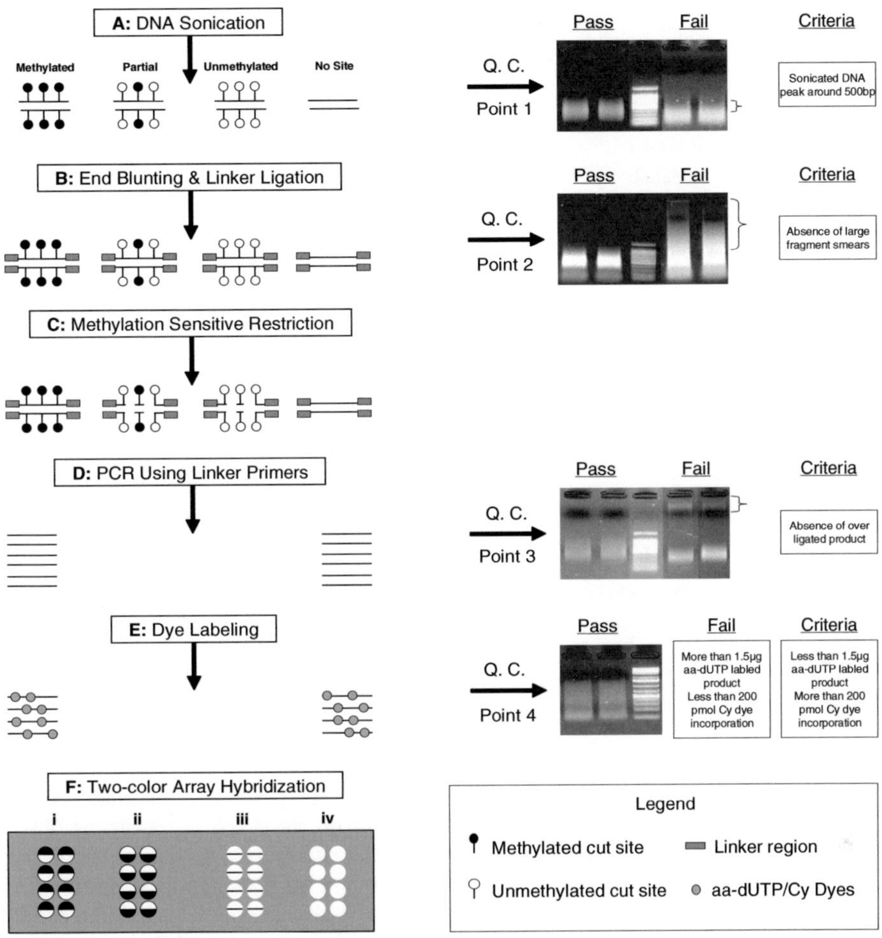

Fig. 8.1. Overview of the DMH protocol. *Left half*: Schematic diagram of the DMH protocol showing sonicated fragments that are methylated, partially methylated, unmethylated, or have no-cut sites. **(A)** Genomic DNA is fragmented to ~500–800 bp in length. **(B)** Ends of sonicated DNA are repaired and adapters are ligated to the repaired ends. **(C)** Linker-ligated fragments are digested with methylation-sensitive restriction enzymes such as *Hin*P1I and *Hpa*II (Note: digestion requires both the presence of the restriction recognition sites and the absence of mC). **(D)** Fragments that are retained after enzyme restrictions as well as fragments without restriction sites between the linkers are amplified by linker-mediated PCR (Note: Fragments with unmethylated restriction sites will not be amplified). **(E)** Aminoallyl dUTPs (aa-dUTPs) are incorporated into the amplified fragments using a high-concentration Klenow fragment. Cy dyes are subsequently chemically coupled to the aa-dUTPs and the sample is ready for array hybridization. **(F)** Two-color array hybridization outcomes: (i) test sample is hypermethylated in comparison to the control sample (pseudo-red); (ii) test sample is hypomethylated (pseudo-green); (iii) both samples have an equal methylation status (pseudo-yellow); (iv) unmethylated regions or poor probe/target interaction (hybridization signals in both channels below background signals). *Right half*: DMH quality controls tracking points to ensure sample reproducibility. Representative gel images characterizing samples with either "pass" or "fail" likelihoods on critical points along the DMH protocol. The bracketed regions highlight the telltale signs of a failed sample.

data-analysis protocols (**Fig. 8.2**) described in this chapter are provided for ease of use as initial data processing. Groups with expertise in computational and data modeling are encouraged to augment these pipelines with their unique data-analysis algorithms as there are many other protocols suitable for this data type.

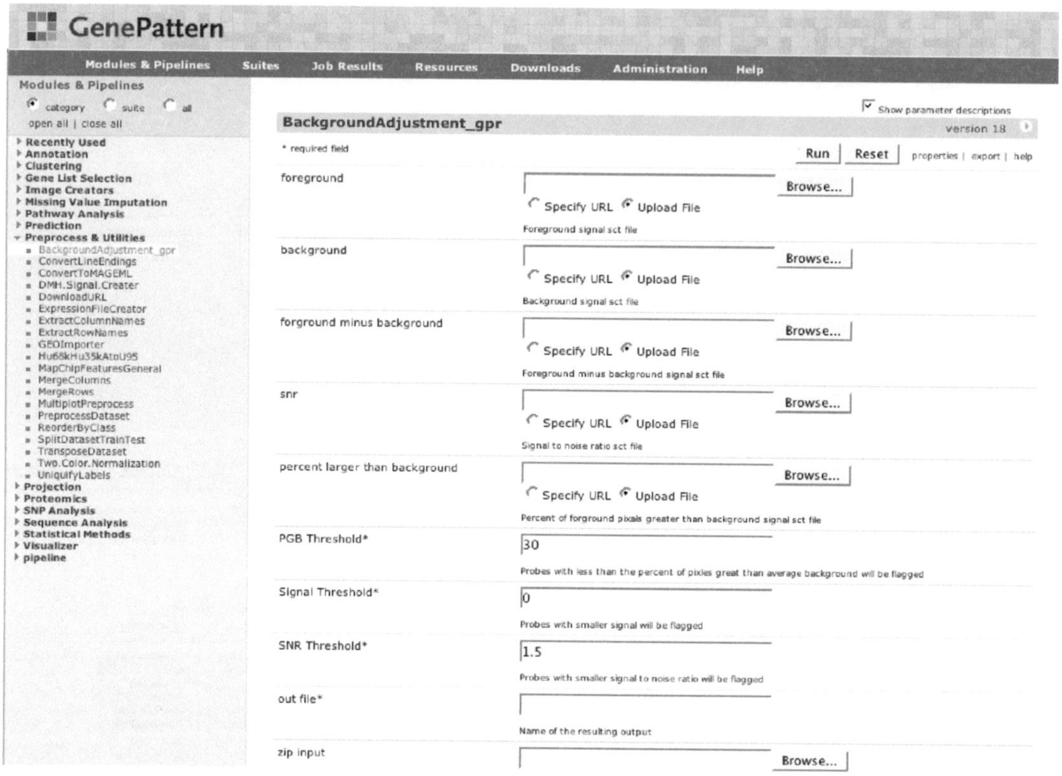

Fig. 8.2. A GenePattern screen shot showing the DMH data preprocessing modules and data analysis modules created by the OSU-ICBP computational scientists. Several modules for processing and analyzing DMH data have been developed for the web application interface GenePattern. The GenePattern application can be accessed at http://www.bisr.osumc.edu:8080/gp/. GenePattern can also be downloaded from http://www.broad.mit.edu/cancer/software/genepattern/ and installed locally. To run the DMH modules locally, they must be installed into GenePattern from http://www.skull.med.ohio-state.edu/tools/files/GenePattern_Modules/ using the *install module from zip* function in GenePattern. The image scanning process using Axon scanner will produce a total of six files. Of importance to subsequent data analysis are the gpr files. In order to employ the preprocessing and analysis methods developed in GenePattern, it is necessary to zip all the gpr files into a single archive.

2. Materials

2.1. Genomic DNA Isolation

1. Genomic DNA isolation: QIAamp DNA Mini Kit (Qiagen, Valencia, CA).
2. DNA concentration: ND-1000 Spectrophotometer (NanoDrop Technologies, Wilmington, DE).

2.2. DNA Fragmentation

1. DNA fragmentation is performed with a Bioruptor 200 (Diagenode SA., Liege, Belgium).

2.3. DMH Amplicon Preparation

1. Blunt-end ligation:
   a. T4 DNA Polymerase (New England Biolabs (NEB), Ipswich, MA, 3 U/μL).
   b. 100 mM dNTP mix.
2. Blunt-end reaction clean-up:
   a. Zymo DNA Clean & Concentrator-5 columns (Zymo Research Corp., Orange, CA).
3. Linker sequences:
   a. JW102 (5′-GCG GTG ACC CGG GAG ATC TGC ATT C-3′).
   b. JW103 (5′-GAA TTC AGA TC-3′).
4. Linker ligation reaction mix:
   a. T4 Ligase (NEB, 400 U/μL).
   b. PEG-6000 (Sigma-Aldrich, St. Louis, MO).
5. Testing ligation reaction efficiency:
   a. DeepVent$_R$ (exo-) DNA Polymerase (NEB, 2 U/μL).
6. Methylation-sensitive restrictions:
   a. *Hpa*II (5′-C↓CGG-3′, NEB, 10 U/μL).
   b. *Hin*P1I (5′-G↓CGC-3′, NEB, 10 U/μL).
7. Linker-mediated PCR:
   a. Deep Vent$_R$ (exo-) DNA polymerase (NEB, 2 U/μL).
8. Amino-allyl dNUP incorporation:
   a. BioPrime labeling kit (Invitrogen, Carlsbad, CA, 18094-011).
   b. Aminoallyl-dUTP (Fermentas, Glen Burnie, MD, #R1101).
9. Dye coupling reaction:
   a. Cy-Dye post-labeling reactive dye (12 single-use Cy5 and 12 single-use Cy3 dye packs; Amersham Health Inc., Princeton, NJ, RPN5661).
   b. Sodium carbonate.
   c. 3 M sodium acetate, pH 5.2 (Sigma-Aldrich).

2.4. Microarray Hybridization

1. Array hybridization:
   a. Human CpG Island ChIP-on-chip Microarray Set (Agilent Technology, Santa Clara, CA, G4492A).
   b. Gasket slides (Agilent Technology, G2534).
   c. Oligo aCGH/ChIP-chip Hybridization Kit (Agilent Techology, #5188).
   d. Hybridization oven (Agilent Technology, G2545A).
   e. Hybridization chamber (Agilent Technology, G2534A).
   f. Human Cot-1 DNA (Invitrogen, 15279-011).
2. Array washing:
   a. Stabilization and Drying Solution (Agilent Technology, 5185-5979).
   b. AccuGENE 20 × SSPE buffer (Lonza, Rockford, ME, 51214).

   c. Sarcosine.
   d. Slide rack and slide tank.
   e. Rotating/rocking platform.
3. Data acquisition:
   a. GenePix 4000B Microarray Scanner (Molecular Devices Corp., Sunnyvale, CA).
   b. GenePix Pro 6.0 (Molecular Devices Corp.).

2.5. Methylation Data Processing, Visualization, and Analyses

1. Gene pattern software (http//www.bisr.osumc.edu:8080/gp/) (**Fig. 8.2**). Many statistical methods are available to analyze high-throughput data and most can be applied to DMH data with some adjustments.

3. Methods

3.1. Genomic DNA Isolation

1. Genomic DNA is isolated using the QIAamp DNA Mini Kit according to manufacturer's direction.
2. The quality of genomic DNA is ascertained by running 60–80 ng DNA on a 0.7% agarose gel. High-quality DNA has the following characteristics:
   a. high MW DNA.
   b. absence of DNA degradation (manifest as DNA smearing).
   c. absence of residual RNA (appear as low MW spot).
3. DNA concentration is determined using a ND-1000 Spectrophotometer. An accurate estimation of DNA concentration is required to ensure a successful linker-ligation step.

3.2. DNA Fragmentation

The DNA fragmentation process is an important step. Many factors can affect the reproducibility of creating a pool of similar size fragments when samples are started on different days (*see* **Note 1**).

1. The Bioruptor needs to be precooled with ice-water mixture. Prior to the beginning of the sonication process, remove all the ice particles using a strainer and add a predetermined amount of fresh ice. Bring the water level to a preset mark.
2. Operate the Bioruptor with rest periods interspersed between sonication cycles. For good-quality genomic DNA, we will sonicate the samples for eight sets of 30 s "On" and 30 s "Off". We replenish ice after the first two cycles and replace both fresh ice and ice-cold water after fourth cycle.
3. The fragment size is determined by gel electrophoresis and additional sonication cycles may be added to obtain the desired fragmentation smear (*see* **Note 1**).

3.3. DMH Amplicon Preparation

1. Take 100–200 ng of sonicated DNA and repair the sonicated ends by adding 3 μL of T4 DNA Polymerase, 4 μL of 2 mM dNTPs, 2.5 μL of 10 × BSA, and 5 μL of 10 × NEB buffer #2 in a final volume of 50 μL. Incubate this reaction mixture at 37°C in a water bath for 2.5 h. Purify the reaction mixture with Zymo DNA purification columns and elute the purified product with 29.5 μL of water.
2. Anneal linker adapters by heating equal molar amounts (stock solutions: 100 μM; amount per sample: 0.125 nmol of each linker) of JW102 and JW103 oligonucleotides at 95–100°C for 5 min and allow the mixture to cool to room temperature gradually. Use freshly prepared linker adapters for every experiment.
3. Combine DNA from step 1 with 2.5 μL (0.25 nmol of annealed linkers) containing of pre-annealed linker adapters, 1 μL of 10 mM ATP, 2.5 μL of 50% PEG-6000, 4 μL of 10 × T4 ligase buffer, and 0.5 μL of T4 ligase in a final volume of 40 μL. All reagents must be kept on ice during the assembly of the reaction mixture. The ligation reaction mixture is held at 14°C for 2 h, preferably using a thermocycler.
4. To evaluate the efficiency of the linker-ligation step, perform a ligation test PCR analysis. 1 μL of the ligation product from step 3 will be added to a PCR mixture containing 0.2 μL of 10 μM JW102, 0.4 μL of 10 mM dNTP, 2 μL of 10 × Buffer, and 0.4 μL of DeepVent polymerase in a final volume of 20 μL. Amplification conditions are as follows: 55°C for 2 min, 72°C for 5 min, 95°C for 2 min followed by a total of 17 cycles of 55°C for 30 s and 72°C for 1 min and the last cycle of 55°C for 30 s and 72°C for 10 min. PCR products are separated on a 1.5% agarose gel. Over-ligated samples will appear as high MW smears or bands not originally present in the sonicated DNA (**Fig. 8.1**). If too much material is lost in the preceding clean-up steps, the smear will be extremely faint, and sometimes a high MW band will also be evident right below the well. Both scenarios warrant restarting the DMH protocol from step 1 (*see* **Note 2**).
5. Samples with ligation test PCR products that closely resemble the smear pattern of the starting material can go on to the next step. There are many enzymes suitable for interrogating the methylation status of the linker-ligated fragments. To decrease the likelihood of incomplete digestion by any one enzyme, we use two methylation-sensitive enzymes for this interrogation (*Hpa*II and *Hin*P1I; 20 units each and 2 h per restriction step). The restrictions are carried out sequentially and the restricted product is purified with Zymo column and eluted with 40 μL of water.

6. The final linker-mediated PCR step will utilize purified restricted fragments (suggested range: 4–10 μL from step 5 as dictated by the intensity of the ligation test PCR smear) as DNA templates. The PCR mix consists of 20 μL of 10 × ThermoPol Buffer, 4 μL of 10 mM dNTP, 2 μL of 10 μM JW102, and 4 μL of Deep Vent DNA polymerase in a final volume of 200 μL. The reaction mixture is aliquot into four PCR tubes (50 μL per tube). Amplification condition is the same as those described in step 4 except that a total of 25 cycles is used in this step. The PCR products are combined and purified with a Qiaquick column. The purified products are eluted twice with 40 μL of water with an expected yield of 0.8–1.5 μg DNA. The smear pattern of the PCR products is determined to ensure it resembles the starting material (**Fig. 8.1**, *see* **Note 3**) for additional information.
7. Fluorescent dyes (Cy 5 and Cy 3) are indirectly coupled to the PCR products through the aminoallyl dNTP incorporation step. 600 ng of PCR product in 68 μL of water is combined with 60 μL of BioPrime 2.5 × random primers and denatured at 95°C for 5 min. The mixture is then placed on ice for 3 min. 15 μL of 10 × dNTP (2 mM dATP, dCTP and dGTP, 0.35 mM TTP), 4 μL of 10 mM aminoallyl-dUTP, and 3 μL of Klenow (40 U/μL) are added to the cooled mixture. The reaction mixture is incubated at 37°C for 6 h and purified with a Qiaquick column. The purified aminoallyl-labeled products are eluted twice with 40 μL of water. We expect the final DNA yield to be between 6 μg and 10 μg (*see* **Note 4**).
8. The fluorescent dye-coupling step should be carried out in dim light (absence of direct fluorescent lighting). Products from the previous step are savant dried and resuspended in 3 μL of water. Cy 5 dye (one single-use dye pack) is reconstituted in 3 μL of 0.1 M sodium carbonate buffer (pH 9.0) and is combined with the test sample. The dye-PCR mix is combined and mixed at 0.5 h intervals for 3.5 h. The Cy 3 dye (one single-use dye pack) is similarly resuspended and combined with the control sample, but it is only mixed for 1.5 h (stagger the starting time of these two mixtures so that the two samples are ready at the same time). At the end of incubation period, add 35 μL of 100 mM sodium acetate (pH 5.2) and 35 μL of water into each reaction mixture. Purify each sample with a Qiaquick column and elute the dye-coupled samples in 80 μL of water. Obtain absorbance at 260 nm for DNA concentration, at 550 nm for Cy3 incorporation and at 650 nm for Cy5 incorporation.

3.4. Microarray Hybridization/Washing

1. Combine volumes of fluorescently labeled samples containing 300 pmol of each Cy dyes. Add 20 μg of Cot-1 DNA, 50 μL of Agilent blocking buffer, 250 μL of Agilent hybridization

buffer to the sample, and bring the total volume to 500 μL. Denature the entire mixture at 95°C for 3 min and hold the mixture at 40°C for 30 min.

2. Assemble the following hybridization accessories to permit uninterrupted hybridization workflow: hybridization chamber (chamber base, chamber top, clamp, and chamber thumbscrew), hybridization gasket slide and CpG island microarray. Add the labeled samples to the gasket slide placed in the chamber base. Invert the microarray slide (with the printed side down) with care over the gasket slide. Place the chamber top over the slide assembly and engage the clamp and chamber thumbscrew to secure the hybridization chamber. The thumbscrew should be tightened a quarter turn after the loosely tighten stage. The chamber is then placed in 65°C hybridization oven set at rotating speed of 10 rpm for 16–20 h.
3. Prewarm the Agilent Stabilization and Drying Solution at 37°C to completely dissolve the dye-stabilizing salts. Gently pry open the gasket slide from the microarray in a solution containing 300 mL 20 × SSPE buffer containing 1 mL of 5% sarcosine/L. Place the microarray in a slide rack and wash in a fresh tank of SSPE/sarcosine buffer with rocking motion for 5 min. Wash the microarray for another 5 min in water containing 3 mL 20 × SSPE buffer/L. Submerge slide in the Stabilization and Drying Solution and gently rock back and forth for 1 min. Pull slide out of the solution in a slow and controlled manner. Place in light-protected slide box and scan immediately.

3.5. Methylation Data Collection and Processing

1. Photomultiplier tube (PMT) settings are adjusted so that the overall Cy5 and Cy3 signals from each hybridized slides are balanced. The "014791_D_20070207.gal" file provided by Agilent Technology is used as the accompanying annotation file for probes present on the 244 K microarray.
2. Data files from each study are archived together into a single zip file.
3. The GenePattern **DMH.Signal.Creator** module is used to read the data files into the appropriate GenePattern file format (sct).The **DMH.Signal.Creator** is a perl script which takes a zipped archive as input and returns files containing the desired columns from each gpr file in the zipped archive.
   a. *Input zipfile*: The location of the zipped archive you wish to upload.
   b. *Input columnfile*: An optional argument indicating the location of the txt file with the names of columns that should be extracted from each gpr file. If this is not provided, the columns to be used in the next step (BackgroundAdjustment_gpr) will be extracted.

c. *Output file*: An optional string used in the naming scheme for files containing the extracted columns.
d. *Output file format*: Indicates which file format the data should be saved (either SCT or GCT). The recommended analysis modules for DMH data require the data to be in the SCT format.
e. *Output zipfile*: An optional string used to name the zipped archive of the resulting files. If left blank, the files will not be zipped in the same file.

4. The GenePattern "**BackgroundAdjustment_gpr**" module (**Fig. 8.2**) is a perl script that is used to flag problematic probes, such as low hybridization signals or hybridization signals from regions plagued with artifacts. The zip file produced by the "DMH.Signal.Creator" module is the appropriate input for this module.
   a. *Foreground*: Location in the sct file containing the foreground signals. (No need to specify this if a zip file from DMH.Signal.Creator is used).
   b. *Background*: Location in the sct file containing the background signals. (No need to specify this if a zip file from DMH.Signal.Creator is used).
   c. *Foreground minus background*: Location in the sct file containing the foreground signal minus the background signal. (No need to specify this if a zip file from DMH.Signal.Creator is used).
   d. *SNR*: Location of the file containing the signal to noise ratio of each hybridization experiment. (No need to specify this if a zip file from DMH.Signal.Creator is used.)
   e. *Percent larger than background*: Location of the file containing the percent larger than background information. (No need to specify this if a zip file from DMH.Signal.Creator is used).
   f. *PGB Threshold*: Location of the file containing the percent of foreground pixels greater than background signal in the sct file. (No need to specify this if a zip file from DMH. Signal Creator is used).
   g. *Signal Threshold*: Probes with signals less than the average background pixels will be flagged.
   h. *SNR Threshold*: Probes with low signal-to-noise ratio will be flagged.
   i. *Out file*: Name of the resulting output file.
   j. *Zip input*: Optional – Archive containing foreground, background, foreground minus background, SNR, and percent larger than background files. This zip file can be generated using the DMH.Signal.Creator module.
5. The GenePattern module **Two.Color.Normalization** consists of an R script that provides three intraslide normalization schemes: *loess, rank invariant loess, and median adjustment*;

and five inter-slide normalization schemes: *scale, quantile, Aquantile, Rquantile, and Gquantile*. For *loess* and *rank invariant loess*, the locally weighted nonparametric scatter plot smoother loess fits the correlation between average intensity of probe signals and ratio of signal intensities; the resulting residual is the normalized ratio which can be transformed back to signal intensities for the Cy3 and Cy5 channels *(13)*. The rank invariant loess fits a rank invariant subset of the data while loess fits the entire data set. The median adjustment simply subtracts the median ratio from each probe *(14)*. The goal of quantile normalization is to transform the data so that each data set will have the same distribution *(15)*. The *scale* method assumes that the arrays are linearly related and rescales each sample with an appropriate scalar *(16)*. The five quantile methods adjust signals to ensure identical distributions: *quantile* ensures signals across arrays and channels have the same distribution; *Aquantile* ensures that the average intensities are identically distributed across arrays; *Rquantile* and *Gquantile* ensure that Cy5 and Cy3 channels, respectively, have the same distributions across arrays *(17)*.

a. *Foreground file*: Location of the file containing the sct file with the array signals.
b. *Intraslide normalization*: To select for one of the following within-slide normalization scheme: loess, rank invariant loess, or median adjustment.
c. *Interslide normalization*: To select for one of the following between-slide normalization scheme: scale, quantile, Aquantile, Rquantile, and Gquantile.
d. *Iter*: Parameter passed to both loess intraslide normalization methods: determines the number of iteration steps used in the fitting process.
e. *Bin*: Parameter passed to both loess methods: determines the bin size for fitting the data.
f. *Span*: Parameter passed to both loess intraslide normalization methods: determines the span for fitting the data.
g. *Max set*: Parameter passed to the rank invariant intraslide normalization loess method: determines the maximum size of the rank invariant subset.
h. *Out file*: Name of the resulting output file.
i. *Gct*: Desired output format. If gct file format is selected, then normalized log ratio of Cy3 to Cy5 will be returned. If sct file format is selected, then normalized Cy3 and Cy5 signal will be returned.

3.6. Methylation Data Visualization and Analyses

1. The "DMH.QC" module in GenePattern will generate images to assess the quality of the scanned image, the range of

probe-signal intensities, and the effectiveness of the data preprocessing steps.

a. **Variability and signal-intensity range of replicated control probes**: Twelve replicates of 20 control sequences whose signal intensities are known to span the entire signal spectrum are printed on the array. A box plot of the log signal intensity of these 20 probes is used to reveal potential spatial artifacts due to slide quality and post-hybridization wash stringency issues. As there should be little variability in signal intensities for replicate probes, these controls are ideal reporters for spatial-specific artifacts that often exhibit as zonal color gradation, either from side to side or from top to bottom.
b. **Background intensities**: MA plots of Negative, Reserve Negatives (SM), Biological Negative (NC1), and Dark Corner (DCP) control probes provide a snapshot of probes whose signal intensities should be below the detectable background intensities.
c. **Dye Bias**: The raw intensity MA plot as well as the QC plots for Cy3, Cy5, and log ratio will reveal the extent of signal bias between the two channels. In theory, there should be no global trend (other than variation around zero) when we plot the log intensity ratio versus the average log intensities (i.e., MA plot). As the probes are randomly dispersed on the array, there should be no discernable pattern in the three spatial QC plots.
d. **Effectiveness of the normalization protocols**: Comparison of the pre- and postnormalized MA plots will reveal the effectiveness of the preprocessing step in removing dye effects as a result of sample preparation and scanning bias. In addition, the density plots for the two separate channels will delineate how the overall array signals are affected by the preprocessing step.
e. **Distribution of flagged probes**: Spatial plot of probes that are either flagged by the preprocessing step to contain hybridization artifacts or have no discernable hybridization signals allows us to assess the quality of the scanned image. It is important that we should observe no discernable pattern in this plot.

2. Methylated DNA fragments (those surviving the restriction of methylation-sensitive enzymes) are expected to be significantly longer than the arrayed probes (ranges from 45mers to 60mers); therefore, the signal intensities of probes in a contiguous chromosomal region are not independent, as fluorescent-labeled fragments will most likely hybridize to several neighboring probes. Therefore we apply a simple kernel-smoothing function termed M-score to integrate the probe-level information within a sliding window to provide

a regional methylation profile. The M-score is included in GenePattern and is used to transform probe-level information to regional information to identify genes with differential promoter methylation. The concept behind this module is presented in **Fig. 8.3**. In brief, the M-score of each probe is established by computing a relative methylation score using probes that are 500-bp upstream and 500-bp downstream from the respective probe. The first step in this approach is to rank probes according to their normalized log ratios. An arbitrary cutoff is set (here we set the cutoffs at the top and bottom 25th percentile) so that only probes with considerable changes contribute to the M-score. The next step involves the transformation of probe information to regional methylation profile. The equation $M\text{-}score = ((\#probe_{log\ upper25th} - \#probe_{log\ lower25th}) / total\ probes\ in\ 1\text{-}kb\ window)$ functions to smooth out peaks and valleys in a specific region so that an overall hyper- or hypomethylation event can become more evident. Other statistical methods are available for analyzing high-throughput data and most are amendable for the analysis of DMH data (*see* **Notes 5** and **6**).

3. Regions identified by the M-score to possess a hyper- or hypomethylation profile will be compiled and the normalized

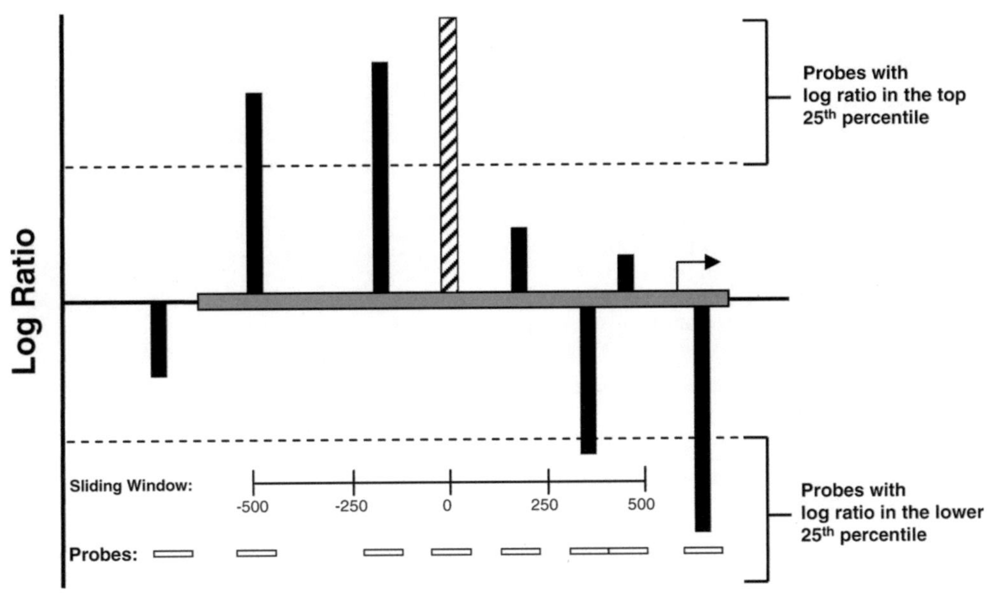

Fig. 8.3. Schematics for M-score analysis. M-score is a computational approach to combine probe-level information to create a regional methylation profile. The M-score of each probe is established by computing a relative methylation score using probes that are 500-bp upstream and 500-bp downstream from the probe itself. In this figure, the striped bar represents the probe of interest and the grey horizontal bar represents a CpG island. The horizontal dashed lines encompass the middle 50th percentile of all the log intensity ratios. *Black bars* signify the normalized log intensity ratios of probes derived from the scanned microarray image. In this schematics, the M-score for the probe of interest would be 0.33 or (3–1)/6.

log ratios in these regions will be visualized in a SMUDGE plot. The "SmudgePlot" module provides a straightforward approach to visualize regional methylation profile of a large number of samples simultaneously in the context of genomic characteristics (i.e., transcription start site, probe location, and density relative to CpG islands, repetitive elements, and methylation-sensitive restriction sites, etc). This method plots the averaged signal along a desired region in a stacked manner so that the methylation signature of multiple samples can be visualized simultaneously (**Fig. 8.4**). In brief, the signal intensity of all probes within a window of 500 bp is averaged providing a signal summary of the given window. This signal summary is calculated for each consecutive 500-bp section in the region of interest. The signal summaries are then visually

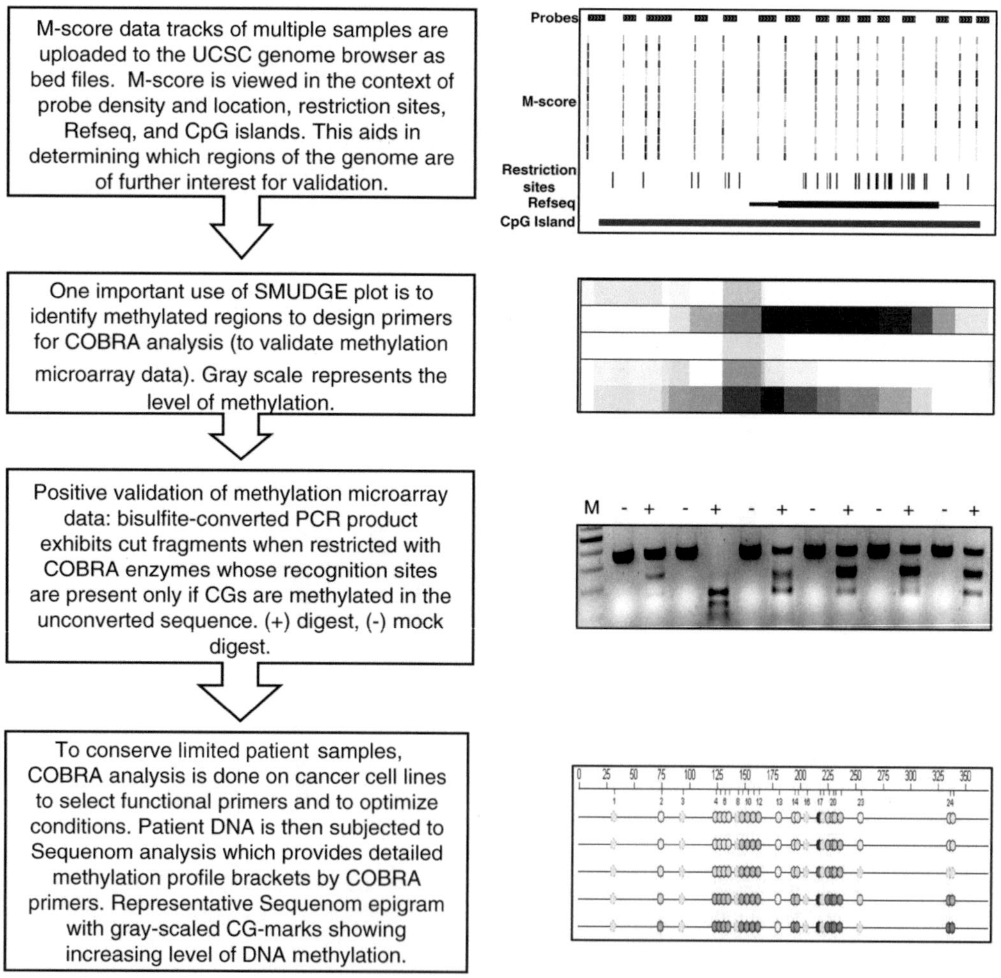

Fig. 8.4. Data visualization and validation. *Left*: Flow chart of DMH data visualization and data validation. Although these processes can be achieved by many means, approaches described here have worked well in our lab. *Right*: Representative images generated at each of these steps.

presented in a gray scale (with black and grey indicating varying degree of hypermethylation to white indicating no differential methylation). The method is amendable to visualize data presented either as log ratios or as intensities from the two separate channels.

a. *Intensity file*: A file containing the probe intensities from all samples of interest. The file must be in either sct or gct format.

b. *Regions*: Genomic region of interest for data visualization (e.g., chr1:915873-918506, chrX:56784965-56786213). One must provide either genomic-region information (*Regions*) or a file containing such information (*Region file*).

c. *Region file*: A text file containing one identifier per line that specifies the desired region to be visualized. If desired, a name for each region (separated from the genomic region information by a tab) may be included to annotate each of the plots (e.g., chr1:915873-918506 SAMD11).

d. *Genome Build*: This information specifies the UCSC human genome built used for data visualization. The default built is hg18.

e. *Zip*: Whether or not to zip the resulting images into a single archive.

3.7. Hypermethylated Loci as Resources for Uncovering Cancer-Specific Markers

1. The M-score approach described here is a simple data analysis workflow readily accessible through the GenePattern platform. This is a good starting point for research groups that do not have the support of a biostatistician/bioinformatician to assist with detailed methylation microarray data analysis. If more computation resources are available, the appropriate steps to utilize data-modeling approaches should be used that account for confounding factors associated with oligonucleotide-based DNA methylation microarray data. These approaches are commonly used for the analysis of ChIP-on-chip data *(18, 19)*.
2. Hypermethylated loci uncovered using the M-score approach (without the benefit of modeling-based data analysis) are a good resource for the identification of cancer-specific methylation markers. Since these loci are selected without correction for any confounding biases associated with the DMH protocol, we expect to encounter a number of false positive findings. We circumvent this by validating the methylation status of selected loci in a panel of relevant cell lines. Validation methods we have successfully utilized include COBRA (COmbined Bisulfite Restriction Analysis, *see* **Chapter 19**) and real-time-based methylation-specific PCR (*see* **Chapters 23** and **24**). We customarily generate two sets of primers to three regions of each CpG island. A validation rate between 30% and 40% is achieved

when performing validation on 30–50 CpG islands. These validated loci then become candidate genes for further analyses using detailed approaches such as MassARRAY (*see* **Chapter 16**) and bisulfite-genome sequencing (*see* **Chapter 14**).

4. Notes

1. DNA sonication appears to be a straightforward step, but it requires close adherence to predetermined parameters to generate DNA with consistent fragmentation size. Some of the parameters are as follows: samples should be placed in Eppendorf tubes from the same manufacturer and the same batch number; tight control of ice:water ratio in the sonication vessel; careful timing of "sonicating" and "resting" periods; standardizing all steps in smear-size determination (e.g., use only freshly prepared agarose gel as its strength increases with increasing storage time); high accuracy in DNA quantity and quality determination, and always use freshly sonicated samples for the DMH protocol. Samples with average smear size considerably lower than 500 bp will generate little linker-mediated PCR products due to loss of small fragments as these are not retained by purification columns. Therefore, QC Point 1 (**Fig. 8.1**) in the DMH protocol is the fragmentation size of the sonicated samples.
2. Samples with very degraded DNA, oversonicated samples, and samples with a starting amount of DNA deviating significantly from 0.1 μg to 0.2 μg will result in too many ligated products (high-intensity smear of the appropriate smear size) or overligated products (high smearing indicative of end-to-end joining of ligated products). Over abundance of ligated products and over-ligated samples will result in samples with skewed representation of the genome. Therefore, QC Point 2 (**Fig. 8.1**) in the DMH protocol is the ligation test PCR gel image.
3. Borderline samples barely meeting QC Points 1 and 2 will produce too much or too little LM-PCR products (0.8–1.5 μg). On the low end of the range, one risks the production of an insufficient quantity of dye-labeled DNA for array hybridization. On the high end of the range, one risks distortion of the genome due to overamplification. Therefore, QC Point 3 (**Fig. 8.1**) in the DMH protocol is the amount of LM-PCR products.
4. Borderline samples barely meeting QC Points 1–3 may lead to too much (>10 μg DNA) or insufficient aminoallyl dUTP-labeled products (<6 μg DNA). Insufficient DNA at this stage will result in low Cy dye signals after microarray

hybridization. Too much aminoallyl-dUTP-labeled products will reduce the ability to gauge differential methylation present in the test (Cy5-labeled sample) and the control (Cy3-labeled sample) samples. Therefore, QC Point 4 (**Fig. 8.1**) in the DMH protocol is the amount of aminoallyl dUTP-labeled samples.

5. Analysis of M-score:
1. Use bootstrapping to empirically estimate significant threshold for the M-score:
   a. Resample observed data 1000 times with replacement, generating 1000 randomly sampled data sets $S_1, \ldots, S_{1000}$.
   b. For each sample set S_i, compute the M-scores.
   c. For a significance level t, say $t = 0.01$, determine the threshold T_i for which t-percent of the M-scores for sample S_i are above T_i
   d. Use T equal to the mean across all T_i's as the threshold.
2. If the interest lies in the detection of concurrent methylation across samples, probes can be ranked with respect to the number of samples with M-score above the threshold T.
3. If the interest lies in the identification of DNA methylation-based signatures that distinguish between two groups, a Fisher's exact test can be used to test the association between samples of each probe (i.e., separate the data into the two categories above or below threshold and from sample 1 or 2). The resulting score can then be used to rank the probes with respect to their ability to distinguish between the two groups.
6. The data-preprocessing protocols such as M-score and normalization methods described in this book chapter are still quite rudimentary in that they do not account for the hybridization signal intensities as a result of the GC content of the microarray probes and the number of methylation-sensitive restriction sites in and around these probes. Algorithms designed to take these factors into account are currently under development.

Acknowledgements

We thank Dr. Christopher Hopkins (Agilent Technology) for advice on the design and handling of the Agilent Human CpG Island ChIP-on-chip Microarray Set, Drs. Ramana Davuluri, Hao Sun, Greg Singer, Joseph Liu, and Yi-Wen Huang (The Ohio State University) for technical support. This work was supported in part by the National Cancer Institute grants CA110475 and U54 CA11300, and by funds from The Ohio State University

Comprehensive Cancer Center-Arthur G. James Cancer Hospital and Richard J. Solove Research Institute. DP is partially supported by an NCI Training Grant T32-CA106196-03.

References

1. Huang, T. H.-M., Laux, D. E., Hamlin, B. C., et al. (1997) Identification of DNA methylation markers for human breast carcinoma using the methylation-sensitive restriction fingerprinting technique. *Cancer Res* 57, 1030–1034.
2. Yan, P. S., Perry, M. R., Laux, D. E., et al. (2000) CpG island arrays: an application toward deciphering epigenetic signatures of breast cancer. *Clin Cancer Res* 6, 1432–1438.
3. Yan, P. S., Chen, C.-M., Shi, H., et al. (2001) Dissecting complex epigenetic alterations in breast cancer using CpG island microarrays. *Cancer Res* 61, 8375–8380.
4. Wei, S. H., Chen, C.-M., Strathdee, G., et al. (2002) Methylation microarray analysis of late-stage ovarian carcinomas distinguishes progression-free survival in patients and identifies candidate epigenetic markers. *Clin Cancer Res* 8, 2246–2252.
5. Shi, H., Wei, S. H., Leu, Y.-W., et al. (2003) Triple analysis of the cancer epigenome: an integrated microarray system for analyzing gene expression, DNA methylation, and his tone acetylating. *Cancer Rees* 63, 2164–2171.
6. Yan, P., Rahmatpanah, F., Shi, H., et al. (2003) Differential distribution of DNA methylation within the *RASSF1*A CpG island in breast cancer. *Cancer Res* 63, 6178–6186.
7. Leu, Y. W., Yan, P. S., Jin, V. X., et al. (2004) Loss of estrogen receptor signaling triggers epigenetic silencing of downstream targets in breast cancer. *Cancer Res* 64, 8184–8192.
8. Waha, A., Rodrigues, F. J., Waha, A., et al. (2004) Separation of glioma subtypes based on differential methylation conglomerates by CpG microarrays. *Cancer Genomics Proteomics* 1, 209–214.
9. van Doorn, R., Zoutman, W. H., Dijkman, D., et al. (2005) Epigenetic profiling of cutaneous T-cell lymphoma: promoter hypermethylation of multiple tumor suppressor genes including *BCL7a, PTPRG,* and *p73. J Clin Oncol* 23, 1–11.
10. Yan, P. S., Venkataramu, C., Ibrahim, A., et al. (2006) Mapping geographical zones of risk in breast cancer by epigenetic biomarkers. *Clin Cancer Res* 12, 6626–6636.
11. Fan, M., Yan, P. S., Hartman-Frey, C., et al. (2006) Diverse gene expression and DNA methylation profiles correlate with differential adaptation of breast cancer cells to the antiestrogens tamoxifen and fulvestrant. *Cancer Res* 66, 11954–11966.
12. Zighelboim, I., Goodfellow, P. J., Schmidt, A., et al. (2007) Differential methylation hybridization array of endometrial cancers reveals two novel cancer specific methylation markers. *Clin Cancer Res* 13, 2882–2889.
13. Yang, Y. H., Dudoit, S., Luu, P., et al. (2001). Normalization for cDNA microarray data, in (Bittner, M. L., Chen, Y., Dorsel, A. N., Dougherty E. R., eds.), *Microarrays: Optical Technologies and Informatics, Proceedings of SPIE,* 4266, pp. 141–152.
14. Smyth, G. K., Speed, T. P. (2003). Normalization of cDNA microarray data. *Methods* 31, 265–273.
15. Bolstad, B. M., Irizarry, R. A., Astrand, M, et al. (2003) A comparison of normalization methods for high-density oligonucleotide array data based on bias and variance. *Bioinformatics* 19, 185–193.
16. Yang, Y. H., Dudoit, S., Luu, P., et al. (2002). Normalization for cDNA microarray data: a robust composite method addressing single and multiple slide systematic variation. *Nucleic Acids Res* 30, e15.
17. Yang, Y. H., Thorne, N. P. (2003). Normalization for two-color cDNA microarray data, in (Goldstein, D. R., ed.), *Science and Statistics: A Festschrift for Terry Speed, IMS Lecture Notes – Monograph Series* 40, pp. 403–418.
18. Johnson, W. E., Li, W., Meyer, C. A., et al. (2006) Model-based analysis of tiling-arrays for ChIP-chip. *Proc Natl Acad Sci USA* 103, 12457–12462.
19. Song, J. S., Johnson, W. E., Zhu, X., et al. (2007) Model-based analysis of two-color arrays (MA2C). *Genome Biol* 8, R178.

Chapter 9

Analysis of DNA Methylation by Amplification of Intermethylated Sites (AIMS)

Mireia Jordà, Jairo Rodríguez, Jordi Frigola, and Miguel A. Peinado

Abstract

DNA methylation is an epigenetic modification that plays a crucial role in the control of gene expression and chromosome structure in plants and mammalian cells. Multiple types of DNA fingerprinting techniques have been developed and applied to investigate DNA methylation profiles in different experimental settings. One of these techniques, the amplification of intermethylated sites (AIMS) is a simple approach appropriate for genome-wide estimates of DNA methylation and the discovery of specific methylated sequences. AIMS is based on the differential enzymatic digestion of genomic DNA with methylation-sensitive and methylation-insensitive isoschizomers followed by restrained PCR amplification of methylated sequences. This method is appropriate to compare large series of samples and the simultaneous identification of hypo- and hypermethylation events. Applications of AIMS include the study of DNA methylation changes in cancer and aging, and the discovery of DNA methylation in a social insect.

Key words: DNA methylation, CpG island, amplification of intermethylated sites (AIMS), isoschizomer, DNA fingerprinting, methylome.

1. Introduction

DNA methylation is an epigenetic modification that plays a crucial role in the control of gene expression and chromosome structure in plants and mammalian cells *(1–3)*. The dynamics of DNA methylation and its involvement in multiple biological and pathological processes demand discovery-based strategies to obtain comprehensive profiles and to reveal specific changes associated with alternative situations. Due to the extent of genomic methylation, there is no need to identify markers a priori and methodologies based on the screening of large sets of randomly selected anonymous markers may render valuable information at both

Jörg Tost (ed.), *DNA Methylation: Methods and Protocols, Second Edition, vol. 507*

DOI 10.1007/978-1-59745-522-0_9 Springerprotocols.com

global and specific levels, favoring the use of DNA fingerprinting-type techniques since early studies *(3–6)*.

One of these techniques, the amplification of intermethylated sites (AIMS) generates easily readable fingerprints representing the DNA methylation profile of the cell *(7)*. AIMS is based on the differential cleavage of isoschizomers with distinct methylation sensitivity (i.e., *Sma*I and *Xma*I). Specific adaptors are ligated to the methylated ends of the digested genomic DNA. The ligated sequences are purified and amplified by PCR using adaptor-specific primers extended at the 3′ end with two to four arbitrarily chosen nucleotide residues to reduce the complexity of the product. To be amplified, a sequence must fulfill two requirements: (i) to contain two closely spaced methylated *Sma*I sites and (ii) to show homology to the nucleotides extended at the 3′ end of the primer. PCR products are resolved in denaturing polyacrylamide-sequencing gels generating fingerprints that consist of multiple anonymous bands that represent the DNA methylome of the cell. These bands are DNA fragments flanked by two methylated sites and can be individually isolated and characterized (**Fig. 9.1**). The usefulness of AIMS lies in two properties of the method: the generation of a high number of sequences representing two close methylated CpGs and the feasibility of reducing the complexity of the product to obtain readable fingerprints. AIMS is

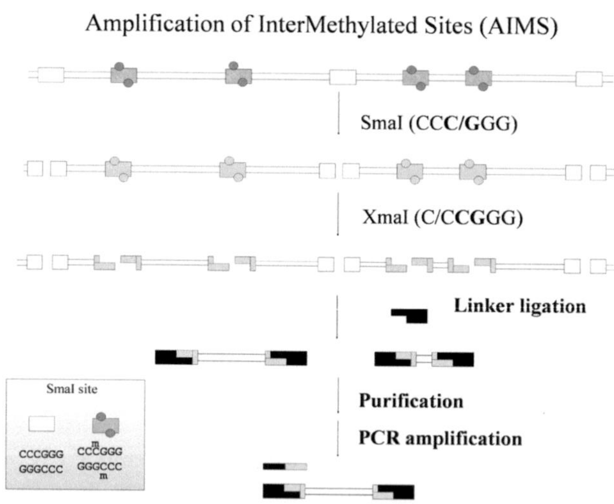

Fig. 9.1. Scheme of the AIMS technique. Genomic DNA is represented by two parallel lines, with seven CCCGGG recognition sites shown as boxes. Nonmethylated (empty boxes) and methylated (filled boxes with dots indicating the methylated cytosines) sites are depicted. In a first step, digestion of unmethylated sites with *Sma*I results in blunt ends, while the remaining methylated sites are cut in a second step with the *Xma*I endonuclease, producing sticky ends. Ligation of an adaptor to the sticky ends and PCR using appropriate primers will result in the amplification of DNA fragments flanked by two methylated *Sma*I sites.

appropriate to screen for methylation in different types of sequences as demonstrated in studies performed in human samples. Specifically, AIMS-represented sequences can be grouped into any of four categories of DNA sequences: those corresponding to CpG islands (contained or located within), non-CpG islands, unknown or anonymous sequences, and sequences generated from repetitive elements such as SINEs, LINEs, LTRs, etc.

This method has been evaluated for reproducibility and sensitivity *(7)*, and allows the identification of hypo- (loss of a band) and hypermethylation (appearance of a new band). Applications of AIMS to investigate epigenetic changes in cancer include the identification of recurrent hypermethylations associated with gene silencing *(8, 9)*, screening for both hypomethylation and hypermethylation in cancer cell lines with altered DNA methylation function *(10, 11)*; and the genome-wide estimation of abnormal DNA methylation in cancers *(12, 13)*. It has been also applied to identify epigenetic differences arising during the lifetime of monozygotic twins *(14)*. Moreover, a modification of AIMS technique has been instrumental to discover functional CpG methylation in the social insect *Apis Mellifera (15, 16)*.

As a screening technique that amplifies multiple sequences, artifacts and irreproducibilities may occur. Therefore, it is always necessary to validate the AIMS results by alternative approaches to obtain accurate determinations at sequence level, of which bisulfite sequencing (**Chapter 14**) is currently the recommended choice.

2. Materials

2.1. DNA Digestion

1. Enzyme: *Sma*I and 10 × buffer #A (Roche Diagnostics GmbH, Mannheim, Germany).
2. Enzyme: *Xma*I, 10 × buffer #4 and BSA (New England Biolabs, Boston, MA).
3. Enzyme: *Hpa*II and 10 × buffer #L (Roche Diagnostics GmbH).
4. Enzyme: *Msp*I and 10 × buffer # 2 (New England Biolabs).

2.2. Fill-in 5′ Ends

1. DNA polymerase I, Large Klenow Fragment (New England Biolabs).
2. 100 mM dNTP set (GE Healthcare, Buckinghamshire, UK). dNTPs are diluted at 500 μM in water and stored in aliquots at −20°C.

2.3. Adaptor Ligation

Oligonucleotides MCA-Blue and Blue are used after digestion with *Sma*I/*Xma*I, oligonucleotides ADPT B-GC, and ADPT B are used after digestion with *Hpa*II/*Msp*I.

1. Oligonucleotide MCA-Blue 100 μM: 5′-CCGGTCAGAGCTTTGCGAAT-3′
2. Oligonucleotide Blue 100 μM: 5′-ATTCGCAAAGCTCTGA-3′
3. Oligonucleotide ADPT B 100 μM: 5′-ATTCGCAAAGCTCTGA-3′
4. Oligonucleotide ADPT B-GC 100 μM: 5′-P-CGTCAGAGCTTTGCGAAT-3′
5. T4 DNA ligase and 10 × buffer (New England Biolabs).

2.4. Sample Purification

1. Illustra™ DNA and Gel Band Purification Kit (GE Healthcare).

2.5. PCR Amplification

1. 100 mM dNTP set (GE Healthcare). dNTPs are diluted at 1 mM in water and stored in aliquots at −20°C.
2. Taq DNA polymerase and 10 × PCR buffer (Roche Diagnostics GmbH).
3. Magnesium chloride ($MgCl_2$) is resuspended at 100 mM in water, filtered with 0.22 μm filters, and stored in aliquots at −20°C.
4. [α-^{33}P]dATP (GE Healthcare, Buckinghamshire, UK).
5. Primers 100 μM: different set of primers corresponding to the Blue primer sequence extended with the CCGGG (overhanging end, *Xma*I digestion) plus three to four arbitrarily chosen nucleotides are used:

Set A: Blue-CCGGG-CTA + Blue-CCGGG-TGG
Set B: Blue-CCGGG-CTG + Blue-CCGGG-TGG
Set C: Blue-CCGGG-CGCG + Blue-CCGGG-CAAC

6. Primers 100 μM: different primers or set of primers corresponding to the ADPT B sequence primer extended with the CGG (overhanging end, *Msp*I digestion) plus two arbitrarily chosen nucleotides are used:

Set A: ADPT B-CGG-AT + ADPT B-CGG-TC
Set B: ADPT B-CGG-TT
Set C: ADPT B-CGG-AA

2.6. Polyacrylamide Denaturing Urea Gel Electrophoresis Preparation

1. Electrophoresis system model S2 (Gibco BRL)
2. TBE (5 ×): 450 mM Tris base, 450 mM boric acid, 1 mM EDTA(free acid), pH 8.0.
3. Urea.
4. 40% acrylamide:bis–acrylamide solution (29:1) (National Diagnostics). Caution: This is a neurotoxin when unpolymerized.
5. *N,N,N,N′*-Tetramethyl-ethylenediamine (TEMED).
6. Ammonium persulfate (*see* **Note 1**): prepare a 10% solution in water and freeze aliquots at −20°C.
7. Sigmacote (Sigma–Aldrich).

8. DLB buffer: 0.09% bromophenol blue, 0.09% xylene cyanol, 10 mM NaOH, 93% deionized formamide. Store at 4°C.
9. Molecular weight marker: ΦX174 DNA / HaeIII (Fermentas).

3. Methods

AIMS is suitable for the study of DNA methylation in genomes rich in CpG sites and a high degree of methylation. This results in a number of methylated *Sma*I/*Xma*I sites flanked by the adaptor and close enough to be amplified by PCR. In the case of genomes poor in CpG sites and/or DNA methylation (like *Apis Mellifera*), methylated *Sma*I/*Xma*I sites are rare and a low number of amplifiable products are expected. In this case, the AIMS method can be applied with some modifications, concretely replacing the restriction endonucleases by alternative endonucleases recognizing a more frequent site, like the *Hpa*II/*Msp*I isoschizomers *(15, 16)*.

3.1. Preparation of Samples

1. DNA is obtained by conventional phenol–chloroform extraction. Check concentration (we work with 100 ng/μL DNA concentration) and purity before proceeding with the digestions (*see* **Note 2**).

3.2. AIMS for Genomes with High Content of Methylated CpGs

3.2.1. SmaI Digestion

1. Digest 1 μg of DNA with 1 U of the methylation-sensitive restriction endonuclease *Sma*I for 16 h at 25°C, which cleaves leaving blunt ends (CCC/GGG). The reaction mix composition is: 1.5 μL 10 × buffer , 0.1 μL *Sma*I (10 U/μL), and 10 μL DNA (100 ng/μL) in a final reaction volume of 15 μL.

3.2.2. XmaI Digestion

1. Subsequently, digest DNA with 5 U of the non-methylation-sensitive restriction endonuclease *Xma*I for 6 h at 37°C, which cleaves leaving sticky ends (C/CCGGG). Add to the previous digestion the following: 0.5 μL *Xma*I (10 U/μL), 0.5 μL 20 × BSA, 1 μL 10 × buffer, and 8 μl bidistilled water. Final reaction volume is 25 μL.

3.2.3. Adaptor Ligation

1. Prepare adaptors by incubation of the same volume of Blue and MCA-Blue oligonucleotides (100 μM) for 2 min at 65°C followed by cooling at room temperature for 30–60 min (*see* **Note 3**).
2. Ligate DNA to 2 nM of adaptor using T4 DNA ligase at 16°C for 16 h. Add to the double digestion the following: 20 μL adaptor (100 μM), 2 μL T4 DNA ligase (400 U/μL), 8 μL

10 × buffer, and 25 μL bidistilled water. Final reaction volume is 80 μL.

3.2.4. Sample Purification

1. Purify the ligation product with the Illustra™ DNA and Gel Band Purification Kit and elute in 250 μL of bidistilled water. Final DNA concentration is ~4 ng/μL (assuming recuperation of total DNA from the columns). Aliquot and keep at −20°C.

3.2.5. PCR Amplification

1. Perform PCR using 3 μL of each ligated DNA (~12 ng) as template in a 25 μL volume containing 1.1 μM of each primer, 2 U Taq polymerase, 125 μM dNTPs, 1.5 mM $MgCl_2$, 1 μCi [α-^{33}P]dATP, and PCR buffer (*see* **Note 4**).
2. PCR with primer set A and B consists of 30 two-step cycles (15 s at 94°C and 1 min 15 s at 74°C). PCR with primer set C consists of 30 three-step cycles (15 s at 94°C, 45 s at 68°C, and 1 min at 72°C). PCR cycles are preceded by denaturation for 1 min at 95°C and ended with an extension step of 5 min at 72°C (*see* **Note 5**).

3.2.6. Polyacrylamide Denaturing Urea Gel Electrophoresis

These instructions assume the use of the electrophoresis sequencing gels system model S2 (Gibco BRL). It is critical to clean the glass plates with distilled water and 70% ethanol, and treat with Sigmacote, which forms a neutral, hydrophobic, microscopically thin film to avoid the gel adhering to the glass, just before use. The two glass plates are placed horizontally separated by 0.15 mm spacers and blocked with metal clamps.

1. Prepare a 8 M urea, 6% polyacrylamide denaturing electrophoresis gel by mixing 28.8 g urea, 12 mL 5 × TBE and 12 mL bidistilled water. Dissolve by heating at 65°C and adjust volume with water up to 51 mL. Cool the mix at room temperature and add 9 mL 40% acrylamide/bis solution (*see* **Note 6**), 15 μL TEMED, and 600 μL ammonium persulfate solution. Quickly pour the gel mix between the two glass plates, taking care to avoid and eliminate air bubbles (*see* **Note 7**). Immediately lay horizontally, insert the comb with the flat side inside the gel and hold in place by metal clamps. It should polymerize within 1 h, but it is recommended to leave it for 3 h or even overnight. After polymerization, remove the clamps and place the gel in the vertical electrophoresis unit.
2. Prepare the running buffer diluting 5 × TBE with bidistilled water and add it to the upper and lower chambers of the gel unit.
3. Carefully remove the comb and use a 20-mL syringe fitted with a 22-gauge needle to wash the front with running buffer. Then invert the comb and insert it again between the two glass plates with the jagged side touching the acrylamide gel edge to form the wells.

4. Flush the sample wells with a syringe containing running buffer and load denaturing loading buffer (DLB) in alternate wells (one empty/one full) to check they are well formed and not interconnected (*see* **Note 8**).
5. Prerun about 30–45 min at 55 W.
6. Dilute PCR products 1:4 in DLB buffer and denature by heating at 95°C for 3 min.
7. Wash the wells again to eliminate urea and load 3 μL of each sample.
8. Run at 55 W for 5 h (*see* **Note 9**).

3.2.7. Sample Visualization

1. After electrophoresis, remove the running buffer and pry the gel plates apart. Then, blot the gel to a sheet of 3 MM Whatman paper, cover with plastic wrap, and dry under vacuum at 85°C.
2. Expose to an X-ray film at room temperature for 3–6 days or use an imaging system such as PhosphorImager. One example of the resulting fingerprints is shown in **Fig. 9.2**.

3.3. AIMS for Genomes with Low Content of Methylated CpGs

For genomes poor in CpG methylation it is more effective to use isoschizomers that recognize a shorter sequence, such as *Hpa*II/*Msp*I (CCGG), which is more frequent than *Sma*I/*Xma*I site (CCCGGG). As *Hpa*II and *Msp*I enzymes cleave leaving the same ends, it is very important to block the *Hpa*II ends in order to differentiate between unmethylated and methylated ends. Blocking is carried out with the large Klenow Fragment.

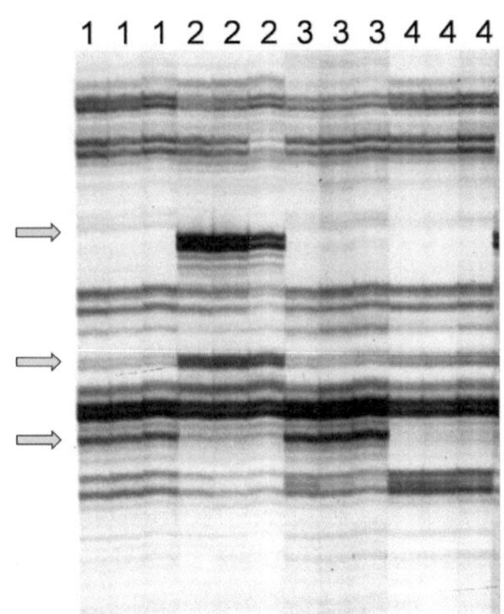

Fig. 9.2. AIMS of four human DNA samples. Analysis was performed in triplicate. Arrows indicate bands with differential display that correspond to hypermethylations (presence of the band) and hypomethylations (loss of the band). Bands of interest can be excised from the gel, re-amplified, cloned, and sequenced. The methylation status should be confirmed by alternative approaches, such as bisulfite sequencing.

3.3.1. HpaII Digestion

1. Digest 1 μg of DNA with 10 U of the methylation-sensitive restriction endonuclease *Hpa*II for 16 h at 37°C, which cuts leaving sticky ends (C/CGG). The reaction mix composition is: 1.5 μL 10 × buffer, 1 μL *Hpa*II (10 U/μL), and 10 μL DNA (100 ng/μL) in a final reaction volume of 15 μL.

3.3.2. Fill-in HpaII Ends

1. Fill-in *Hpa*II ends by incubation with 2 U of the DNA polymerase I, Large Klenow Fragment for 15 min at 37°C. Add to the previous digestion the following: 1 μL Klenow (2 U/μL), and 1.5 μL dNTPs (500 μM). The final volume is 17.5 μL. Klenow is active in many restriction enzyme buffers, so do not add new buffer. Stop the reaction heating at 75°C for 10 min (*see* **Note 10**).

3.3.3. MspI Digestion

1. Subsequently, digest DNA with 10 U of the non-methylation-sensitive restriction endonuclease *Msp*I for 6 h at 37°C, which cleaves leaving sticky ends (C/CGG). Add to the previous digestion the following: 0.5 μL *Msp*I (20 U), 0.75 μL 10 × buffer, and 6.25 μL bidistilled water. The final reaction volume is 25 μL.

3.3.4. Adaptor Ligation

1. The rest of the protocol is the same used for conventional AIMS (starting from **Section 3.2.3**) but with different adaptors (ADPT B and ADPT B-GC) and primers (ADPT sets), so PCR conditions are different. Therefore, replace step in **Section 3.2.5** with step in **Section 3.3.5**.

3.3.5. PCR Amplification

1. Perform PCR using 4 μL of each ligated DNA (~16 ng) as template in a 25 μL volume containing 1.1 μM of each primer, 2 U Taq polymerase, 125 μM dNTPs, 1.5 mM $MgCl_2$, 1 μCi [α-^{33}P]dATP, and PCR buffer (*see* **Note 4**).
2. The PCR (independent of the primer set) consists of 35 three-step cycles (1 min at 94°C, 45 s at 63°C, and 1 min 30 s at 72°C). PCR cycles are preceded by denaturation for 2 min at 95°C and ended with an extension step of 5 min at 72°C (*see* **Note 5**).

3.4. Isolation and Sequencing of DNA Fragments

To identify and to confirm the amplified sequences, bands can be excised from dried polyacrylamide gels, and DNA re-eluted in H_2O (80°C, 10 min). A detailed description of band isolation and validation may be found elsewhere *(17, 18)*. PCR with the same primers and conditions used in the AIMS experiment is performed to amplify the fragment. PCR products are cloned into plasmid vectors using the TA cloning kit (Invitrogen, Groningen, The Netherlands) and sequenced using standard procedures (**Chapter 14**).

4. Notes

1. Ammonium persulfate is best stored at room temperature in a dessicator.
2. DNA purity and quality are of significant importance, since contaminated or degraded DNA may result in aberrant methylation patterns.
3. A stock volume of adaptors can be prepared, aliquoted, and stored at −20°C.
4. If it is necessary, a higher amount of [α-^{33}P]dATP can be used, always adjusting with water to 22 μL.
5. Other nucleotides can be added to the 3′ end of the primers to obtain more fingerprints.
6. If the urea gel mix is not cool enough when acrylamide-bis is added, it can precipitate.
7. In case bubbles are formed, use a radiography film cut with a hook shape to eliminate them.
8. If DLB transfers to the adjacent wells, press the comb more into the gel, and load DLB again after having run the gel for a few minutes. On the other hand, avoid sticking the comb too far into the polyacrylamide gel, as this will cause interwell marked lines and band *bending* of the fingerprints.
9. The running time can vary depending on the size of the bands of the fingerprint. It is therefore advisable to run the samples first in a small nondenaturing acrylamide gel, which only allows detecting a smear, but will help to determine the size of the bands of the fingerprint.
10. It is very important to have a control of the blocking step, such as DNA digested with *Hpa*II, incubated with Klenow, and ligated with the adaptors. In this case, no specific PCR amplification is expected.

Acknowledgements

The research was supported by the Spanish Ministry of Education and Science (SAF2006/351 and Consolider-Ingenio 2010 CSD2006-49).

References

1. Jones, P. A., Baylin, S. B. (2002) The fundamental role of epigenetic events in cancer. *Nat Rev Genet* 3, 415–428.
2. Jaenisch, R., Bird, A. (2003) Epigenetic regulation of gene expression: how the genome integrates intrinsic and environmental signals. *Nat Genet* 33 Suppl, 245–254.
3. Feinberg, A. P., Tycko, B. (2004) The history of cancer epigenetics. *Nat Rev Cancer* 4, 143–153.
4. Cottrell, S. E. (2004) Molecular diagnostic applications of DNA methylation technology. *Clin Biochem* 37, 595–604.

5. Fazzari, M. J., Greally, J. M. (2004) Epigenomics: beyond CpG islands. *Nat Rev Genet* 5, 446–455.
6. Esteller, M. (2007) Cancer epigenomics: DNA methylomes and histone-modification maps. *Nat Rev Genet* 6, 286–298.
7. Frigola, J., Ribas, M., Risques, R. A., et al. (2002) Methylome profiling of cancer cells by amplification of inter- methylated sites (AIMS). *Nucleic Acids Res* 30, e28.
8. Frigola, J., Mu noz, M., Clark, S. J., et al. (2005) Hypermethylation of the prostacyclin synthase (PTGIS) promoter is a frequent event in colorectal cancer and associated with aneuploidy. *Oncogene* 24, 7320–7326.
9. Frigola J, Song, J., Stirzaker, C., et al. (2006) Epigenetic remodeling in colorectal cancer results in coordinate gene suppression across an entire chromosome band. *Nat Genet* 38, 540–549.
10. Paz, M. F., Wei, S., Cigudosa, J. C., et al. (2003) Genetic unmasking of epigenetically silenced tumor suppressor genes in colon cancer cells deficient in DNA methyltransferases. *Hum Mol Genet* 12, 2209–2219.
11. Sadikovic, B., Haines, T. R., Butcher, D. T., et al. (2004) Chemically induced DNA hypomethylation in breast carcinoma cells detected by the amplification of intermethylated sites. *Breast Cancer Res* 6, R329-337.
12. Frigola, J., Sole, X., Paz, M. F., et al. (2005) Differential DNA hypermethylation and hypomethylation signatures in colorectal cancer. *Hum Mol Genet* 14, 319–326.
13. Rodriguez, J., Frigola, J., Vendrell, E., et al. (2006) Chromosomal instability correlates with genome-wide DNA demethylation in human primary colorectal cancers. *Cancer Res* 66, 8462–8468.
14. Fraga, M. F., Ballestar, E., Paz, M. F., et al. (2005) Epigenetic differences arise during the lifetime of monozygotic twins.*Proc Natl Acad Sci USA* 102, 10604–10609.
15. Honeybee Genome Sequence Consortium. (2006) Insights into social insects from the genome of the honeybee *Apis mellifera*. *Nature* 443, 931–949.
16. Wang, Y., Jorda, M., Jones, P. L., et al. (2006) Functional CpG methylation system in a social insect. *Science* 314, 645–647.
17. Perucho, M., Welsh, J., Peinado, M. A., et al. (1995) Fingerprinting of DNA and RNA by arbitrarily primed polymerase chain reaction: applications in cancer research. *Methods Enzymol* 254, 275–290.
18. Arribas, R., Tortola, S., Welsh, J., et al. (1997) Analysis of tumor-specific genetic alterations by Arbitrarily Primed PCR, in (Micheli, M. R., and Bova, R., eds.), *Fingerprinting Methods Based on Arbitrarily Primed PCR*, pp. 177–185, Heidelberg, Springer, Germany.

Chapter 10

Methylation-Sensitive Representational Difference Analysis (MS-RDA)

Toshikazu Ushijima and Satoshi Yamashita

Abstract

Methylation-sensitive representational difference analysis (MS-RDA) is a genome subtraction method that isolates DNA fragments differentially methylated between two genomes. It can be performed in any organism, even in those for which no microarray products are available. An important characteristic of MS-RDA is that it enriches unmethylated CpG-rich regions of the genome (amplicon), most of which are unique sequences. DNA fragments differentially methylated between two DNA samples will be present in one amplicon, but not in the other. The difference can be identified by RDA. Most technical difficulties reside in the RDA procedure, and many fine techniques are necessary for a successful application of this powerful technology.

Key words: MS-RDA, genome-wide screening, epigenome, DNA methylation, epigenetics.

1. Introduction

Methylation-sensitive representational difference analysis (MS-RDA) is a genome subtraction method that isolates DNA fragments differentially methylated between two genomes *(1–3)*. In an era when microarray analyses are extensively performed, the advantage of MS-RDA is that it can be performed to analyze any organism, even those for which no microarrays are available. Even if microarrays are available for the species of interest, MS-RDA does not require any fancy instruments, such as expensive scanners. If an investigator is familiar with the RDA procedure itself, MS-RDA can be easily performed.

For MS-RDA, genomic DNA is digested with a methylation-sensitive restriction enzyme, and DNA fragments are restricted

Jörg Tost (ed.), *DNA Methylation: Methods and Protocols, Second Edition, vol. 507*

DOI 10.1007/978-1-59745-522-0_10 Springerprotocols.com

into sizes suitable for PCR that are subsequently amplified into an "amplicon" by PCR using a universal adaptor and primer (**Fig. 10.1**). The amplicon is enriched with DNA fragments derived from unmethylated genomic regions with frequent occurrence of the recognition sites of the restriction enzyme. Preparation of an amplicon reduces the complexity of the genome, and genomic subtraction can be efficiently performed using two amplicons from two different genomes. DNA fragments present only in the "tester" sample, but not in the "driver" sample, can be isolated by the following genomic subtraction RDA (**Fig. 10.2**).

Any conventional method for genome-wide screening of differences in DNA methylation needs to prepare a representation

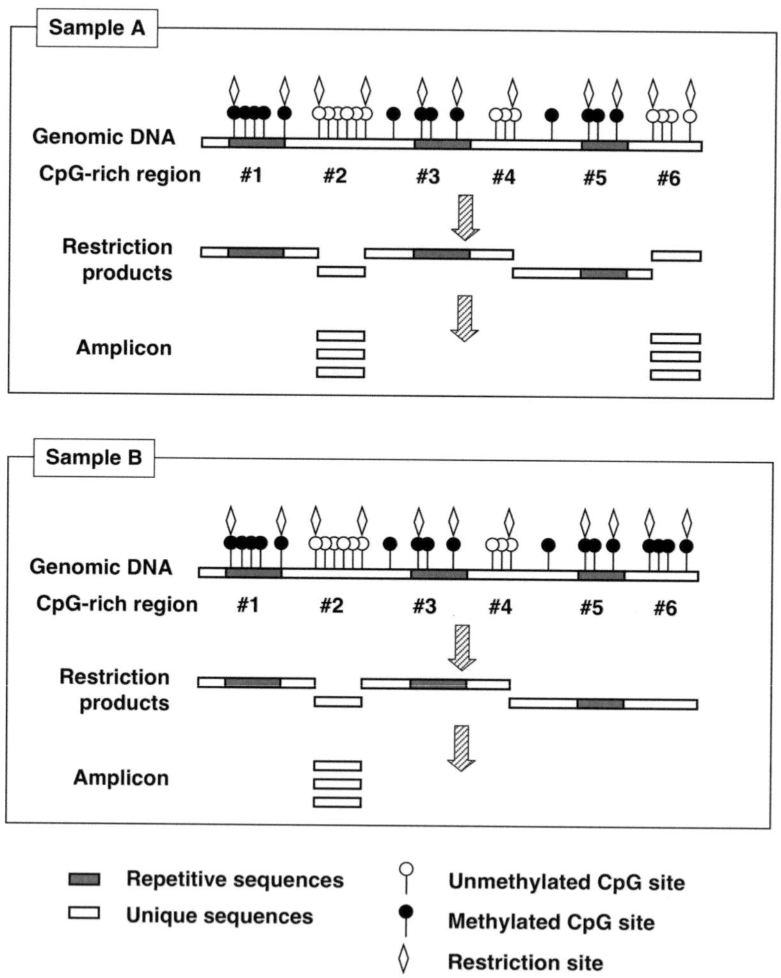

Fig. 10.1. Preparation of amplicon in MS-RDA. Tester and driver amplicons are prepared from samples A and B, respectively. The samples are digested by a methylation-sensitive restriction enzyme, such as *Hpa*II, *Sac*II, or *Nar*I. DNA fragments suitable for PCR are derived from unmethylated CpG-rich regions and amplified into an amplicon by PCR using a universal adaptor.

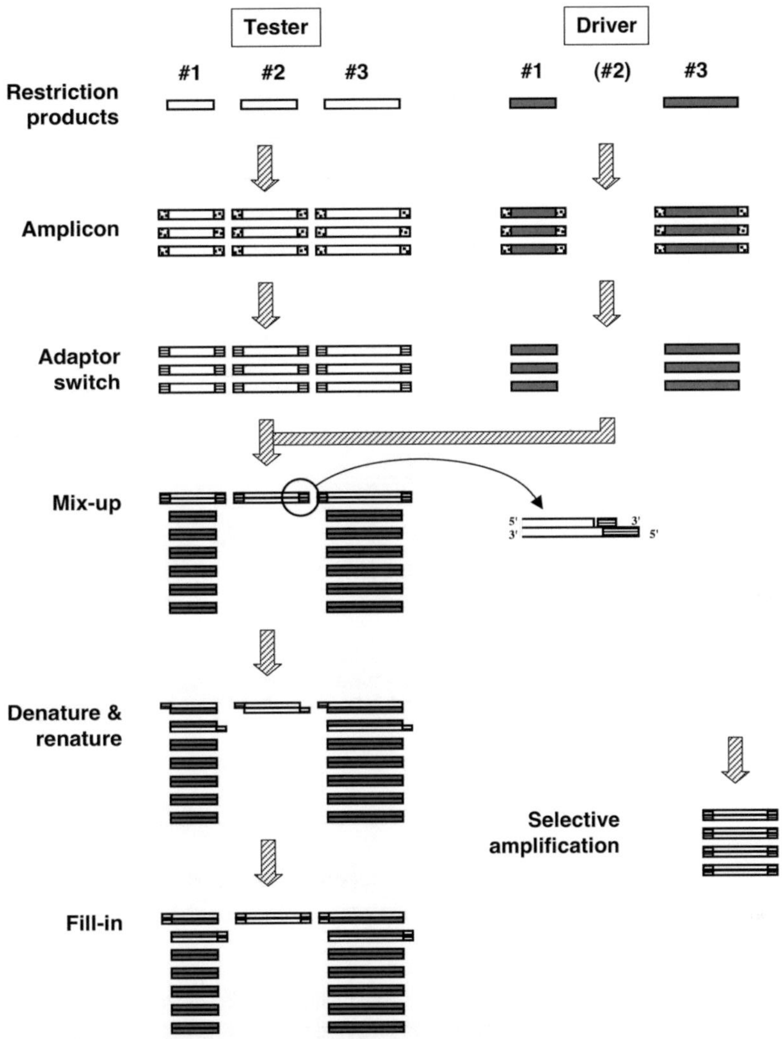

Fig. 10.2. Procedure of RDA. The overall procedure of RDA. Amplicons are prepared from restriction products of the tester and driver DNA (also shown in **Fig. 10.1**). Only for the tester, a new adaptor is ligated. A small amount of the tester is mixed with the driver amplicon, and the mixture is denatured by heat and undergoes re-annealing under stringent conditions. If a DNA fragment in the tester had its counterpart in the driver (fragments #1 and #3), it will re-anneal with a fragment in the driver because the driver is present in large excess. If a DNA fragment does not have its counterpart in the driver (fragment #2), it will re-anneal with a tester fragment. Only such DNA fragments have an adaptor on both ends and can be amplified exponentially in the following selective amplification.

of the genome as the entire genome spans 3×10^9 bp and is too large to be comprehensively analyzed by any conventional method and even by most microarray techniques. The representation that is used for the screening can be prepared by various means such as subtraction (MS-RDA, methylated CpG island amplification-RDA, etc.) or two-dimensional gel-electrophoresis (restriction landmark genomic scanning; RLGS,

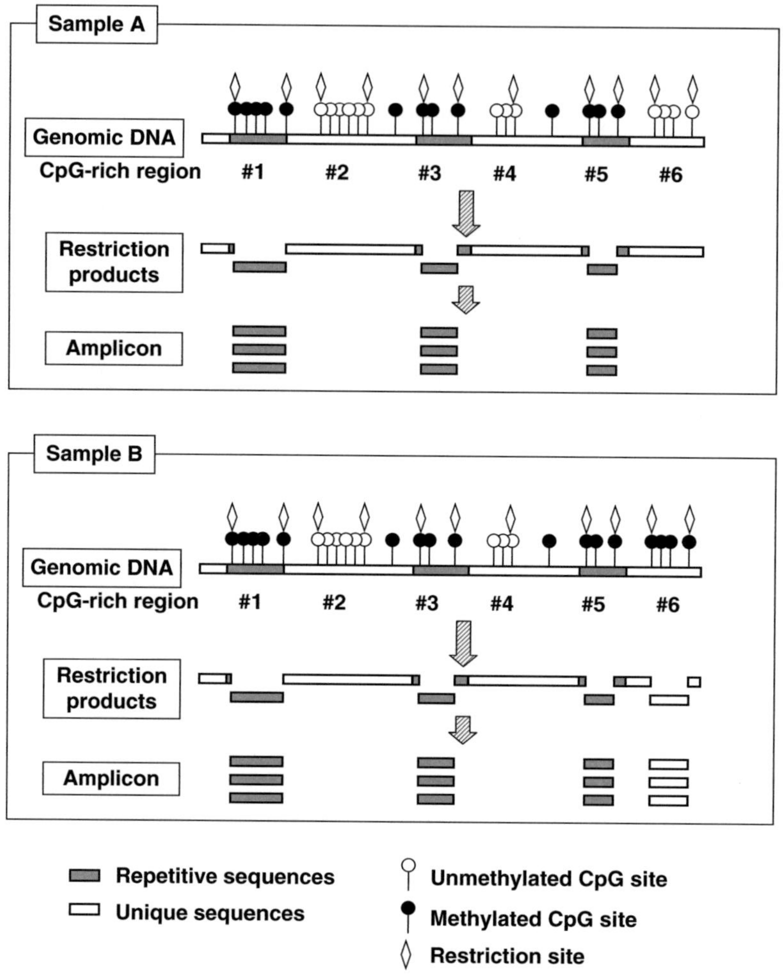

Fig. 10.3. Preparation of amplicons from methylated regions of the genome. Amplicons can also be prepared from methylated regions of the genome. However, if such regions are amplified, the vast majority is derived from repetitive sequences and not suitable for genome subtraction.

Chapter 11) *(3)*. To prepare a representation of the genome, either the unmethylated or methylated regions of the genome can be extracted. Since more than 40% of the human genome is made of repetitive sequences that are CpG rich and highly methylated, a representation constructed from methylated regions of the genome will contain a significant amount of repetitive sequences (**Fig. 10.3**). Repetitive sequences are resistant to removal by genome subtraction and prevent efficient isolation of differentially methylated unique sequences. MS-RDA prepares its representation from unmethylated regions of the genome, and avoids thereby the interference from repetitive sequences.

MS-RDA has been applied to isolate DNA fragments methylated in human lung cancers, gastric cancers, breast cancers, pancreatic cancers, neuroblastomas, and melanomas *(4–10)*. A tumor-suppressor gene was identified in gastric cancers *(11)*, and a promising prognostic marker was identified in neuroblastomas *(9, 12)*.

2. Materials

2.1. Regular Reagents Used in Molecular Biology

1. 5 M NaCl,
2. Neutralized, water-saturated phenol and chloroform,
3. 3 M NaOAc,
4. 10 M NH_4OAc,
5. Agarose and NuSieve GTG agarose,
6. 0.5 × TBE: 5 mM Tris, 0.45 mM boric acid, and 1 mM EDTA, and
7. Competent cells.

2.2. Oligonucleotides

1. Oligonucleotides to prepare adaptors for the *Hpa*II series:
   RHpa24: 5′-AGC ACT CTC CAG CCT CTC ACC GAC-3′,
   RHpa11: 5′-CGG TCG GTG AG-3′,
   JHpa24: 5′-ACC GAC GTC GAC TAT CCA TGA AAC-3′,
   JHpa11: 5′-CGG TTT CAT GG-3′,
   NHpa24: 5′-AGG CAA CTG TGC TAT CCG AGG GAC-3′,
   NHpa11: 5′-CGG TCC CTC GG-3′.
2. Oligonucleotides to prepare adaptors for the *Sac*II series:
   RSac26: 5′-AGC ACT CTC CAG CCT CTC ACG ACC GC-3′,
   RSac9: 5′-GGT CGT GAG-3′,
   JSac26: 5′-ACC GAC GTC GAC TAT CCA TGA ACC GC-3′,
   JSac9: 5′-GGT TCA TGG-3′,
   NSac26: 5′-AGG CAA CTG TGC TAT CCG AGG ACC GC-3′, and
   NSac9: 5′-GGT CCT CGG-3′.
3. Oligonucleotides to prepare adaptors for the *Nar*I series:
   RNar24: 5′-AGC ACT CTC CAG GCA CTC ACC AGG-3′,
   RNar11: 5′-CGC CTG GTG AG-3′,
   JNar24: 5′-ACC GAC GTC GAC TAT CCA TGA AGG-3′,
   JNar11: 5′-CGC CTT CAT GG-3′,
   NNar24: 5′-AGG CAA CTG TGC TAT CCG AGG AGG-3′, and
   NNar11: 5′-CGC CTC CTC GG-3.
4. Oligonucleotides used as PCR primers.
   For the *Hpa*II series, use RHpa24, JHpa24, and NHpa24.
   For the *Sac*II series, use RSac24: 5′-AGC ACT CTC CAG CCT CTC ACG ACC-3′,
   JSac24: 5′-ACC GAC GTC GAC TAT CCA TGA ACC-3′ and

NSac24: 5′-AGG CAA CTG TGC TAT CCG AGG ACC-3′.
For the *Nar*I series, use RNar24, JNar24, and NNar24.

2.3. Amplicon Preparation

1. Methylation-sensitive restriction enzymes *Hpa*II, *Sac*II, and *Nar*I (New England Biolabs, Beverly, MA), and the methylation-insensitive restriction enzyme, *Msp*I (New England Biolabs).
2. T4 ligase (New England Biolabs), and T4 ligase buffer (10 × accompanying T4 ligase). The T4 ligase buffer contains ATP and should be thawed on ice.
3. Taq polymerase. Hot Start Taq polymerases activated by heating at 95°C are "not" suitable.
4. dNTP each 2 mM or 2.5 mM.
5. 10 × PCR buffer III: 67 mM Tris-HCl (pH 8.8), 4 mM $MgCl_2$, 16 mM $(NH_4)_2SO_4$, 10 mM β-mercaptoethanol, and 100 μg/mL bovine serum albumin (BSA). After mixing these reagents, filter with an Acrodisk (0.2 μm), aliquot into several 1-mL stock tubes, and keep at −20°C.
6. 5 M betaine. After dissolution, filter with an Acrodisk (0.2 μm), aliquot into several 1 mL stock tubes, and keep at −20°C.
7. Chroma Spin + TE-200 column (Clontech K1325).

2.4. Competitive Hybridization and Selective Amplification

1. 3 × EE solution: 30 mM EPPS (4-(2-Hydroxyethyl)-1-piperazinepropanesulfonic acid, pH 8.0), and 3 mM EDTA. EPPS is known to have a consistent pH at high temperatures *(13)*, and is stable for years once the pH is adjusted.
2. Mung bean nuclease (New England Biolabs).

2.5. Cloning of the MS-RDA Product

1. TA cloning kit using for example pGEM-T Easy (Promega).

3. Methods

The difficulty of MS-RDA mainly resides in the technical complexity of RDA. Since there are many steps in RDA, it is evident that even if 80% efficiency is achieved in each step, the final yield will go down to 11% after 10 steps. If 90% efficiency is achieved, it will be 35%; and at 95% efficiency, it will be 60%. Therefore, it is essential to maximize the efficiency at each step by maintaining DNA fragments that should be maintained and by eliminating those that should be eliminated.

The following procedures describe the *Hpa*II series of MS-RDA. Except for the optimal ratio of the mixture of the tester and driver amplicons, the *Sac*II series and *Nar*I series can be performed in the same manner using the appropriate adaptors (**Section 2.2**).

3.1. Amplicon Preparation

1. Genomic DNA (10 μg) is digested twice with an excess amount (100 U) of a methylation-sensitive restriction enzyme such as *Hpa*II, *Sac*II, or *Nar*I (*see* **Note 1**). The digestion product is extracted twice with phenol, twice with chloroform, and precipitated with ethanol using NH_4OAc. Dissolve the pellet in 20 μL of TE and quantify the solution.
2. Prepare adaptors for PCR amplification in parallel to step 1. To prepare 100 μM RHpa adaptor solution, dilute both RHpa24 and RHpa11 oligonucleotides with TE to final a concentration of 100 μM. Heat the mixture at 70°C for 5 min and cool down to 10°C. Prepare JHpa and NHpa adaptors in a similar manner (*see* **Note2**).
3. Ligate the RHpa adaptor to the purified *Hpa*II-restricted DNA. The ligation mixture (30 μL) contains 500 pmol of RHpa adaptor, 1 μg of the restricted DNA, 1 × T4 ligase buffer, and 800 U of T4 ligase. Keep the mixture at 16°C overnight (*see* **Note 3**).
4. Prepare a PCR mix (400 μL per sample) containing 3 μL of the ligation mixture, 1 × PCR buffer III, 300 μM dNTPs, 1 M betaine (*see* **Note 4**), and 1 μM RHpa24 primer. Prepare one 500 μL Eppendorf tube for the tester sample, and 10 tubes for the driver sample (*see* **Note 5**).
5. Start the PCR reaction by heating tubes at 72°C and addition of 3 μL of Taq polymerase (5 U/μL) to each tube (*see* **Note 6**). Then, perform 20 cycles of PCR (95°C for 1 min and 72°C for 3 min), and check the degree of amplification by running 10 μL of the PCR solution in a 0.9% agarose gel. If the PCR product (smear) has an intensity comparable to the DNA marker (250 ng/lane), the amplification is sufficient (**Fig. 10.4**). If not, add two to ten more cycles.
6. When sufficient amplification is achieved (*see* **Note 7**), purify the PCR product by two phenol extractions, two chloroform extractions, and ethanol precipitation with NH_4OAc. Dissolve the precipitate in 30 μL (tester) and 300–400 μL (driver) of TE, and quantify the samples (*see* **Note 8**).
7. Digest 20 μg of tester amplicon and 200 μg of driver amplicon with 100 U and 1000 U of *Msp*I, respectively. By running 1 μg of the digested DNA in a 3% NuSieveGTG agarose gel, confirm that the adaptor is completely restricted (**Fig. 10.5**).
8. Inactivate the restriction enzyme by extracting the digestion solution with phenol. Remove the digested adaptor by applying the extracted solution to a Chroma Spin + TE-200 column (one column for the tester, and four columns for the driver). For the tester, only the first elute (E1) should be used for the following steps. For the driver, the amplicon remaining in the

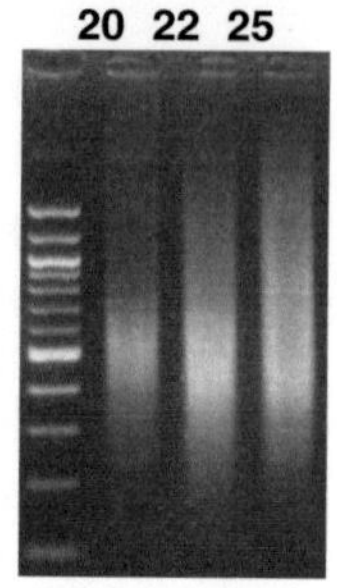

Fig. 10.4. Typical preparation of an amplicon. With 20 cycles of PCR the smear was not intense enough, and two cycles were added (22 cycles). If a further three cycles were added (25 cycles), the smear extended into a high molecular weight and was considered to have a significant amplification bias.

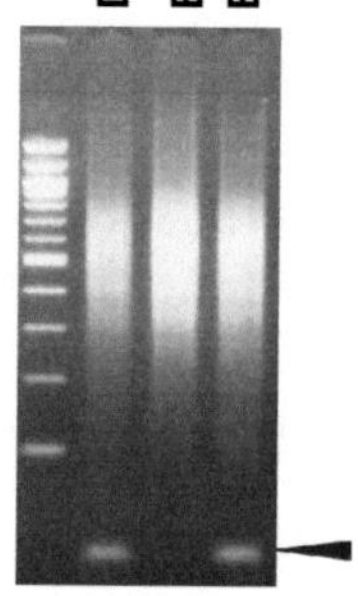

Fig. 10.5. Elimination of the digested adaptor. The digested amplicon showed a strong band for the 24-bp adaptor (shown by an arrow), and the band has completely disappeared in E1. However, the band still appeared in E2, and this E2 was again column purified and used as a driver.

column can be washed out (E2) by adding another TE for elution (*see* **Note 9**). By running 5 μL of E1 (and E2) in 3% NuSieveGTG agarose gel, confirm that the digested adaptor is at least for E1 completely removed (**Fig. 10.5**).

3.2. Competitive Hybridization

1. Ligate a new adaptor (JHpa adaptor) only to the tester amplicon E1. The ligation solution (30 μL) contains 500 pmol of JHpa adaptor, 200 ng of the tester amplicon (E1), 1 × ligation buffer, and 800 U of T4 ligase. Keep the solution at 16°C overnight.
2. Add 70 μL of TE to the ligation solution, and extract the solution once with phenol. Mix the extracted solution (100 μL) with 40 μg of the driver amplicon (E1 and E2). Extract the mixture once with phenol, twice with chloroform, and precipitate it with NaOAc (*see* **Note 10**).

3. Centrifuge the tube and rinse the pellet with 70% ethanol. Dry the pellet to a degree that it is dry but still retains water (*see* **Note 11**). Dissolve the pellet in 4 μL of 3 × EE solution (*see* **Note 12**).
4. Denaturation and re-annealing are successively performed. Place one drop of mineral oil onto the 4 μL 3 × EE solution in tube C. Denature the DNA mixture by heating at 96°C for 10 min in a thermal cycler. Add 1 μL of 5 M NaCl, paying attention so that the tube is "not" cooled during the handling (*see* **Note 13**), and keep the tube at 67°C for 16–24 h for re-annealing.

3.3. Selective Amplification

1. The denatured and re-annealed mixture is diluted by adding 45 μL of 1 M NaCl preheated at 67°C. To prevent evaporation, 1 M NaCl should be prepared in a large volume (400–500 μL). Great care should be taken that the mixture does not cool down.
2. Prepare a PCR solution (387 μL for now, but final 400 μL) that contains 1 × PCR III buffer, 300 μM dNTP, 1 M betaine, and 15 U of Taq polymerase; keep the solution at 72°C.
3. To the PCR solution, add 5 μL of the diluted mixture (step 1) at 72°C avoiding cool down of the mixture (*see* **Note 14**). This step is necessary to fill in 3′ ends of re-annealed products.
4. Add 8 μL of JHpa24 primer (50 μM) to the PCR solution at the first 95°C step, and perform 10 cycles of PCR (95°C for 1 min and 70°C for 3 min). After 10 cycles of PCR, keep the PCR solution at 72°C. Do not cool it down.
5. Take the tubes out of the thermal cycler and add immediately 40 μL of 10 × mung bean buffer to the PCR tube (*see* **Note 15**). Then add 10 μL of mung bean nuclease (10 U/μL). Keep the tube at 30°C for 30 min.
6. Extract the solution twice with phenol, twice with chloroform, and precipitate it with ethanol using NH_4OAc (*see* **Note 16**). Dissolve the pellet in 30 μL of TE.
7. Prepare a PCR solution that contains 1 × PCR III buffer, 300 μM dNTPs, 15 U Taq polymerase, 1 M betaine, and 3 μL of the above solution (step 6) (*see* **Note 17**). Heat the tube to 95°C, and then add 8 μL of JHpa24 primer (*see* **Note 18**). Perform 20 cycles of PCR (95°C for 1 min and 70°C for 3 min).
8. Check the degree of PCR amplification by running 10 μL of the PCR solution in a 2% NuSieve agarose gel (**Fig. 10.6**). If the amplification is not sufficient, add 2–10 cycles of PCR.
9. Extract the PCR solution twice with phenol, twice with chloroform, and precipitate it with ethanol using NH_4OAc. Dissolve the pellet in 50 μL of TE, and quantify the solution. This is the product of the first cycle of competitive hybridization (C1).

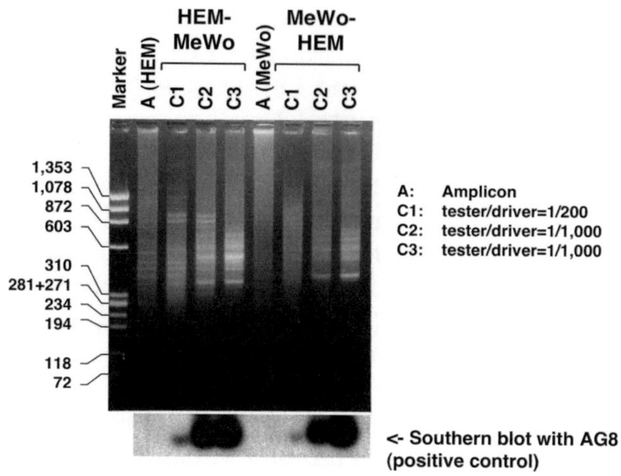

Fig. 10.6. A typical course of MS-RDA. Amplicons were prepared from a normal human embryonic melanocyte (HEM) and a melanoma cell line (MeWo). Two series of MS-RDA were performed using HEM as the tester (HEM – MeWo) and MeWo as the tester (MeWo – HEM). In C1, some differentially methylated fragments seem to be enriched. In C2 and C3, differentially methylated fragments became clearly visible by ethidium bromide staining of the entire DNA. Appropriate enrichment of differentially methylated fragments was confirmed by adding an unmethylated and methylated DNA fragment (AG8) to the tester and driver, respectively, at a concentration of one copy per haploid genome, and by observing its amplification by Southern blot analysis.

3.4. The Second Cycle

1. Digest 5–20 μg of C1 with 50–200 U of *Msp*I to switch the JHpa adaptor to the NHpa adaptor.
2. In a similar manner to the adaptor switch from RHpa to JHpa (step 8 of **Section 3.1** amplicon preparation), remove the digested JHpa adaptor, and confirm that the removal was complete.
3. To the C1 whose adaptor was removed (C1/MspI/E1), ligate the NHpa adaptor in a 30 μL solution that contains 500 pmol of the NHpa adaptor, 200 ng of the C1/MspI/E1, 1 × T4 ligase buffer, and 800 U of T4 ligase. Keep the solution at 16°C overnight.
4. Add 70 μL of TE to the ligation solution, and extract the solution once with phenol. Mix 40 ng (20 μL of the diluted solution) of the diluted ligation solution with 40 μg of driver amplicon (E1 and E2). Extract the mixture once with phenol, twice with chloroform, and precipitate it with ethanol using NaOAc.
5. Perform competitive hybridization and selective amplification as described for C1 in steps 3 and 4 of **Section 3.2** (competitive hybridization) and steps 1–9 of **Section 3.3** (selective amplification), except that the NHpa24 primer is used (*see* **Note 19**). This will produce C2. For C2, 20 cycles are usually enough for the second PCR as described in step 7

of **Section 3.3** (selective amplification). Confirm that selective amplification took place by running C2 in a 2% NuSieve agarose (**Fig. 10.6**).

3.5. The Third Cycle (OPTIONAL)

1. If selective amplification is not satisfactory in C2, a third cycle could be performed. In this case, switch the adaptor from NHpa back to JHpa, and perform competitive hybridization and selective amplification as described above (*see* **Note 20**).

3.6. Analysis of the MS-RDA Product

1. Clone the product of the last selective amplification (C2 or C3) into pGEM-T Easy vectors (*see* **Note 21**). After transformation into competent cells, pick up 96 clones and sequence them (*see* **Note 22**, *see also* **Chapter 14**). The cloned fragments have the final adaptor on their ends, and their sequence could be searched in genome databases.

4. Notes

1. This step is the only step that detects the DNA methylation status of the various genomic regions in the tester and driver. Therefore, the digestion by a methylation-sensitive restriction enzyme must be complete. Complete digestion can be confirmed by electrophoresing an aliquot of the digestion solution in a 0.9% agarose gel and observing the smear pattern. Also, some DNA samples are contaminated with RNA, and, if so, RNA must be eliminated during the digestion by RNase treatment. This will allow an appropriate amount of genomic DNA to be used for the following steps to prepare an amplicon.
2. Once an adaptor is prepared, it can be stored at −20°C for years.
3. The ligation time can be shortened down to 4 h. Since ATP is necessary for the ligation reaction, 10 × T4 ligase buffer should be handled on ice.
4. Betaine is known to facilitate amplification of G+C-rich sequences *(14)*.
5. These can be prepared at room temperature, while some of the later steps need careful temperature control. The ligation solution can be used without any purification.
6. To prevent nonspecific amplification, Taq polymerase should be added at a high temperature. A hot-start using an engineered-Taq polymerase, such as AmpliTaq Gold® (Applied Biosystems), is not suitable because the 3′ ends of the ligation products must be filled in at the initial incubation step at 72°C (**Fig. 10.2**).

7. For the following steps, a sufficient quantity of amplicon should be prepared. However, if too many cycles of PCR are performed, the amplification bias among DNA fragments will become significant. Therefore, the number of PCR cycles should be kept to a minimum within the range that yields a sufficient quantity of amplicon.
8. Typically, the driver amplicon yields 200–400 μg of amplified DNA.
9. Since 40 μg of driver amplicon without the adaptor are necessary for a competitive hybridization, 80–120 μg of the driver amplicon are necessary for a complete MS-RDA procedure. Use of E2 is sometimes inevitable to secure sufficient amounts of driver amplicon without the adaptor. However, for the tester, only E1 should be used because any residual adaptor can be ligated again and impairs the efficiency of competitive hybridization and selective amplification.
10. Since the efficiency of competitive hybridization and re-annealing is critical for the success of the experiment, the mixture is purified completely.
11. As with usual ethanol precipitation, the pellet should be dried to an appropriate degree. If it is too dry, its dissolution will become very difficult. If it contains much ethanol, the volume becomes larger and the efficiency of re-annealing will be impaired.
12. Ethanol precipitation is typically performed in a 1.5 mL Eppendorf tube, and the next step will be performed in a 0.5 mL Eppendorf tube. To dissolve the pellet of 40 μg of DNA (in tube A) completely:
    (a) take two 0.5 mL tubes (B, C),
    (b) put 5 μL of 3 × EE in tube B,
    (c) move 2 μL of 3 × EE from tube B to the pellet in tube A,
    (d) vortex tube A for more than 1 min paying attention that the solution is on the pellet, and spin it down (the solution is very sticky),
    (e) move the 2(+α) μL of the solution in tube A to tube C,
    (f) add fresh 2 μL of 3 × EE from tube B to tube A without changing the tip,
    (g) vortex tube A for more than 1 min, and spin it down,
    (h) move the 2 μL in tube A to tube C without changing the tip, and
    (i) vortex tube C, and spin it down.
13. The 5 M NaCl can be preheated, but care should be taken to avoid evaporation. Wear gloves to avoid burns while handling a hot tube. Cooling down of the tube accelerates nonspecific annealing, and will impair the efficiency of RDA.

14. The remaining solution (45 μL) from step 1 is usually kept at 67°C until the success of step 8 is confirmed. If the pellet is lost at step 6, it can be started over from this point.
15. While adding the mung bean buffer to the PCR solution, its temperature goes down to 30–40°C. Do not cool it down too much.
16. The pellet is very tiny, and the highest care must be taken that it is not lost. This step is one of the most difficult steps in the entire RDA procedure.
17. At this step, for the first time, all the DNA molecules in the tube are completely double stranded because all the single-stranded DNA molecules were digested by mung bean nuclease. Therefore, taking care of the temperature is not necessary during preparation of the PCR solution.
18. Only for this PCR, hot-start PCR can be used, if addition of a primer at 95°C is troublesome. The entire solution can be prepared at room temperature using an engineered-Taq polymerase, such as AmpliTaq Gold®.
19. The product of the first 10 cycles of PCR of the second competitive hybridization is usually much more abundant than the one in the first competitive hybridization, and the risk of loss is lower.
20. An addition of a fourth cycle does not improve the experiment. If selective amplification is not obtained in the third cycle, the quality of the initial samples, selection of the initial samples, and technical errors should be considered.
21. Any TA cloning vector is fine.
22. The number of clones sequenced is dependent upon the diversity of the final product. If redundant clones are observed by sequencing, sequencing can be suspended at that time point. If only nonredundant clones are observed, more clones should be sequenced.

Acknowledgements

The protocol is improved through the support of a Grant-in-Aid for the Third-Term Comprehensive Cancer Control Strategy from the Ministry of Health, Labour and Welfare, Japan.

References

1. Ushijima, T., Morimura, K., Hosoya, Y., et al. (1997) Establishment of methylation-sensitive representational difference analysis and isolation of hypo- and hypermethylated genomic fragments in mouse liver tumors. *Proc Natl Acad Sci USA* 94, 2284–2289.
2. Kaneda, A., Takai, D., Kaminishi, M., et al. (2003) Methylation-sensitive representational

difference analysis and its application to cancer research. *Ann NY Acad Sci* 983, 131–141.
3. Ushijima, T. (2005) Detection and interpretation of altered methylation patterns in cancer cells. *Nat Rev Cancer* 5, 223–231.
4. Takai, D., Yagi, Y., Wakazono, K., et al. (2001) Silencing of *HTR1B* and reduced expression of *EDN1* in human lung cancers, revealed by methylation-sensitive representational difference analysis. *Oncogene* 20, 7505–7513.
5. Kaneda, A., Kaminishi, M., Nakanishi, Y., et al. (2002) Reduced expression of the *insulin-induced protein 1* and *p41 Arp2/3 complex* genes in human gastric cancers. *Int J Cancer* 100, 57–62.
6. Miyamoto, K., Asada, K., Fukutomi, T., et al. (2003) Methylation-associated silencing of heparan sulfate D-glucosaminyl 3-*O*-sulfotransferase-2 (3-OST-2) in human breast, colon, lung and pancreatic cancers. *Oncogene* 22, 274–280.
7. Miyamoto, K., Fukutomi, T., Akashi-Tanaka, S., et al. (2005) Identification of 20 genes aberrantly methylated in human breast cancers. *Int J Cancer* 116, 407–414.
8. Hagihara, A., Miyamoto, K., Furuta, J., et al. (2004) Identification of 27 5′ CpG islands aberrantly methylated and 13 genes silenced in human pancreatic cancers. *Oncogene* 23, 8705–8710.
9. Abe, M., Ohira, M., Kaneda, A., et al. (2005) CpG island methylator phenotype is a strong determinant of poor prognosis in neuroblastomas. *Cancer Res* 65, 828–834.
10. Furuta, J., Nobeyama, Y., Umebayashi, Y., et al. (2006) Silencing of peroxiredoxin 2 and aberrant methylation of 33 CpG islands in putative promoter regions in human malignant melanomas. *Cancer Res* 66, 6080–6086.
11. Kaneda, A., Wakazono, K., Tsukamoto, T., et al. (2004) *Lysyl oxidase* is a tumor suppressor gene inactivated by methylation and loss of heterozygosity in human gastric cancers. *Cancer Res* 64, 6410–6415.
12. Abe, M., Westermann, F., Nakagawara, A., et al. (2007) Marked and independent prognostic significance of the CpG island methylator phenotype in neuroblastomas. *Cancer Lett* 247, 253–258.
13. Lisitsyn, N., Lisitsyn, N., Wigler, M. (1993) Cloning the differences between two complex genomes. *Science* 259, 946–951.
14. Henke, W., Herdel, K., Jung, K., et al. (1997) Betaine improves the PCR amplification of GC-rich DNA sequences. *Nucleic Acids Res* 25, 3957–3958.

Chapter 11

Restriction Landmark Genomic Scanning: Analysis of CpG Islands in Genomes by 2D Gel Electrophoresis

Joseph F. Costello, Dominic J. Smiraglia, Chibo Hong, and Christoph Plass

Abstract

Restriction landmark genomic scanning (RLGS) is a method that provides a quantitative genetic and epigenetic (cytosine methylation) assessment of thousands of CpG islands in a single gel without prior knowledge of gene sequence. The method is based on two-dimensional separation of radiolabeled genomic DNA into nearly 2,000 discrete fragments that have a high probability of containing gene sequences. Genomic DNA is digested with an infrequently cutting restriction enzyme, such as *Not*I or *Asc*I, radiolabeled at the cleaved ends, digested with a second restriction enzyme, and then electrophoresed through a narrow, 60-cm-long agarose tube-shaped gel. The DNA in the tube gel is then digested by a third, more frequently cutting restriction enzyme and electrophoresed, in a direction perpendicular to the first separation, through a 5% nondenaturing polyacrylamide gel, and the gel is autoradiographed. Radiolabeled *Not*I or *Asc*I sites are frequently used as "landmarks" because *Not*I or *Asc*I cannot cleave methylated sites and since an estimated 89% and 83% of the recognition sites, respectively, are found within CpG islands. Using a methylation-sensitive enzyme, the technique has been termed RLGS-M. The resulting RLGS profile displays both the copy number and methylation status of the CpG islands. Integrated with high-resolution gene copy-number analyses, RLGS enables one to define genetic or epigenetic alteration in cells. These profiles are highly reproducible and are therefore amenable to inter- and intraindividual DNA sample comparisons. RLGS was the first of many technologies to allow large-scale DNA methylation analysis of CpG islands.

Key words: RLGS, genomics, epigenomics, CpG islands, methylation, cancer, tissue-specific, restriction digestion, two-dimensional gel electrophoresis.

1. Introduction

Restriction landmark genomic scanning (RLGS) provides a quantitative genetic and epigenetic (cytosine methylation) assessment of thousands of gene-associated CpG islands in a single gel *(1)*.

Jörg Tost (ed.), *DNA Methylation: Methods and Protocols, Second Edition, vol. 507*

DOI 10.1007/978-1-59745-522-0_11 Springerprotocols.com

The method is based on two-dimensional separation of radiolabeled genomic DNA into nearly 2,000 discrete fragments, which have a high probability of containing gene and/or promoter sequences.

Genomic DNA is digested with an infrequently cutting restriction enzyme such as *Not*I, radiolabeled at the cleaved ends, digested with a second restriction enzyme and then electrophoresed through a narrow, 60-cm-long agarose tube-shaped gel. The DNA in the tube gel is then digested by a third, more frequently cutting restriction enzyme and electrophoresed, in a direction perpendicular to the first separation, through a 5% non-denaturing polyacrylamide gel, and the gel is autoradiographed. Radiolabeled *Not*I sites or *Asc*I sites are frequently used as "landmarks" since the majority of *Not*I or *Asc*I sites are within CpG islands *(2)*. The resulting RLGS profile displays the copy number of the CpG islands. These profiles are highly reproducible and are therefore amenable to inter- and intraindividual DNA sample comparisons. Using RLGS, the genome of mammals can be scanned at approximately 1 Mb intervals *(1)*.

To increase the number of fragments analyzed by RLGS, gel conditions can be altered *(3)* (allowing up to 6,000 fragments to be analyzed) or the DNA samples can be processed with a different series of enzymes (*see* **Table 11.1**). The choice of a particular "landmark" enzyme is critical since this site determines the bias of the displayed fragments. To maintain a strong bias for CpG islands, landmark enzymes such as *Not*I, *Asc*I *(4)*, *Bss*HII, or *Eag*I are generally used. Alternatively, a different second and/or third restriction enzyme may be used along with same landmark enzyme to display a different subset of fragments.

Table 11.1
Examples of additional enzyme combinations

Enzyme combination	Reference
*Not*I / *Eco*RV / *Hinf*I	*(1)*
*Not*I / *Pvu*II / *Pst*I	*(16)*
*Not*I / *Pst*I / *Pvu*II	*(31)*
*Not*I / *Bam*HI / *Hinf*I	*(31)*
*Bss*HII / *Pvu*II / *Pst*I	*(31)*
*Bss*HII / *Bam*HI / *Eco*RI	*(31)*
*Bss*HII / *Bam*HI / *Eco*RV	*(31)*
*Asc*I / *Eco*RV / *Hinf*I	*(4)*

RLGS fragments may be analyzed and quantified using one of several computer software programs having automated spot detection algorithms (http://www.expasy.ch/ch2d/2d-index.html). These include Conime (free download at http://www.cse.ohio-state.edu/research/graphics/conime/), Phoretix 2D Professional from Nonlinear Dynamics, Melanie III and PDQuest, which are available from BIO-RAD, and Bio-Image which is available from Genomic Solutions.

If the sequence of an RLGS fragment is not known, their sequence may be predicted with an *in silico* method using a sequenced genome *(5–8)* or obtained directly from an arrayed genomic library created from the same subset of fragments displayed on the RLGS profile *(4, 9, 10)*. One *in silico* method uses a custom PERL script to generate RLGS digestion fragments from the human genome sequence *(6, 7)*, similar to an informatics tool termed Virtual Genome Scan *(11)*. *In silico* restriction fragments are then matched with actual RLGS fragments on the basis of their two-dimensional fragment size and chromosome of origin *(12)* and confirmed with PCR. Alternatively, RLGS fragments are cloned from an arrayed genomic library. Arrayed libraries have been made from *Not*I/*Eco*RV fragments as well as *Asc*I/*Eco*RV fragments *(4)* and are useful when gels created by the same enzyme combinations are used. Although highly successful, this method is limited by the amount of effort needed to generate the library resource of each genome of interest and each enzyme combination of interest *(4, 8, 10)*. Direct cloning is technically more difficult due to the small amount of DNA in the gel, but may be applied to all enzyme combinations.

Differences between RLGS profiles have been used to identify important genes involved in normal cellular processes and in disease states. Two novel imprinted genes, one encoding a ribonucleoprotein auxiliary factor and the second encoding $Cdc25^{Mm}$, were isolated using this approach *(13, 14)*. It has been possible to obtain an estimate of the total number of imprinted genes in the genome by determining the proportion of DNA fragments on RLGS profiles which display a potentially imprinted pattern *(13)*. These genomic loci were identified as having a parent-of-origin-specific methylation pattern and indicated that a 50% change in the intensity of a single-copy DNA fragment was readily detectable in an RLGS profile. Such a reduction is also apparent in X-chromosome-specific fragments derived from either males or females, since there is methylation-related inactivation of one X chromosome in the latter. Similarly, comparison of profiles from normal individuals to those from Down's syndrome patients has revealed a proportional increase in the intensity of many chromosome 21-specific loci, as well as several chromosome 21 CpG islands, which were methylated on one copy of chromosome 21 and potentially represent an attenuation mechanism allowing for viability of a trisomy chromosome 21 fetus *(15)*. This approach has identified previously known tumor-suppressor genes as targets of aberrant methylation in cancer, such as *Itga4* (α4-integrin) *(16)*, and *Igfbp7 (17)* as well as novel tumor-suppressor gene candidates such as *TCF21 (18)*, *SLC5A8 (19–22)*, *ID4 (23)*, *BMP3B (24)*, *SOCS1 (25)*, and *WNK2 (26)*. RLGS analysis of tumors has also identified amplified oncogenes, such as CDK6, and others *(27–31)*. In a study of 98 primary human tumors, nonrandom and tumor-type-specific methylation patterns were discovered by systematic RLGS analyses of more than 1,000 CpG islands per tumor, in total including more than 100,000 methylation measurements *(32)*. Normal genetic variation among related individuals has also been observed using RLGS *(33)*. When integrated with high-resolution deletion maps from microarray-based comparative genomic hybridization (array CGH), RLGS can be used to define genetic or epigenetic alteration in cells *(5, 6)*.

The chromosomal origins of the majority of DNA fragments displayed on *Not*I-based RLGS profiles from human DNA have been mapped by chromosome-assigned RLGS (CA-RLGS) *(12)*. CA-RLGS profiles were generated from flow-sorted human chromosomes and then each individual chromosome-specific profile (except chromosomes 9–12 which were not separable from each other by flow sorting) was integrated into a total genomic DNA profile.

RLGS was shown to be a useful tool for the development of genetic maps *(34)* and loss of heterozygosity *(35)*. A detailed linkage map of 1,045 DNA fragments displayed on RLGS profiles and specific loci within each chromosome has been generated from profiles derived from a panel of recombinant inbred mouse strains *(34)*. Strain-specific differences in restriction sites lead to either the loss or gain of a labeled fragment on an RLGS profile. Similar to other approaches using RFLP probes in Southern blots or polymorphic microsatellite markers in PCR assays, a single RLGS profile displays several hundred polymorphic markers in intra- and interspecific back crosses, interspecific congenic strains, and recombinant inbred strains *(36–38)*. Previous sequence information is not required to establish a genetic map using RLGS. These high-density genetic maps and standard positional cloning techniques have been used to identify genes of interest. In a study by Okazaki et al. RLGS was used for the identification of a large number of loci in a defined genetic interval containing the mouse reeler locus on chromosome 5 *(38)*. Reeler backcross progeny were generated and typed for markers flanking the rl locus. DNAs from homozygotes and heterozygotes at the reeler locus were pooled separately and analyzed with three different restriction enzyme combinations that displayed a total of 8,856 fragments. Thirty-one new markers spanning approximately 8.5 cM were then used to establish a high-density genetic map surrounding the reeler locus and, in combination with positional cloning, led to the identification of the reeler gene *(39)*.

2. Materials

2.1. Restriction Landmark Genomic Scanning Procedure

2.1.1. Isolation of Genomic DNA

1. Liquid nitrogen.
2. Mortar and pestle, heavy-duty aluminum foil, and hammer.
3. Dialysis tubing: 3/4 in. × 25 ft (Gibco BRL).
4. Dialysis clips/closures (Spectra/Por).
5. 100% ethanol.
6. Proteinase K.
7. RNAse A (Boehringer Mannheim).
8. Lysis buffer: 10 mM Tris–HCl (pH 8.0), 150 mM EDTA (pH 8.0), and 1% sarkosyl.
9. PCI: phenol:chloroform:isoamylalcohol (in the ratio 25:24:1).
10. 10 mM Tris–HCl, pH 8.0.

2.1.2. Enzymatic Processing of Genomic DNA

1. Wide-bore pipette tips.
2. Klenow DNA polymerase I.
3. Sequenase ver. 2.0 (13 U/ μL, USB/Amersham).
4. *Not*I (10 U/ μL), *Eco*RV (10 U/ μL), *Hinf*I (70 U/ μL) (Promega or New England Biolabs).
5. [α-^{32}P]-dGTP (20 mCi/mL, 6,000 Ci/mmol; New England Nuclear).
6. [α-^{32}P]-dCTP (10 mCi/mL, 3,000 Ci/mmol; Amersham).
7. 10 × Buffer 1: 500 mM Tris–HCl (pH 7.4), 100 mM $MgCl_2$, 1 M NaCl, 10 mM DTT (store at −20°C).
8. 20 × Buffer 2: 3 M NaCl, 0.2% Triton X-100, 0.2% BSA (store at −20°C).
9. Blocking buffer: 1 μL 10 × buffer 1, 0.1 μL 1 M DTT, 0.4 μL each of 10 μM dGTPαS (Pharmacia), 10 μM ddATP, 10 μM ddTTP, and 0.2 μL 10 μM dCTPαS (Pharmacia). Make stock and store in aliquots at −20°C.
10. Second enzyme digestion buffer: 1 μL 1 mM ddGTP, 1 μL 1 mM ddCTP, 4.4 μL ddH_2O, and 1.2 μL 100 mM $MgCl_2$.
11. 6 × loading dye (first-dimension): 0.25% bromophenol blue (BPB), 0.25% xylene cyanol (XC), and 15% Ficoll type 400.

2.1.3. Sample Quantitation

1. Salmon sperm DNA.
2. 20% Trichloracetic acid.
3. Whatman GF/F filter.
4. Scintillation counter.

2.1.4. First-Dimension Gel Set-Up and Electrophoresis

1. First-dimension gel apparatus (Biocraft, Tokyo, Japan).
2. PFA-grade teflon tubing for the first-dimension gel (2.4 mm i.d., 3.0 mm o.d., 10 M – sufficient for at least 16 gels) (PFA 11 thin wall, natural; American Plastic, Columbus Ohio).
3. Glass tubes (4 mm i.d., 5 mm o.d., 60 cm) with a tapering at the top end extending over 1.2 cm to a final dimension of 3 mm i.d.).
4. Two-way stopcocks (4–8).
5. Flexible Tygon tubing (3/16 in. i.d., 1/4 in. o.d., VWR).
6. Seakem GTG agarose (do not substitute).
7. 20 × Boyer's buffer: 1 M Tris–HCl, 360 mM NaCl, 400 mM sodium acetate, 40 mM EDTA, pH 8.0 autoclave. Note: Boyer's buffer is used at 2 × concentration for the gel and running buffer.
8. First-dimension gel (0.8%): 0.48 g agarose, 60 mL 2 × Boyer's buffer.

2.1.5. In-Gel Digestion

1. Digest tubing: PFA grade teflon, 9, thin-wall, natural (2.7 mm i.d. and approximately 3.3 mm o.d.; American Plastic, Columbus Ohio).
2. 10-mL syringe.
3. 10 × Buffer K: 200 mM Tris–HCl, 100 mM $MgCl_2$, 1 M NaCl, pH 7.4, autoclave.
4. *Hinf*I (70 U/μL; Promega or New England Biolabs).
5. BSA.

2.1.6. Second-Dimension Electrophoresis

1. Second-dimension gel apparatus (Biocraft Research Landmark Genome Scan System, Tokyo, Japan).
2. Five glass plates, eight spacers, one Plexiglas sheet (C.B.S. Scientific).
3. Flexible tubing (1/8 in. i.d., 3/16 in. o.d., VWR).
4. Three-way stopcock.
5. 10- and 60-mL syringes, 26-gauge needle.
6. "Plastic" tape (Scotch brand, do not substitute).
7. 10 × TBE: 900 mM Tris–borate, 20 mM EDTA, pH 8.3.
8. 10 × TE: 100 mM Tris–HCl, 10 mM EDTA, pH 8.0.
9. Connecting agarose: 1 × TBE, pH 8.3, 0.8% Seakem GTG agarose.
10. Second-dimension loading dye: 1 × TE, pH 8.0, 0.25% bromophenol blue, and 0.25% xylene cyanol.
11. 5% nondenaturing polyacrylamide gel: 1 × TBE, pH 8.3, 96.9 g acrylamide, 3.3 g bis-acrylamide, 1.3 g ammonium persulfate (APS) in a total volume of 2 L. Add 700 μL TEMED before pouring the gel. This volume is sufficient for four gels (one second-dimension apparatus).
12. Isopropanol or water-saturated butanol.
13. Whatmann paper (three for each gel, 13.5 in. × 16.5 in.).

14. Saran wrap.
15. X-ray film (BioMax MS).

2.2. Analysis of RLGS Profiles

1. Computer software program with automated spot-detection algorithms (e.g., Phoretix 2D from Nonlinear Dynamics and Melanie III from Bio-Rad).
2. For visual analysis of overlaid autoradiographs: transparent acetate sheets or used 14 × 17 X-ray film which have been cleared of emulsion by washing in 5% bleach.

2.3. Identification of RLGS Fragments

1. Elution Buffer: 0.5 M ammonium acetate, 1 mM EDTA.
2. PCI—phenol:chloroform: isoamylalcohol (in the ratio 25:24:1).
3. Scalpel.
4. 100% Ethanol.
5. Glycogen (20 mg/mL).
6. 1 × TE: 10 mM Tris–HCl, 1 mM EDTA, pH 8.0.
7. PCR reagents: Taq polymerase, dNTPs, primers designed based on predicted fragment sequences.

3. Methods

3.1. Restriction Landmark Genomic Scanning

3.1.1. Isolation of Genomic DNA

1. Expel 100–300 mg of the frozen tissue (*see* **Note 1**), wrap it in aluminum foil, and quickly break into pieces with a hammer. Keep the foil/tissue cold by submerging it in liquid nitrogen.
2. Transfer the tissue to a mortar (prechilled in liquid nitrogen) and grind to a powder with a prechilled pestle. Transfer the tissue powder into a 50 mL tube and store in dry ice until all samples are processed.
3. Add 5–25 mL of lysis buffer and proteinase K (approximately 0.01–0.1 mg/mL final concentration in lysis buffer). Mix gently using a glass rod or 1 mL disposable pipette. Incubate at 55°C for 20 min, mixing very gently every 5 min.
4. Cool the lysed samples on ice for 10 min. Add an equal volume of PCI. Rotate tubes gently for 30–60 min.
5. Centrifuge for 30 min at 1,400*g* and transfer the DNA (aqueous phase) to a 50-mL tube using a wide-bore pipette. Repeat the PCI extraction twice.
6. Transfer the DNA into dialysis tubing and dialyze against 4 L of 10 mM Tris, pH 8.0 for 2 h. Transfer the tubing into fresh 10 mM Tris and dialyze overnight at room temperature. Dialyze in fresh 10 mM Tris for an additional 2 h.
7. Transfer the DNA to 50-mL tubes and add RNase A to a final concentration of 1 μg/mL. Incubate at 37°C for 2 h.

8. Add 2.5 volumes 100% ethanol to the DNA and rotate gently for 30–60 min.
9. Centrifuge for 30 min at 1,400*g* to collect DNA. For smaller amounts of tissue, it may be necessary to collect the DNA by centrifugation at 2,000*g*.
10. Discard the ethanol and briefly air-dry the pellet. Do not overdry the pellet.
11. Resuspend the DNA to a final concentration of approximately 1 μg/μL (may take several days at 4°C). To check for DNA degradation, electrophorese a small amount of the DNA on a standard 0.8% agarose gel. DNA isolated in this manner has an average size of 200–300 kb and should be transferred with positive displacement pipettes.

3.1.2. Enzymatic Processing of Genomic DNA

1. In a 1.5-mL tube, add 7 μL of genomic DNA (0.2–1.0 μg/μL), 2.5 μL of blocking buffer, and 0.5 μL of Klenow DNA polymerase I (5 U/μL), mix thoroughly by stirring, and incubate reaction at 37°C for 20 min. (Transfer the DNA with a wide-bore pipette tip. Do not pipette to mix (*see* **Note 2**). Using a master mix increases the uniformity among samples.
2. Incubate the reaction at 65°C for 30 min to inactivate the polymerase. Cool the reaction on ice for 2 min and centrifuge briefly at high speed.
3. Add to the sample 8 μL of 2.5 × buffer 2, 2 μL of *Not*I (10 U/μL), and mix thoroughly by stirring. Incubate at 37°C for 2 h.
4. Add 0.3 μL of 1 M DTT, 1 μL of [α-^{32}P]-dGTP, 1 μL of [α-^{32}P]-dCTP, and 0.1 μL of Sequenase ver. 2.0 (13 U/μL) to the sample (use a master mix). Mix thoroughly by stirring and incubate at 37°C for 30 min (*see* **Note 3**).
5. Add to the sample 7.6 μL of the second digestion buffer and 2 μL of *Eco*RV (10 U/μL). Mix by stirring and incubate at 37°C for 1 h. Cool on ice and add 7 μL of 6 × first-dimension loading dye. To confirm that the DNA was digested to completion, check 2.5 μL of the reaction on a standard 0.8% agarose gel.

3.1.3. Determining Amount of Sample to Load on First-Dimension Gel

1. Mix 2 μL of each sample with 100 μL of salmon sperm DNA (500 μg/mL) and add 100 μL of 20% trichloroacetic acid (TCA). Incubate for 10 min on ice. Do not centrifuge. Mix by stirring.
2. Filter each sample through a Whatman GF/F filter. Wash each filter twice with 10 mL of 20% TCA.
3. Measure the dpms on each filter using a scintillation counter. For mouse genomic DNA, loading 130,000 dpm is sufficient. For human genomic DNA, 450,000 dpm should be

loaded. These values may vary with the quality of the DNA (*see* **Notes 4** and **5**).

3.1.4. First-Dimension Gel Set-Up and Electrophoresis

1. Using a sharp razor, cut one end of the Teflon tubing at an angle to make a bevel. Feed the beveled end into the glass rod of the gel holder until it protrudes slightly from the tapered end. Using a hemostat, pull the beveled end up through the tapered end of the glass rod until it protrudes 2–4 cm. Cut the tubing horizontally at the same end, leaving a 2-mm protrusion (this is the top of the gel holder). Cut the opposite end horizontally to leave a 2-cm protrusion from the glass rod. Invert the gel holder and press the top protruding end firmly against a hot metal surface (metal spatula heated by a Bunsen burner) to fold the edges of the Teflon outward onto the rim of the glass support, making sure to avoid folding the edges inward and sealing the tubing. Pull a rubber stopper with cored center over the top end of the gel holder until it is just past the taper of the glass rod. It is essential that all tubing is clean and free of liquid and particulate matter. The Teflon tubing should be rinsed by suctioning through ddH_2O and then dried by continued application of the vacuum.
2. Attach a two-way stopcock to a 10-mL syringe and then to the gel holder via 2–3 cm of flexible tubing. Adjust the stopcock valve to the open position.
3. In a clean 200-mL glass bottle, add 60 mL 2 × Boyer's buffer and 0.48 g Seakem GTG agarose (0.8%). Weigh the solution. Microwave until the agarose is dissolved, stopping occasionally to swirl the contents and to avoid boil over. Weigh the solution and add ddH_2O to return to the starting weight. Equilibrate the gel solution to 55°C in a water bath.
4. With the stopcock valve in the open position, lower the protruding Teflon tube into the molten agarose solution. Suction the gel solution into the gel holder until the gel solution has reached 1–2 cm from the top of the gel holder and then close the stopcock valve. Keeping the gel upright, suspend the gel from a ring stand. Add a drop of gel solution to the bottom of each Teflon tube to allow for the slight gel shrinkage during the drying period. Allow the gel to solidify for a minimum of 20 min (60 mL gel solution is sufficient for eight gels).
5. Open the stopcock valve and remove the syringe and connecting tubes from each gel. After adding 2 × Boyer's buffer to the bottom of the first-dimension apparatus (to approximately 5 cm from the bottom), lower the gels into the first-dimension gel apparatus, seating the rubber stopper firmly into the appropriate holes in the top portion of the apparatus. Fill the top chamber with 2 × Boyer's buffer until the tops of the gels are submerged. Remove air bubbles from the

space between the top of the gel holder and the top of the gel. The sample will not run properly if the gel or the loading well contains bubbles or particulate matter.

6. Load an appropriate amount of sample onto each gel. Electrophorese at 110 V for 4 h, then increase to 230 V for approximately 20 h (or until the bromophenol blue (BPB) dye has reached 10 cm from the bottom of the lower buffer chamber).

3.1.5. In-Gel Digest

1. Remove buffer and gel holders from the first-dimension apparatus. Extrude the gel into a pan containing 1 × buffer K by forcing the gel out through the bottom of the gel holder. This is accomplished using a 1-mL syringe fitted with a pipette tip and filled with buffer K. Firmly insert the tip into the top of the gel holder and depress the plunger until the gel begins to come out through the bottom of the gel holder. Carefully replace the 1-mL syringe with a 5-mL syringe, depress the plunger until the entire gel is expelled. With a razor, cut a bevel in the low-molecular weight end of the gel and cut horizontally at the high-molecular weight end so that the gel is approximately 43 cm in length. The gel length is now the same as the width of the second-dimension gel.
2. Place each gel into a separate 50-mL tube containing 40 mL of 1 × buffer K. Incubate for 10 min at room temperature. Carefully pour off the buffer and incubate in 1 × buffer K for an additional 10 min. (The gel may be transferred by carefully looping it onto gloved fingers.)
3. Carefully pour the buffer K and gel into a pan containing fresh buffer K. Using a 10-mL syringe attached to restriction digest tubing (via a 1–2 cm segment of flexible tubing), suction the gel into the digest tubing, low molecular weight (beveled) end first. The gel is suctioned into the digest tubing by placing the end of tubing in line with the beveled end of the gel and pulling the syringe plunger. Be careful to stop once the gel has completely entered the tubing. Carefully position the tubing vertically, with the syringe at the bottom. Suction any remaining buffer from the tubing into the syringe. Do not continue suctioning if the gel blocks the syringe opening. If this occurs, depress plunger gently and force gel away from syringe opening. Detach the syringe, expel buffer, and reattach.
4. In a clean tube, make a 1.6 mL mix of 1 × restriction enzyme buffer, 0.1% BSA, and 750 U of the restriction enzyme *HinfI*. Place the open end of the digest tubing into the tube containing restriction digestion solution, now holding the syringe end up, apply suction until a small amount of digestion solution appears in the syringe. Carefully remove the digest

tubing and orient both ends upward in a U-shape. Remove the syringe and attach the two ends of the tubing to form a closed circle. Place in a moist chamber and incubate at 37°C for 2 h. Preparing a master mix of the digest solution works well. Avoid bubbles in the digest tubing as these may interfere with complete digestion. In general, the restriction enzyme is in excess and incomplete digests are very rare.

3.1.6. Second-Dimension Electrophoresis

1. Assemble the second-dimension gel apparatus. All glass plates should be cleaned thoroughly and wiped with 95% ethanol. The nonbeveled face of each plate should be coated with Gel-slick or Sigmacote (only once every 10 uses). Lay the back of the apparatus horizontally on a table top with the upper buffer chamber hanging over the table edge. Insert the two small clear plastic blocks at the bottom corners of each apparatus. Place a glass plate in the apparatus, beveled edge facing upward and near the upper buffer chamber, followed by two spacers, one along each side. Add glass plates and spacers in this manner until the fifth plate has been added. After the third plate, slide flexible Tygon tubing down the side channel of the apparatus, with a bevel cut in the leading end of the tubing. Cut the other end, leaving approximately 10 cm protruding from the apparatus. Position the front part of the apparatus by aligning the screw holes of the front and back parts. Secure with the Teflon screws. Seal the oblong oval "windows" at the lower, front face with Plastic tape. Stand the apparatus upright in the lower buffer chamber.
2. Using a three-way stopcock, attach the gel apparatus tubing in series with a 2-L reservoir and attach a 60-mL syringe to the remaining stopcock outlet. The tubing should be attached to the 2-L reservoir through a bottom drain (a 2-L graduated cylinder works well). Secure the reservoir above the gel apparatus to allow gravity flow. Adjust the stopcock valve to allow liquid to flow between the 2-L reservoir and the 60-mL syringe. Once the TEMED has been added, pour the acrylamide solution into the 2-L reservoir. Pull the syringe plunger down to the 50-mL mark. Depress the plunger to push the air out of the upper tubing. Once all air has been removed, adjust the valve so that all three ports are open. Acrylamide will flow into the apparatus, filling all four gels simultaneously from the bottom upward. Stop the flow when the level reaches 3 mm from the top edge of the glass plates. Allow the solution to settle for 2–3 min. If the acrylamide level drops, resume flow briefly. Immediately add 1 mL of isopropyl alcohol or water-saturated butanol along the top edge of each gel. After the valve leading to the gel apparatus has been closed, detach the syringe and reservoir. Once the acrylamide has polymerized, the gels may be stored overnight by adding 1 × TBE to the upper reservoir.

Just before use, rinse wells thoroughly with water and then dry gently with a tissue.

3. Gently separate the ends of the digest tubing and extrude the first-dimension gel into a pan containing 1 × TBE, pH 8.3. The gel may slide out by gravity or may require gentle liquid pressure.
4. Transfer the gel to a 50-mL tube containing 40 mL 1 × TBE, pH 8.3. Incubate for 10 min at room temperature, replace with fresh TBE, and incubate for an additional 10 min.
5. Place each first-dimension gel in a horizontal position across the beveled edge of each glass plate. Once all gels are in place, fill the space between the agarose gel and the top of each polyacrylamide gel with molten 0.8% agarose (equilibrated to 55°C). Use a 26-gauge needle attached to a 10-mL syringe to add the connecting agarose. Be sure to avoid bubbles between the first- and second-dimension gels. Allow connecting agarose to solidify for 5 min, add 250 μL second-dimension loading dye along the length of each gel. Add 1 × TBE, pH 8.3 to the upper and lower buffer chambers and electrophorese at 100 V for 1 h, then increase to 150 V for approximately 24 h (or until the bromophenol blue (BPB) reaches the bottom of the gel).
6. Remove buffers and disassemble apparatus. Lift each gel from the plates by overlaying with Whatmann paper cut to size for autoradiographic or phosphorimager cassettes. Trace the perimeter of the paper with edge of a plastic ruler, removing excess gel. Carefully peel back and lift Whatmann paper and gel. Place gel side up on second Whatmann paper. Overlay with saran wrap, add third Whatmann paper to top and fold edges of saran wrap over top Whatmann. In the same orientation, place in a gel drier for 1 h at 80°C while applying a vacuum. It is critical that the entire gel is completely dried. Remove lower and upper Whatman paper, fold Saran Wrap under remaining paper and expose to X-ray film (**Fig. 11.1**).

3.2. Analysis of RLGS Profiles

Visual assessment is performed by overlaying two RLGS profiles on a light box and comparing relative intensities of fragments. Overlaying "master" profiles (profiles used as a standard for comparison) with clear acetate sheets allows one to mark differences or similarities between multiple profiles and to generate cumulative data sets. This is also a convenient form in which to retain a usable record of the analysis results, and is a standard that has allowed the sharing of results between labs. To allow uniform comparisons of RLGS profiles from different samples and different laboratories, each fragment has been given a three-variable designation (Y coordinate, X coordinate, and fragment number) *(32)*. A publicly accessible website containing these designations is available (http://www.cse.ohio-state.edu/graphics/conime).

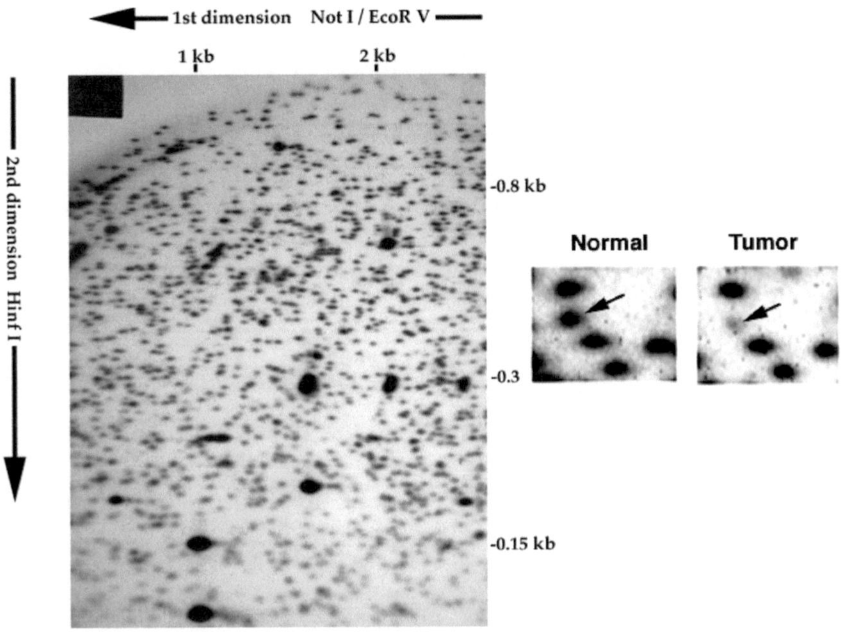

Fig. 11.1. RLGS profile of human brain tumor DNA using the enzyme combination *Not*I / *Eco*RV / *Hinf*I.

Alternatively, software systems developed for analysis of two-dimensional protein gels are useful for automated analysis of RLGS profiles (*see* **Note 6**.).

3.3. Identification of RLGS Fragments

Recent work has determined the identity of 800 RLGS fragments from the *Not*I–*Eco*RV–*Hinf*I combination *(8)*. Thus, there is very little need for spot identification. However, if a spot of interest is not within this list, or if a new enzyme combination is used to increase the genome coverage, the following elution protocol should be useful.

3.3.1. Elution of an Individual Spot DNA from the RLGS Gel

1. Remove the piece of X-ray film and Saran Wrap from the excised gel slice. Wet the gel slice with 20 μL of elution buffer (0.5 M ammonium acetate, 1 mM EDTA). Gently lift the rehydrated gel off of the paper.
2. Cut the gel slice into smaller pieces with a scalpel. Add 150 μL elution buffer to the wetted gel slice. Place in a shaking incubator at 37°C overnight.
3. Centrifuge and transfer supernatant to a clean tube. Rinse acrylamide with 50 μL elution buffer and combine with supernatant.
4. Add 1 μL of glycogen (20 μg/μL) and 2.5 volumes of 100% ethanol, chill for several hours or overnight at –80°C.
5. Centrifuge at 4°C for 30 min, remove ethanol, and dry pellet. Resuspend in 10 μL TE.

3.3.2. Prediction and Confirmation of RLGS Fragments with In Silico RLGS

A strategy has been developed to identify the nucleotide sequences of the majority of the RLGS fragments using a sequenced genome assembly.

1. Run the custom PERL program *(6, 8)* that performs restriction digests using combined enzymes, such as *Not*I/*Eco*RV (first dimension) and *Not*I/*Hinf*I (second dimension) to generate RLGS digestion fragments from a sequenced genome assembly, similar to an informatics tool termed Virtual Genome Scan *(11)*.
2. *In silico* restriction fragments are matched with actual RLGS fragments on the basis of their two-dimensional fragment size and chromosome of origin *(8, 12)*.
3. PCR primers are designed based on the putative matched sequence.
4. Confirm the prediction by PCR. Confirmation of a match requires single PCR products of the correct size from the eluted, matched fragment and from total genomic DNA, and no PCR product or consistently minimal product from a nearby but unrelated fragment and from water. Once the identity and specific chromosomal localization of a fragment is known, it is possible to integrate the methylation data from RLGS with copy-number data, gene function, and regulation (*see* **Note 7**).

4. Notes

1. DNA quality is a critical parameter for generating high-quality RLGS profiles. Small amounts of degraded DNA can cause a diffuse background. Therefore, tissue and cell pellets should be snap frozen in liquid nitrogen and stored at −80°C prior to isolation.
2. Genomic DNA is first blocked at sheared sites by the addition of dideoxynucleotides and sulfur-substituted nucleotides. The DNA is digested with an infrequent cutting restriction enzyme end labeled at the restriction sites, and then digested with a second restriction enzyme. To prevent nonspecific shearing of the DNA during this procedure, wide-bore pipette tips should be used for transferring the DNA, and all reactions must be mixed by stirring, rather than pipetting.
3. The protocol describes in detail the procedure and solutions for the most frequently used restriction enzyme set (*Not*I, *Eco*RV, and *Hinf*I), although many different methylation-sensitive and -insensitive restriction enzymes may be used (*see* **Table 11.1**).
4. A critical factor in the first-dimension electrophoresis is loading the correct amount of labeled DNA on the gel. Loading

more than 1.5 μg may cause smearing of high molecular weight DNA fragments. Loading less than 1 μg may result in a less intense profile, but longer film exposure times may compensate for this. Generally, a short "test exposure" for 1 day helps to estimate the full exposure time needed. In addition to the quality of DNA, the sample volume loaded on the first-dimension gel correlates roughly with quality of the profile. Sample volumes >10 μL occasionally result in a smeared appearance of the high molecular weight DNA fragments, although the exact cause of this technical problem is unknown.

5. For visual determination of the concentration of the DNA in each sample, co-electrophorese control restriction enzyme digested DNA in the range of 0.2–1 μg on a standard 0.8% agarose gel. Approximate the microliter amount of the sample that will contain 1.5 μg. This is the maximum amount that should be loaded on the first-dimension gel. Alternatively, the amount of sample to be loaded on the first-dimension gel can be determined by loading a constant amount of incorporated radiolabel. If an important area of the profile has streaks rather than discrete spots, but low signal strength, the specific activity of the labeled material may be poor. This is generally due to the DNA being too concentrated. Ideally, the DNA will be viscous, but not clumpy.
6. Using direct visual assessment of profiles has proven very reliable. There is a nearly 100% concordance between alterations of DNA fragments detected in RLGS profiles and their subsequent validation by other methods, such as Southern blotting and bisulfite-based methods. Comparisons by computer or by visual assessment are facilitated by the fact that RLGS profiles from different tissues or from different individuals are identical at the majority of loci. Several computer-assisted analysis systems which were originally designed for analysis of 2-D protein gels have been developed for the 2-D DNA analysis *(40, 41)*
7. One of the disadvantages of using RLGS for global methylation analysis of CpG islands, at least in the analysis of tumor tissue, is that a loss of a fragment from an RLGS profile could be due to either deletion or methylation. Once the fragment is identified, along with the genetic data, such as copy-number change of the same locus, the loss of the RLGS fragment can be ascribed to deletion, methylation or both *(6)*. The identification of DNA fragments which display a tumor-specific increase in intensity has led to two very different findings which are indistinguishable until the DNA fragments are identified and tested further. The increased intensity in some cases corresponded to gene amplification *(27)*, while for others this identified DNA fragments from repetitive sequences

that had become demethylated in a tumor-specific manner *(42)*. Recently the *in silico* program (vRLGS) has been significantly improved to increase the accuracy of prediction. Not only the size of the fragment is included in the calculations, but also the sequence characteristics, such as CpG content *(8)*. Mapping RLGS fragments *in silico* is widely applicable to any sequenced genome and enzyme combination *(6, 8)*. These maps will facilitate a more comprehensive assessment of DNA methylation and genetic alteration and provide a new capability for genome-wide analyses of methylation and gene expression.

Acknowledgement

This work was supported in part by grants from NIH, The Brain Tumor Society, and the Accelerated Brain Cancer Cure to JFC.

References

1. Hatada, I., Hayashizaki, Y., Hirotsune, S., et al. (1991) A genomic scanning method for higher organisms using restriction sites as landmarks. *Proc Natl Acad Sci USA* 88, 9523–9527.
2. Lindsay, S., Bird, A. P. (1987) Use of restriction enzymes to detect potential gene-sequences in mammalian DNA. *Nature* 327, 336–338.
3. Imoto, H., Hirotsune, S., Muramatsu, M., et al. (1994) Direct determination of NotI cleavage sites in the genomic DNA of adult mouse kidney and human trophoblast using whole-range restriction landmark genomic scanning. *DNA Res* 1, 239–243.
4. Dai, Z., Weichenhan, D., Wu, Y. Z., et al. (2002) An AscI boundary library for the studies of genetic and epigenetic alterations in CpG islands. *Genome Res* 12, 1591–1598.
5. Hong, C., Bollen, A. W., Costello, J. F. (2003) The contribution of genetic and epigenetic mechanisms to gene silencing in oligodendrogliomas. *Cancer Res* 63, 7600–7605.
6. Zardo, G., Tiirikainen, M. I., Hong, C., et al. (2002) Integrated genomic and epigenomic analyses pinpoint biallelic gene inactivation in tumors. *Nat Genet* 32, 453–458.
7. Matsuyama, T., Kimura, M. T., Koike, K., et al. (2003) Global methylation screening in the Arabidopsis thaliana and Mus musculus genome: applications of virtual image restriction landmark genomic scanning (Vi-RLGS). *Nucleic Acids Res* 31, 4490–4496.
8. Smiraglia, D. J., Kazhiyur-Mannar, R., Oakes, C. C., et al. (2007) Restriction Landmark Genomic Scanning (RLGS) spot identification by second generation virtual RLGS in multiple genomes with multiple enzyme combinations. *BMC Genomics* 8, 446.
9. Plass, C., Weichenhan, D., Catanese, J., et al. (1997) An arrayed human not I-EcoRV boundary library as a tool for RLGS spot analysis. *DNA Res* 4, 253–255.
10. Smiraglia, D. J., Frühwald, M. C., Costello, J. F., et al. (1999) A new tool for the rapid cloning of amplified and hypermethylated human DNA sequences from restriction landmark genomic scanning gels. *Genomics* 58, 254–262.
11. Rouillard, J. M., Erson, A. E., Kuick, R., et al. (2001) Virtual genome scan: a tool for restriction landmark-based scanning of the human genome. *Genome Res* 11, 1453–1459.
12. Yoshikawa, H., de la Monte, S., Nagai, H., et al. (1996) Chromosomal assignment of human genomic NotI restriction fragments in a two-dimensional electrophoresis profile. *Genomics* 31, 28–35.
13. Hayashizaki, Y., Shibata, H., Hirotsune, S., et al. (1994) Identification of an imprinted U2af binding protein related sequence on mouse chromosome 11 using the RLGS method. *Nat Genet* 6, 33–40.
14. Plass, C., Shibata, H., Kalcheva, I., et al. (1996) Identification of Grf1 on mouse chromosome 9 as an imprinted gene by RLGS-M. *Nat Genet* 14, 106–109.

15. Kuromitsu, J., Yamashita, H., Kataoka, H., et al. (1997) A unique downregulation of h2-calponin gene expression in Down syndrome: a possible attenuation mechanism for fetal survival by methylation at the CpG island in the trisomic chromosome 21. *Mol Cell Biol* 17, 707–712.
16. Akama, T. O., Okazaki, Y., Ito, M., et al. (1997) Restriction landmark genomic scanning (RLGS-M)-based genome-wide scanning of mouse liver tumors for alterations in DNA methylation status. *Cancer Res* 57, 3294–3299.
17. Komatsu, S., Okazaki, Y., Tateno, M., et al. (2000) Methylation and downregulated expression of mac25/insulin-like growth factor binding protein-7 is associated with liver tumorigenesis in SV40T/t antigen transgenic mice, screened by restriction landmark genomic scanning for methylation (RLGS-M). *Biochem Biophys Res Commun* 267, 109–117.
18. Smith, L. T., Lin, M., Brena, R. M., et al. (2006) Epigenetic regulation of the tumor suppressor gene TCF21 on 6q23-q24 in lung and head and neck cancer. *Proc Natl Acad Sci USA* 103, 982–987.
19. Li, H., Myeroff, L., Smiraglia, D., et al. (2003) SLC5A8, a sodium transporter, is a tumor suppressor gene silenced by methylation in human colon aberrant crypt foci and cancers. *Proc Natl Acad Sci USA* 100, 8412–8417.
20. Hong, C., Maunakea, A., Jun, P., et al. (2005) Shared epigenetic mechanisms in human and mouse gliomas inactivate expression of the growth suppressor SLC5A8. *Cancer Res* 65, 3617–3623.
21. Ueno, M., Toyota, M., Akino, K., et al. (2004) Aberrant methylation and histone deacetylation associated with silencing of SLC5A8 in gastric cancer. *Tumour Biol* 25, 134–140.
22. Thangaraju, M., Gopal, E., Martin, P. M., et al. (2006) SLC5A8 triggers tumor cell apoptosis through pyruvate-dependent inhibition of histone deacetylases. *Cancer Res* 66, 11560–11564.
23. Yu, L., Liu, C., Vandeusen, J., et al. (2005) Global assessment of promoter methylation in a mouse model of cancer identifies ID4 as a putative tumor-suppressor gene in human leukemia. *Nat Genet* 37, 265–274.
24. Dai, Z., Lakshmanan, R. R., Zhu, W. G., et al. (2001) Global methylation profiling of lung cancer identifies novel methylated genes. *Neoplasia* 3, 314–323.
25. Yoshikawa, H., Matsubara, K., Qian, G. S., et al. (2001) SOCS-1, a negative regulator of the JAK/STAT pathway, is silenced by methylation in human hepatocellular carcinoma and shows growth-suppression activity. *Nat Genet* 28, 29–35.
26. Hong, C., Moorefield, K. S., Jun, P., et al. (2007) Epigenome scans and cancer genome sequencing converge on WNK2, a kinase-independent suppressor of cell growth. *Proc Natl Acad Sci USA* 104, 10974–10979.
27. Costello, J. F., Plass, C., Arap, W., et al. (1997) Cyclin-dependent kinase 6 (CDK6) amplification in human gliomas identified using two-dimensional separation of genomic DNA. *Cancer Res* 57, 1250–1254.
28. Dai, Z., Zhu, W. G., Morrison, C. D., et al. (2003) A comprehensive search for DNA amplification in lung cancer identifies inhibitors of apoptosis cIAP1 and cIAP2 as candidate oncogenes. *Hum Mol Genet* 12, 791–801.
29. Frühwald, M. C., O'Dorisio, M. S., Rush, L. J., et al. (2000) Gene amplification in PNETs/medulloblastomas: mapping of a novel amplified gene within the MYCN amplicon. *J Med Genet* 37, 501–509.
30. Hughes, S. J., Glover, T. W., Zhu, X. X., et al. (1998) A novel amplicon at 8p22-23 results in overexpression of cathepsin B in esophageal adenocarcinoma. *Proc Natl Acad Sci USA* 95, 12410–12415.
31. Smiraglia, D. J., Frühwald, M. C., Costello, J. F., et al. (1999) A new tool for the rapid cloning of amplified and hypermethylated human DNA sequences from restriction landmark genome scanning gels. *Genomics* 58, 254–262.
32. Costello, J. F., Frühwald, M. C., Smiraglia, D. J., et al. (2000) Aberrant CpG-island methylation has non-random and tumour-type-specific patterns. *Nat Genet* 24, 132–138.
33. Kuick, R., Asakawa, J., Neel, J. V., et al. (1996) Studies of the inheritance of human ribosomal DNA variants detected in two-dimensional separations of genomic restriction fragments. *Genetics* 144, 307–316.
34. Hayashizaki, Y., Hirotsune, S., Okazaki, Y., et al. (1994) A genetic linkage map of the mouse using restriction landmark genomic scanning (RLGS). *Genetics* 138, 1207–1238.
35. Ohsumi, T., Okazaki, Y., Okuizumi, H., et al. (1995) Loss of heterozygosity in chromosomes 1, 5, 7 and 13 in mouse hepatoma detected by systematic genome-wide scanning using RLGS genetic map. *Biochem Biophys Res Commun* 212, 632–639.
36. Okuizumi, H., Okazaki, Y., Ohsumi, T., et al. (1995) Genetic mapping of restriction landmark genomic scanning loci in the mouse. *Electrophoresis* 16, 233–240.
37. Okuizumi, H., Okazaki, Y., Ohsumi, T., et al. (1995) A single gel analysis of 575 dominant

and codominant restriction landmark genomic scanning loci in mice interspecific backcross progeny. *Electrophoresis* 16, 253–260.

38. Okazaki, Y., Hirose, K., Hirotsune, S., et al. (1995) Direct detection and isolation of restriction landmark genomic scanning (RLGS) spot DNA markers tightly linked to a specific trait by using the RLGS spot-bombing method. *Proc Natl Acad Sci USA* 92, 5610–5614.
39. Hirotsune, S., Takahara, T., Sasaki, N., et al. (1995) The reeler gene encodes a protein with an EGF-like motif expressed by pioneer neurons. *Nat Genet* 10, 77–83.
40. Kuick, R. D., Skolnick, M. M., Hanash, S. M., et al. (1991) A two-dimensional electrophoresis-related laboratory information processing system: spot matching. *Electrophoresis* 12, 736–746.
41. Sugahara, Y., Akiyoshi, S., Okazaki, Y., et al. (1998) An automatic image analysis system for RLGS films. *Mamm Genome* 9, 643–651.
42. Miwa, W., Yashima, K., Sekine, T., et al. (1995) Demethylation of a repetitive DNA sequence in human cancers. *Electrophoresis* 16, 227–232.

Chapter 12

GoldenGate® Assay for DNA Methylation Profiling

Marina Bibikova and Jian-Bing Fan

Abstract

We describe a highly reproducible and multiplexed method, the GoldenGate assay for methylation, for high-throughput quantitative measurements of DNA methylation. It can analyze up to 1,536 targeted CpG sites in 96 samples simultaneously, using only 250 ng of genomic DNA. The method is akin to a "genotyping" of bisulfite-converted DNA. Assay probes can be designed to interrogate the Watson strand, the Crick strand, or both strands at each CpG site. Assay end products are processed using Illumina universal bead arrays. As a result, gene or CpG sets can be refined iteratively—no custom arrays need to be developed. This method allows the detection of as little as 2.5% methylation, and can distinguish 17% difference in absolute methylation level between samples. The method is highly reproducible and compares very well with other common methods of methylation detection, such as methylation-specific PCR and bisulfite sequencing. The Illumina GoldenGate Methylation technology should prove useful for DNA methylation analyses in large populations, with potential application to biomarker discovery and validation.

Key words: DNA methylation, epigenetics, BeadArray, GoldenGate assay, CpG island, biomarker.

1. Introduction

DNA methylation is widespread and plays a critical role in the regulation of gene expression in development, differentiation, and diseases, such as multiple sclerosis, diabetes, schizophrenia, aging, and cancer *(1)*. The ability to access the epigenomic information for a large number of genes or the entire genome should greatly facilitate the understanding of the nature of gene regulation in cells, and epigenomic mechanism of interactions between cells and the environment *(2)*. It should also have significance for studies of human epigenetic disorders and assisted reproduction. Microarray-based DNA methylation profiling

Jörg Tost (ed.), *DNA Methylation: Methods and Protocols, Second Edition, vol. 507*

DOI 10.1007/978-1-59745-522-0_12 Springerprotocols.com

technologies have been developed to meet this goal. These methods can be categorized into three main classes based on how the methylation status is interrogated: (1) discrimination of bisulfite-induced C to T transition; (2) cleavage of genomic DNA by methylation-sensitive restriction enzymes; and (3) immunoprecipitation with methyl-binding protein or antibodies against methylated cytosines. Each of these methods has its own limitations. For example, methylation-sensitive restriction enzymes cannot interrogate every CpG site, while the immunoprecipitation method cannot provide methylation information at single-base resolution for any targeted sequence. For bisulfite-based approaches, the challenges lie in the reduced genome complexity after bisulfite conversion of the genomic DNA; target-specific probe selection and hybridization specificity remain as the main technical hurdles.

We have adapted a high-throughput, single-nucleotide polymorphism (SNP)-genotyping system to DNA methylation detection, based on "genotyping" of bisulfite-converted genomic DNA *(3, 4)*. This technology, the GoldenGate assay for methylation, combines a miniaturized bead-based array platform, a high level of assay multiplexing, and scalable automation for sample handling and data processing. Unlike restriction enzyme-based methods, assay probes can be specifically designed for many of the CpG sites in the genome, and assay oligos can be designed to interrogate either the Watson or Crick strand at each CpG site. Different from other direct hybridization approaches, our method incorporates an allele-specific extension and oligonucleotide ligation step, which enables biochemical discrimination and assay specificity while allowing multiplexed profiling of CpG methylation status in several hundred genes.

GoldenGate Assay for methylation interrogates specific CpG dinucleotides with oligos linked to unique address sequences that can hybridize to their complementary strand on the universal arrays *(5)*. The assay requires a relatively short target sequence of about 50 nucleotides for query oligonucleotide annealing. The assay uses one set of universal PCR primers to amplify all of the targets and generates amplicons of ~100 bp. This uniformity results in a relatively unbiased amplification of the "methylated" and "unmethylated" PCR template populations. The assay multiplexes to over 1,500 CpG sites. The technology has been successfully used to analyze methylation profiles of 1,536 CpG sites from 371 genes in cancer cell lines, lung cancers, and normal tissues, and to identify a panel of adenocarcinoma-specific methylation markers *(3, 6)*. It has also been used to assess the epigenetic specificity of the loss of imprinting of the *IGF2* gene in Wilms tumors *(7)* and to identify a unique epigenetic signature for human embryonic stem cells *(8)*. These results effectively demonstrate the utility of Illumina GoldenGate methylation technology for high-throughput methylation profiling of hundreds of genes in large number of samples. It opens up new avenues to large-scale discovery, validation, and clinical application for DNA methylation markers of disease.

2. Materials

The Materials and Methods sections assume that the user has access to the Illumina BeadStation or BeadLab systems and associated reagents and equipment.

2.1. Recommended Kit for Bisulfite Conversion of Genomic DNA

Of the several kits for DNA bisulfite conversion which we have tested, the EZ DNA Methylation Kit from Zymo Research (Catalogue #D5002 or D5004) yielded the highest-quality bisulfite-converted DNA for use in the GoldenGate Assay (*see* **Notes 1** and **2**). Recently, more bisulfite-conversion kits have become available. However, in addition to high conversion rate, sample elution volume recommended by Zymo kit is better compatible with the downstream GoldenGate procedures than other currently available kits, and we strongly recommend it for sample preparation. The bisulfite-converted samples are transferred to the BCD (bisulfite-converted DNA) plate and stored at −20°C until use.

2.2. Illumina-Supplied Reagents for the GoldenGate Assay for Methylation

1. MS1/MM1 (Master mix for DNA activation for single use/multiple use), reagent for DNA activation.
2. PS1 (Precipitation Solution 1) reagent increases efficiency of activated DNA precipitation.
3. RS1 (Resuspension Solution 1) reagent is used for resuspension of activated DNA after precipitation.
4. OMA (Oligos for methylation assay oligo pool), mixture of oligonucleotides designed to query individual CpG dinucleotides in target sequences.
5. OB1 (Oligo-binding buffer 1), oligo-annealing buffer which also contains paramagnetic particles to optimize washing, extension, and ligation steps of the assay.
6. AM1 (Add MEL buffer 1), wash buffer used to remove excess of mis- or unhybridized query oligonucleotides.
7. UB1 (Universal wash buffer 1), wash buffer for several pre-PCR steps.
8. MEL (Master mix for extension and ligation), optimized mixture of enzymes for the extension/ligation step.
9. MMP (Master mix for PCR), PCR master mix which contains fluorescent common primers for the multiplexed ligated oligo templates.

10. IP1 (Inoculate PCR buffer 1), elution buffer for inoculating the PCR reaction with the ligated templates.
11. MPB (Magnetic particle buffer), a suspension of paramagnetic particles used to bind PCR products.
12. UB2 (Universal buffer 2), a wash buffer used in post-PCR processes.
13. MH1 (Make Hyb buffer 1), a neutralization buffer used in making the final sample single stranded for hybridization to the Sentrix® Array Matrix or BeadChip.
14. IS1 (Image SAM buffer 1), a buffer needed to dry bundles and protect fluors before imaging.

Please refer to **Table 12.1** for reagent storage conditions.

Table 12.1
Reagent storage conditions

Reagent	Storage conditions	Shelf life	Comments
MS1/MM1	−20°C	6 months	Aliquot to refreeze
OMA	−20°C	2 years	Can be stored at 4°C up to 2 weeks
PS1	4°C	6 months	
RS1	Room temperature	6 months	
OB1	−20°C	6 months	Does not completely freeze
AM1	4°C	6 months	
UB1	4°C	6 months	
MEL	−20°C	6 months	Aliquot to refreeze
MMP	−20°C	6 months	Aliquot to refreeze after adding DNA polymerase
IP1	−20°C	1 year	
MPB	4°C	6 months	Do not freeze
UB2	RT	6 months	
MH1	RT	6 months	Keep away from light
IS1	Room temperature while dry; −20°C when reconstituted	6 months	Must resuspend at least 1 day in advance
WC1	−20°C	6 months	Must dilute
AC1	Room temperature	6 months	

2.3. Other Reagents Required for the Assay

1. 0.1 M NaOH.
2. TE buffer: 10 mM Tris–HCl, 1 mM EDTA, pH 8.0.

3. DNA polymerase (Clontech BD, 639208, or 639209).
4. Uracil DNA glycosylase (UDG), 1 U/μL (Invitrogen, 18054-015).
5. 95% EtOH (v/v).
6. 2-butanol 99% (Sigma-Aldrich, B85919).
7. PicoGreen DNA quantitation kit (Invitrogen).
8. Serological pipettes (10, 25, and 50 mL).
9. 96-well, 0.2-mL skirted microtiter plates (Bio-Rad, MSP-9601).
10. 0.8-mL storage plate, conical well bottom (ABgene, AB-0765).
11. 96-well, black, flat-bottom Fluotrac 200 plates (Greiner, 655076).
12. 96-well sealing mats, round cap, autoclavable (ABgene, AB-0674).
13. 96-well cap mats, sealing mats, round cap, pierceable, nonautoclavable (ABgene, AB-0566).
14. Heat sealing foil sheets, thermo seal (ABgene, AB-0559).
15. Microtiter plate clear adhesive film, 2mil Sealplate Adhesive Film, Nonsterile (Phenix Research Products, LMT-SEAL-EX).
16. Microseal "A" Film, PCR plate sealing film (Bio-Rad, MSA-5001).
17. Microseal "F" film, aluminum adhesive film (Bio-Rad, MSF-1001).
18. 96-well V-bottom plates, Corning Costar Polypropylene (Fisher Scientific, 07-200-695 or VWR, 29444-102).
19. Multiscreen Filter plates, 0.45 μm, clear, Styrene (Millipore, MAHV-N45 10/50).
20. Nonsterile solution basins (Labcor Products Inc., 730-001 or VWR, 21007-970).
21. Cliniplate 384-well microtiter plates (Thermo Labsystems, 95040000).
22. Omni trays (Nunc, 242811 or VWR, 62409-600).

2.4. Fiber-Optic 96-Array Matrix and Other Array Formats

Illumina BeadArray™ technology is based on the random self-assembly of a bead pool onto a patterned substrate *(9)*. The arrays are assembled into two different formats: the Sentrix Array Matrix (SAM) and the Sentrix BeadChip. In the array matrix, the fiber optic bundles, each containing ~50,000 5-μm fibers, are arranged into an Array of Arrays™ format that is compatible with and can access the wells of a 96-well microtiter plate. The fiber optic bundles in the array matrix assembly are polished flat on both ends. On one end, the cores of each fiber are etched to form nanowells that will accept 3-μm silica beads. Each of the beads has been derivatized with several hundred thousand oligonucleotides of a particular sequence. In the BeadChip format, several microarrays are arranged on silicon slides that have been processed by micro-electro-mechanical systems (MEMS) technology to also have nanowells that support self-assembly of beads. In both formats, an average of over 30 beads of each type is maintained, a strategy that provides the accuracy of multiple measurements and statistical power. Beads are self-assembled in the wells during the manufacturing process, and each array is decoded to determine the map of bead-type positions *(10)*. Combinatorial series of brief hybridizations and rinses are used for decoding that result in a level of accuracy well beyond the requirements of any application. This decoding process also provides a quality-control measure of the function of each bead that is incorporated into the final bead map. BeadArray technology is the only microarray technology with functional QC of every element in the array.

2.5. Array Imaging

The Illumina BeadArray Reader is required for array imaging. Each assay results in a fluorescent signal associated with individual bead types on the array. To read out these signals, we developed the BeadArray Reader, a two-channel, 0.8-μm resolution confocal laser scanner which can simultaneously (via menu-driven software) scan a BeadChip or SAM at two wavelengths, and create an image file for each channel. For Illumina BeadArray Reader instructions, see the Illumina BeadArray Reader User Guide.

3. Methods

3.1. Bisulfite Conversion of Genomic DNA

Complete bisulfite conversion of genomic DNA, which modifies unmethylated cytosines to uracils while leaving methylated cytosines unchanged *(11)*, is the critical step for success in the GoldenGate assay for DNA methylation measurement. We recommend the EZ DNA Methylation kit (Zymo Research) for bisulfite conversion of DNA, using 500 nanogram or more of genomic DNA in one conversion reaction. All buffers and reagents are provided with the kit. The bisulfite conversion of DNA is carried out following manufacturers' recommendations.

3.2. GoldenGate Assay for Methylation Protocols

These protocols are not intended to replace the GoldenGate assay for methylation manual supplied with Illumina systems, but rather give a detailed overview of the process.

3.2.1. Assay Probe Design

The GoldenGate assay for methylation is based on sequence-specific extension and ligation of correctly hybridized query oligos. For each CpG site, four probes are designed: two allele-specific oligos (ASO) and two locus-specific oligos (LSO). Each ASO–LSO pair corresponds to either the methylated or unmethylated state of the CpG site **(Fig. 12.1)**. If other CpG sites are present in close vicinity of the target CpG site, we make the assumption that they have the same methylation status as the

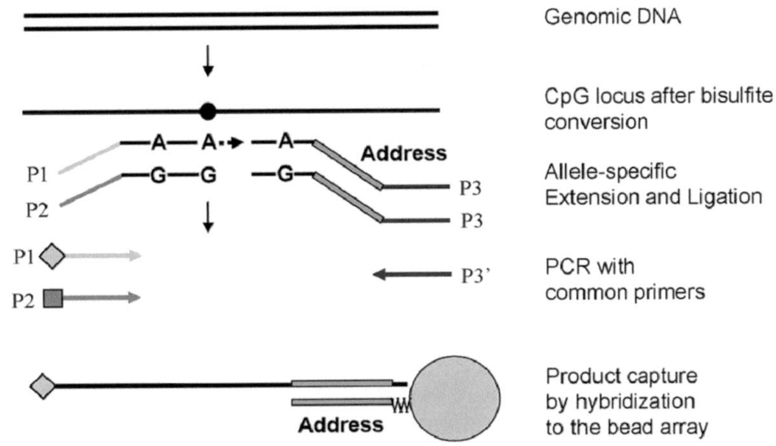

Fig. 12.1. Assay scheme. In the GoldenGate assay for methylation, four oligos are designed to target a specific CpG dinucleotide. Genomic DNA is first treated by sodium bisulfite. The corresponding query oligos bind to the converted DNA, and become extended and ligated enzymatically. The ligated products are then amplified and fluorescently labeled during PCR, and finally detected by hybridization to address sequences on the array. Methylation status of each locus is calculated by taking the ratio of "methylated" signal of the locus to the total locus intensity.

site of interest. This design hypothesis was based on previously reported bisulfite-sequencing results, in which a majority (>90%) of the adjacent CpG sites was shown to be co-methylated or co-demethylated *(12–14)*. This assumption was also confirmed by our own bisulfite-sequencing results. It is worth pointing out that this design strategy is used in methylation-specific PCR primer design (*15*, **Chapter 22**) and other microarray-based DNA methylation analysis *(16)*. While there are many CpG sites within each CpG island, we only select those for which robust assays can designed.

The ASOs consist of two parts: the gene-specific sequence and universal PCR primer sequences, with P1 and P2 at the 5′-end **(Fig. 12.1)**. The downstream oligo consists of three parts: the gene-specific sequence, a unique address sequence which is complementary to one of the 1,536 capture sequences on the array, and a universal PCR primer sequence (P3) at the 3′-end. A single address sequence is uniquely associated with a single target site. This address sequence allows the PCR-amplified products to hybridize to a universal microarray bearing the complementary probe sequences.

3.2.2. Assay Protocol

During this process bisulfite-converted DNA is biotinylated so that it will bind to the paramagnetic beads later on.

3.2.2.1. The Make BCS (Bisulfite-Converted DNA for Single Use) Process for DNA Activation

1. Preheat the heat block to 95°C and allow temperature to stabilize. Turn on the heat sealer to preheat it. Thaw DNA activation reagent (MS1, Illumina) to room temperature, and vortex to fully mix tube contents.

2. Thaw the BCD plate to room temperature, if frozen. We recommend using 250 ng of DNA (before bisulfite conversion) for one assay (*see* **Note 3**).
3. Add 5 μL of MS1 to each well of microtiter plate labelled "BCS". Transfer 5 μL of bisulfite-converted DNA sample to each well of the BCD plate, heat seal with foil, and vortex at 2300 rpm for 20 s. Pulse-centrifuge sealed plate to 250*g* to prevent wells from evaporating during the incubation.
4. Place the BCS plate in the 95°C heat block and close the heat block cover. Incubate the BCS plate at 95°C for 30 min in the preheated heat block.
5. When incubation is complete, remove the plate from the heat block and pulse-centrifuge to 250*g* to collect any condensation.
6. If the experiment will be continued on the same day, set the heat block to 70°C.

Proceed to precipitate the DNA samples to remove any free biotin.

3.2.2.2. The Precip BCS Process for Removal of Free Biotin from Activated DNA

1. Add 5 μL of PS1 reagent to each well of the BCS plate and seal the plate with adhesive film. Pulse-centrifuge the plate to 250*g*, and then vortex the plate at 2300 rpm for 20 s to mix the content.
2. Add 15 μL of 2-propanol to each well of the BCS plate and seal the plate with adhesive film. Vortex the plate at 1600 rpm for 20 s or until the wells are uniformly blue.
3. Centrifuge the sealed plate at 3000*g* for 20 min. Perform the next step immediately to avoid dislodging of the activated DNA pellets. If any delay occurs, re-centrifuge at 3000*g* for 10 min before proceeding.
4. Remove the plate seal and decant the supernatant by quickly inverting each BCS plate and tapping it firmly onto an absorbent pad. Blot off excess fluid. Tap firmly several times over a period of 1 min or until all wells are devoid of liquid. Do not allow supernatant to pour into other wells.
5. Place the inverted BCS plate on a new absorbent pad. Centrifuge to 8*g* for 1 min. Do not centrifuge to more than 8*g*, or the sample will be lost!
6. Set the plate upright and allow to air-dry for 15 min.

Proceed to resuspend the precipitated DNA samples to use in the assay.

3.2.2.3. The Resuspend BCS Process for Dissolving Biotinylated Bisulfite-Converted DNA Pellets and Resuspending Them in Solution

1. Add 10 μL RS1 to each well of the BCS plate.
2. Seal the plate with adhesive film and pulse-centrifuge to 250*g*. Vortex at 2300 rpm for 1 min or until the blue DNA pellets are completely resuspended.
3. Proceed to Make ASE or heat-seal the BCS plate and store at 4°C overnight.

3.2.2.4. The Make ASE (Assay-Specific Extension) Process for Annealing of Query Oligonucleotides to Bisulfite-Converted Activated DNA

1. Preheat heat block to 70°C and allow temperature to stabilize.
2. Remove oligonucleotides for methylation assay (OMA) tube from refrigerator (if frozen, thaw, vortex, and then centrifuge). Thaw oligonucleotide-annealing reagent (OB1) to room temperature and vortex. Do not centrifuge the OB1 tube.
3. Dispense 10 μL of OMA to each well of a new, 96-well, 0.2-mL skirted microtiter plate labelled "ASE". Add 30 μL of well-resuspended OB1 to each well of the same plate.
4. Centrifuge the BCS plate to 250*g* to collect samples at the bottom of the wells. Transfer 10 μL of the biotinylated DNA to the ASE plate containing OMA and OB1 to bring the final volume to 50 μL. Heat seal the plate and vortex briefly at 1600 rpm to mix the content of the wells. Place the ASE plate in the 70°C heat block and immediately reduce the temperature setting to 30°C. This will carry out assay oligonucleotide annealing to the bisulfite converted DNA target by ramping the temperature over approximately 2 h and then holding at 30°C until the next processing step.

3.2.2.5. The Add MEL Process for Assay Oligonucleotide Extension and Ligation

1. Remove the ASE plate from the heat block, reset it to 45°C, and allow temperature to stabilize. Thaw the extension and ligation reagent (MEL) to room temperature.
2. Place the ASE plate with assay oligonucleotides annealed to the bisulfite-converted DNA template on the Illumina-supplied magnetic plate for at least 2 min, or until beads are completely captured. Washing the beads removes excess and mis-hybridized oligonucleotides.
3. After the paramagnetic particles are captured, remove the heat seal from the plate and remove and discard all liquid (~50 μL) from wells, retaining the beads. Add 50 μL of AM1 buffer to each well of the assay plate. Seal the plate with adhesive film and vortex at 1600 rpm for 20 s or until all beads are resuspended.
4. Place the ASE plate on the magnet for at least 2 min, or until beads are completely captured. Remove all AM1 from each well, leaving beads in wells. Repeat addition of 50 μL AM1, vortexing, and removal of buffer.
5. Remove the ASE plate from the magnet and add 50 μL of universal wash buffer (UB1) to each well.
6. Place the ASE plate onto the magnet for at least 2 min, or until beads are completely captured. Remove all UB1 from each well. Repeat addition of 50 μL UB1 and removal of buffer.
7. Add 37 μL of extension and ligation master mix MEL to each well of the ASE plate. Seal plate with adhesive film and vortex at 1600 rpm for 1 min.
8. Incubate the ASE plate on the preheated 45°C heat block for 15 min.

3.2.2.6. The Make PCR and Inoc PCR Processes for Preparing the PCR Mix and Setting up the PCR Reaction

1. Prepare PCR master mix by adding 64 μL of DNA polymerase and 50 μL of UDG to the tube of PCR reagent (MMP). Invert MMP tube several times to mix contents, and aliquot 30 μL of mixture into each well of a new 96-well, 0.2-mL microtiter plate (the PCR plate).
2. Remove the ASE plate from the heat block after the extension and ligation step, and reset the heat block to 95°C.
3. Place the ASE plate on the magnet for at least 2 min, or until beads are captured. Remove clear adhesive film from the assay plate, and remove and discard supernatant (~50 μL) from all wells of the ASE plate, leaving beads in wells. Leave the ASE plate on the magnet and add 50 μL of the universal wash buffer UB1 to each well of the plate.
4. Allow the ASE plate to rest on the magnetic plate for at least 2 min to collect the paramagnetic particles. Remove and discard all supernatant (~50 μL) from all wells of the ASE plate, leaving beads in wells.
5. Add 35 μL of the elution buffer IP1 to each well of the assay plate and seal it with adhesive film. Vortex the plate at 1800 rpm for 1 min, or until all beads are resuspended. Place the plate on the 95°C heat block for 1 min.
6. Remove the ASE plate from heat block and place onto the magnet for at least 2 min, or until the beads have been completely captured. Transfer 30 μL supernatant from the first column of the ASE plate into the first column of the PCR plate. Repeat transfer for the remaining columns using new pipette tips for each column.
7. Seal PCR plate with Microseal "A" PCR plate sealing film. Immediately transfer the plate to the thermal cycler and run the following cycling program: 10 min at 37°C; 34 cycles (35 s at 95°C, 35 s at 56°C, 2 min at 72°C); 10 min at 72°C; and 4°C for 5 min.
8. Proceed immediately to the preparation of PCR product for array hybridization, or seal and store the PCR plate at −20°C.

3.2.2.7. The Make HYB Process to Prepare Samples for Array Hybridization

1. Vortex tube with the suspension of paramagnetic particles (MPB) until beads are completely resuspended.
2. Dispense 20 μL of resuspended MPB into each well of the PCR plate. Mix the beads with PCR product by pipetting up and down, and then transfer the mixed solution to the filter plate. Cover the filter plate with its cover and store at room temperature, protected from light, for 60 min.
3. Place the filter plate containing the bound PCR products onto a new 96-well, V-bottom waste plate using a filter plate adapter. Centrifuge at 1000*g* for 5 min at 25°C.
4. Remove filter plate lid. Add 50 μL of universal wash buffer 2 (UB2) to each well of the filter plate. Dispense slowly so that

the beads are undisturbed. Replace the lid of the filter plate and centrifuge at 1000*g* for 5 min at 25°C.

5. Prepare a new 96-well, V-bottom plate and dispense 30 μL of hybridization buffer (MH1) to all the wells of the new intermediate plate. Place the filter plate onto the intermediate plate such that column A1 of the filter plate matches column A1 of the intermediate plate.
6. Dispense 30 μL of 0.1 M NaOH to all wells of the filter plate. Replace the lid of the filter plate and centrifuge immediately at 1000*g* for 5 min at 25°C. Gently mix the contents of the intermediate plate by moving it from side to side without splashing.
7. Prepare HYB Cliniplate 384-well microtiter plate. First, dispense 30 μL UB2 buffer in every other well (B2, B4, etc., D2, D4, etc.) of the hybridization plate for humidity control. Then transfer 50 μL neutralized hybridization solution from the intermediate plate into appropriate wells (A1, A3, etc., C1, C3, etc.) of hybridization plate. It is helpful to use separate templates for humidity control wells and for hybridization wells to assist in sample dispensing.
8. Seal the plate with clear adhesive film and centrifuge at 3000*g* for 4 min at 25°C. If hybridization is not performed immediately, the HYB plate can be stored at −20°C.

3.3. Hybridization to Bead Arrays

1. Preheat oven to 60°C and allow temperature to equilibrate. If the HYB plate has been frozen, thaw it completely at room temperature in a light-protected drawer. Centrifuge the plate at 3000*g* for 4 min.
2. Prepare two OmniTrays – one with 70 mL UB2, and another with 60 mL NaOH. Carefully place SAM, with the barcode facing up and the fiber bundles facing down, into the UB2 tray. Agitate SAM gently (10 s only) to remove bubbles from bottoms of arrays. After 3 min, move the SAM into NaOH tray and incubate 30 s, then move the SAM back into the UB2 conditioning tray.
   Allow SAM to sit in the UB2 conditioning tray for at least 30 s to neutralize the NaOH.
3. Remove the clear adhesive film from the HYB plate. Insert SAM fiber optic bundles into appropriate wells of the HYB plate using the SAM alignment fixture or a SAM hybridization cartridge. Place the HYB plate/SAM pair into the hybridization oven preheated to 60°C and incubate for 30 min.
4. Hybridization is then conducted under a temperature gradient program from 60°C to 45°C over approximately 12 h.

3.4. Array Imaging

3.4.1. Wash and Dry Array Matrix

1. Reconstitute array drying reagent (IS1) by adding 94 mL of an equal mixture of 95% ethanol and 2-butanol to 6 mL of IS1 to prepare enough reagent to image one array.
2. For each SAM to be imaged, prepare two OmniTrays with 70 mL of UB2, and one with 70 mL of diluted IS1.
3. Carefully separate SAM from HYB plate. Place the SAM into the first UB2 tray and agitate 1 min at room temperature. Transfer SAM to the second UB2 tray and repeat agitation for 1 min at room temperature. Ensure that there are no bubbles on the bottom of the array fibre bundles. These bubbles may prevent certain areas of the fibre bundle from contacting the liquid.
4. Dip the SAM into the IS1 tray. It is important to dip the SAM several times to ensure that the UB2 buffer is completely exchanged. After 5 min, remove SAM from IS1 and place onto an empty OmniTray to air dry for 20 min, fiber bundles up. The dry array is ready for imaging.

3.4.2. Imaging the BeadArray

The arrays are imaged using the BeadArray Reader (Illumina). Image processing and intensity data extraction are performed by the BeadScan software included with the BeadArray Reader.

3.5. Data Collection and Analysis

The BeadStudio software package is included with the Illumina GoldenGate assay for methylation product and is used as a tool for analyzing DNA methylation data from scanned microarray images collected from the Illumina BeadArray Reader. Alternatively, BeadStudio can be used to export the array intensity data for processing by other methylation or statistical analysis programs.

Specifically, BeadStudio executes two types of data analysis:

1. Methylation Analysis – determining DNA methylation levels (beta value) and
2. Differential Methylation Analysis – determining if DNA methylation levels are different between any experimental groups.

Analysis can be performed on individual arrays or on groups of arrays treated as replicates. BeadStudio reports experiment performance based on built-in controls that accompany each experiment (*see* **Notes 6–11**). In addition, BeadStudio provides scatter-plotting and dendrogram tools, facilitating quick, visual means for exploratory analysis (*see* **Note 12**).

4. Notes

1. It is essential to have a robust method for bisulfite conversion of genomic DNA. There are various commercially available

kits for this application, but we found the EZ DNA Methylation kit worked best in our experiments.

2. We recommend using at least 500 ng of genomic DNA for bisulfite conversion. The average DNA recovery after conversion is 70%.
3. DNA quantitation before bisulfite conversion is an important step to ensure that sufficient amount of material is used in the assay to generate high-quality data. We recommend the Invitrogen PicoGreen assay kit for quantitation of DNA samples. The PicoGreen assay can quantitate small DNA volumes, and measures DNA directly. Other techniques may pick up contamination such as small molecules and proteins. We recommend using a fluorometer, as fluorometry provides DNA-specific quantitation, whereas spectrophotometry may be affected by RNA contamination, leading to artificially inflated amounts.
4. We also support customers to select assays to target specific CpG sites within genes or regions of interest to meet specific research needs. Illumina Tech-Support scientists will evaluate customer submissions and work collaboratively with the customer to ensure optimal probe design. Once all of the assays have been evaluated, a list is sent to the customer for final approval before the Oligo Pool for Methylation Assay (OMA) is manufactured. Custom content can be submitted using any of the following formats:
   NCBI's GeneID; Gene symbol (HUGO & RefSeq); RefSeq mRNA accession number with or without version number; RefSeq mRNA GI number; Chromosomal region (RefSeq build, chromosome, pair of coordinates); Sequence.
   Custom methylation is available in panels of 96 and 384–1536 assays per OMA tube. Larger panels can be accommodated with multiple OMA tubes and the appropriate GoldenGate reagents and arrays.
5. Careful experimental design may help to maximize the utility of the GoldenGate assay for methylation. A well-designed experiment should include replicates and adequate numbers of case and controls samples.
6. The GoldenGate assay for methylation has nine controls which help to monitor each step of the assay process, starting from bisulfite conversion to analytical hybridization of the assay to the universal array. Bisulfite conversion controls test for presence of unconverted genomic DNA in assay samples. Primers are designed to the same DNA locus, with one pair targeting converted and the other pair targeting unconverted DNA sequence.
7. The first hybridization controls test the specificity of annealing ASOs with different Tm to the same DNA locus.

8. The allele-specific extension controls test the extension efficiency of properly matched versus mismatched ASOs.
9. The extension gap control tests the efficiency of extending 15 bases from the 3′ end of the allele-specific oligo to the 5′ end of the locus-specific oligo.
10. The hybridization controls monitor the overall intensity of the arrays. In our experience, the hybridization controls should exceed ~10,000 counts in the appropriate channels. If there is an outlier in hybridization control intensities, then normalization may be attempted. If the hybridization controls are near background for any individual array, then there is likely a problem with that array or sample. We have observed this behaviour when UB2 from the humidity control wells splashes into a sample well.
11. The negative controls should be below ~1,000 counts, and the standard deviation of the negative controls should be lower than their intensity. Any outlier samples that show high counts or large standard deviations on the negative controls should be considered with caution. We find that high negative controls can be associated with low DNA input, or losses during bisulfite conversion.
12. A valuable approach to assessing data quality is to cluster the samples using the dendrogram tool in BeadStudio. If there are replicate samples or known sample relationships, outlier samples can be readily detected by their failure to associate with each other in clustering.

References

1. Feinberg, A. P., Tycko, B. (2004) The history of cancer epigenetics. *Nat Rev Cancer* 4, 143–153.
2. Jones, P. A., Martienssen, R. (2005) A blueprint for a Human Epigenome Project: the AACR Human Epigenome Workshop. *Cancer Res* 65, 11241–11246.
3. Bibikova, M., Lin, Z., Zhou, L., et al. (2006) High-throughput DNA methylation profiling using universal bead arrays. *Genome Res* 16, 383–393.
4. Fan, J. B., Oliphant, A., Shen, R., et al. (2003) Highly parallel SNP genotyping. *Cold Spring Harb Symp Quant Biol* 68, 69–78.
5. Fan, J. B., Gunderson, K. L., Bibikova, M., et al. (2006) Illumina universal bead arrays. *Methods Enzymol* 410, 57–73.
6. Lin, Z., Thomas, N. J., Bibikova, M., et al. (2007) DNA methylation markers of surfactant proteins in lung cancer. *Int J Oncol* 31, 181–191.
7. Bjornsson, H. T., Brown, L. J., Fallin, M.D., et al. (2007) Epigenetic specificity of loss of imprinting of the IGF2 gene in Wilms tumors. *J Natl Cancer Inst* 99, 1270–1273.
8. Bibikova, M., Chudin, E., Wu, B., et al. (2006) Human embryonic stem cells have a unique epigenetic signature. *Genome Res* 16, 1075–1083.
9. Barker, D. L., Theriault, D., Che, D., et al. (2003) Self-assembled random arrays: high-performance imaging and genomics applications on a high-density microarray platform, in (Nicolau, D. V., Raghavachari, R., eds.), *Microarrays and Combinatorial Technologies for Biomedical Applications: Design, Fabrication, and Analysis, Proceedings of SPIE*, 4966, pp. 1–11.
10. Gunderson, K. L., Kruglyak, S., Graige, M. S., et al. (2004) Decoding randomly ordered DNA arrays. *Genome Res* 14, 870–877.
11. Wang, R. Y., Gehrke, C. W., Ehrlich, M. (1980) Comparison of bisulfite modification of 5-methyldeoxycytidine and deoxycytidine residues. *Nucleic Acids Res* 8, 4777–4790.

12. Grunau, C., Renault, E., Roizes, G. (2002) DNA Methylation Database "MethDB": a user guide. *J Nutr* 132, 2435S–2439S.
13. Rakyan, V. K., Hildmann, T., Novik, K. L., et al. (2004) DNA methylation profiling of the human major histocompatibility complex: a pilot study for the human epigenome project. *PLoS Biol* 2, e405.
14. Eckhardt, F., Lewin, J., Cortese, R., et al. (2006) DNA methylation profiling of human chromosomes 6, 20 and 22. *Nat Genet* 38, 1378–1385.
15. Herman, J. G., Graff, J. R., Myohanen, S., et al. (1996) Methylation-specific PCR: a novel PCR assay for methylation status of CpG islands. *Proc Natl Acad Sci USA* 93, 9821–9826.
16. Adorjan, P., Distler, J., Lipscher, E., et al. (2002) Tumour class prediction and discovery by microarray-based DNA methylation analysis. *Nucleic Acids Res* 30, e21.

Chapter 13

5′-Azacytidine Expression Arrays

Paul Cairns

Abstract

Epigenetic silencing of a gene can be reversed, resulting in reactivation of expression, by drugs such as the DNA methylation inhibitor 5-Aza-2′-deoxycytidine (5Aza-dC, azacytidine). This drug is added to cell culture media and is incorporated into the new strand during DNA replication in the cell. 5Aza-dC forms a covalent complex with the active sites of the DNA methyltransferase, depleting methyltransferase activity, which results in generalized demethylation. Until recently, global analyses of gene methylation in cancer cells were largely restricted to array or gel-based comparisons of the methylation status of CpG islands between normal and tumor cell DNA. An expression microarray-based screen has the advantage of a more genome-wide analysis with a better gene annotation and, coupled with a reactivation strategy, has the further advantage that it should preferentially identify reexpression of epigenetically silenced genes over methylated CpG islands that do not influence transcription. However, the direct reactivation of methylated genes, as well as secondary effects of azacytidine treatment, can lead to a cascade of deregulation in downstream unmethylated gene expression. A validation strategy is therefore the key for efficient identification of genes methylated in the wild-type cultured tumor cells. An azacytidine-based reactivation approach can only be used on cell lines so validation should include analysis of primary tumors. The potential of this approach for the identification of new hypermethylated genes and pathways has been demonstrated in bladder, colorectal, esophageal, and most other cancer types.

Key words: Azacytidine, expression array, epigenetic reactivation, tumor suppressor gene, CpG island, demethylation, tumor cell lines.

1. Introduction

Tumorigenesis is a multistep process that results from the accumulation and interplay of genetic and epigenetic mutations. The epigenetic alteration of aberrant DNA methylation of CpG islands in the promoter region of genes is well established as a common mechanism for the silencing of tumor suppressor genes in cancer cells *(1)*. Until recently, the majority of genes identified as

Jörg Tost (ed.), *DNA Methylation: Methods and Protocols, Second Edition, vol. 507*

DOI 10.1007/978-1-59745-522-0_13 Springerprotocols.com

hypermethylated with associated loss of expression in cancer had been found by a candidate approach targeting a classical tumor-suppressor gene or a gene of interest to the particular investigator. By definition, a candidate-gene approach has resulted in the examination of a limited number of genes for epigenetic alteration. Many other tumor suppressor and cancer genes important in tumorigenesis likely remain to be identified. The average total number of genes methylated with functional significance in the human tumor cell is presently unknown but a reasonable estimate might be several hundred *(2, 3)*. A global approach to the identification of epigenetically silenced genes in tumor cells can lead to further elucidation of the biology of cancer as well as provide methylation signatures for early detection and for prognostic stratification as well as identify novel targets for therapy.

Initially, global strategies to identify methylated genes in cancer tended to use arrayed CpG island fragments identified through methylation-sensitive restriction enzyme recognition sequences. One issue with such an approach is that many CpG islands are located outside promoter regions and methylation of such islands does not have a functional effect upon transcription *(4)*. Epigenetic silencing of a gene can be reversed, resulting in reactivation of expression, by drugs such as 5-Aza-2′-deoxycytidine (5Aza-dC) which acts through incorporation of the modified nucleotide in the place of cytosine into the newly synthesized strand during DNA replication where it forms a covalent complex with the active sites of the DNA methyltransferase, depleting methyltransferase activity resulting in turn in generalized demethylation *(1)*. A microarray-based screen of genes reexpressed after demethylation treatment uses reactivation of transcription, rather than the presence of a CpG island, as the identifying determinant *(5)*. In addition, commercially available expression microarrays represented more of the genome and were better annotated, for example, by gene name. There have now been several global epigenetic studies highlighting the advantages and issues of this approach. To my knowledge, the first study was of the T24 bladder cancer cell line and produced a list of genes and gene families upregulated after azacytidine treatment *(6)*. Subsequent studies extensively validated upregulated genes *(4, 7)*. These epigenetic reactivation profiles may in part also be array-type dependent, although a number of identical genes has been identified on different arrays and in different laboratories. The profiles are likely also sensitive to different 5Aza-dC doses and treatment times. The amount of methylation and the response obviously can vary for the same gene from one cell line to another and between different genes in the same cell line. One disadvantage of a drug-based reactivation approach is that, unlike CpG island array screening, use is essentially restricted to cell cultures and is not amenable to examination of primary tumors. This latter issue is addressed by inclusion of primary tumor analysis during validation of genes identified as upregulated after azacytidine treatment *(5)*.

A typical approach is to culture several tumor cell lines, representative if possible of histological cell type and tumor stage, of a cancer type and treat with azacytidine over two cell-doubling times. Untreated cells are also grown. Optimization of the azacytidine dose may be necessary. RNA is isolated from both the treated and untreated cells, labeled and hybridized to a human genome expression array. Statistical analysis of the microarray data provides a list of genes ranked by fold upregulation of expression in the azacytidine-treated cells. The list of upregulated genes is interrogated by publicly available database analysis, for example, to select for genes containing a typical CpG island in the promoter region and expressed in the normal cell counterpart to the tumor type or to select known imprinted genes. The subsequent list of selected genes is first validated by bisulfite sequencing of DNA from the untreated cell lines to screen for hypermethylation. Normal cell DNA and primary tumor DNA are then bisulfite sequenced to identify genes with aberrant promoter hypermethylation in cancer cells.

2. Materials

2.1. Azacytidine Treatment

1. Phosphate-buffered saline (PBS) buffer (1 ×), pH 7.2 (Invitrogen, Carlsbad CA): 1.5 mM potassium phosphate monobasic, 155.17 mM NaCl, 2.71 mM sodium phosphate dibasic, and pH 7.2 (pH adjustment with HCl)
2. 5Aza-dC (Sigma, St. Louis, MO) is dissolved in PBS, pH 7.2 (Invitrogen, Carlsbad CA) as a 5-mM stock solution, and stored in aliquots at −80°C. Once thawed, aliquots are discarded (*see* **Note 1**).
3. MTS cytotoxicity assay reagent/cell titer 96 aqueous 1 solution, cell proliferation assay (Promega, Madison, WI) stored at −20°C.

2.2. Cell Culture

1. RPMI (Cellgro, Herndon, VA) or appropriate medium supplemented with 10–15% fetal bovine serum (Hyclone, Logan, UT) stored at −20°C.
2. Penicillin/Streptomycin (Cellgro) 10,000 IU/mL stored as a 100 × solution at −20°C and used in media at 1 × concentration.
3. L-Glutamine (Cellgro) 200 mM 100 × solution stored at −20°C and used in media at 1 × concentration.
4. Trypsin (Difco, Lawrence, Kansas) stored at 0.04% concentration.
5. T75 flasks, 10-cm plates (BD Falcon, San Jose, CA).

2.3. Nucleic Acid Isolation

1. Trizol (Invitrogen) stored at +4°C.
2. DEPC-treated water (Invitrogen) stored at −20°C.
3. Proteinase K buffer (1 ×): 0.075 M NaCl, 0.024 M EDTA pH 8.0 (Invitrogen) autoclaved and stored at room temperature.
4. Proteinase K dissolved at 10 mg/mL in sterile distilled water, aliquoted, and stored at −20°C.
5. Phenol/chloroform.
6. Chloroform/isoamyl alcohol (24:1).
7. 100% Ethanol.
8. Sterile distilled water autoclaved and stored at room temperature.

2.4. Labeling and Hybridization to Array

1. RNeasy Mini Kit (Qiagen, Valencia, CA).
2. RNAse-free DNase I Set (Qiagen).
3. QIAquick PCR purification kit (Qiagen).
4. Fluorescent Direct Label Kit (Agilent Technologies, Santa Clara, CA)
5. Agilent Human Whole Genome Microarray (Agilent Technologies) or microarray of choice.

2.5. Sodium Bisulfite Modification for Sequencing and MSP

These procedures are described in detail in **Chapters 14**, **22**, and **23**.

3. Methods

3.1. Determination of Optimal 5Aza-dC Dose for Array-Based Reactivation

An obvious issue of an epigenetic reactivation and expression array approach is that sensitivity of signal might vary depending upon different baseline expression levels of genes and different types of expression microarray. Furthermore, different genes in the same cell line, or the same gene locus in different cell lines, may be more, or less, strongly epigenetically silenced. It is therefore useful to first determine the optimal 5Aza-dC dosage to a particular tumor cell line for robust detection of transcriptional reactivation on the intended microarray without excessive toxicity to the treated cells.

For many tumor cell lines, it is possible to choose a gene(s) described as hypermethylated with associated loss of expression from the literature as a positive control for the testing of dose and treatment time upon reactivation. We examined the reexpression of five tumor suppressor genes, $p16^{INK4a}$, *MLH1*, *MGMT*, *RARβ2*, and *Timp-3*, well characterized as hypermethylated with associated transcriptional silencing in the SW48 colorectal tumor cell line which has a similar cell-doubling time to many available tumor cell lines, at different doses of 5Aza-dC. RNA was extracted from SW48 cell cultures after treatment with 1 μM,

5 μM, or 10 μM 5Aza-dC and optionally 500 nM trichostatin A (*see* **Note 2**). Reverse transcription-polymerase chain reaction (RT-PCR) was performed, and the cDNA product labeled and hybridized to the expression array. Microarray analysis of upregulation of the five tumor-suppressor genes demonstrated that treatment with 5 μM 5Aza-dC for at least two cell-doubling times resulted in reexpression of the five tumor-suppressor genes with minimal toxicity as assessed by comparison of cell morphology and cell death between untreated and treated cells *(5)*.

Another approach is to first perform an MTS cytotoxicity assay in order to establish appropriate drug concentrations for larger cultures of the particular cell line.

1. Cells are seeded into multiwell plates at a density of 5.0×10^3 cells/well in triplicate.
2. 24 h after seeding, cells are treated with serial dilutions of azacytidine, for example, 0.1, 0.5, 1, 2.5, 5, 10, and 20 μM. Untreated cells serve as a control.
3. 72 h after treatment, cell viability is determined using the MTS in vitro cytotoxicity assay. 10 μL of MTS assay reagent (2 mg/mL) is mixed with 190 μL of the cell media and added to each well.
4. 4 h later, absorbance is read on a multiscan plate reader (Thermo Electron Corp.) at 490 nm.
5. Viability is graphed using Microsoft Excel and is calculated as the percentage of viable cells remaining compared to the untreated control. The option of isolating RNA and testing a gene(s) for reactivation of expression by qRT-PCR is available.

Alternatively, if no dose optimization studies are performed, a good starting regimen is to seed and treat the given tumor cell line with 5 μM of azacytidine as described below.

3.2. Cell Culture

1. The tumor cell lines are split to low density, typically 1×10^6 cells, grown in recommended medium supplemented with 10–15% fetal bovine serum, standard antibiotics and any supplements specified. Set up nine T75 flasks (or 10-cm diameter plates) for drug treatment and six flasks or plates for mock. The three additional plates are for counting treated cells (*see* **Note 3**).
2. The cells are seeded at 1×10^6 density at time 0 h.
3. At 24 h, azacytidine is added at a final concentration of 5 μM or to the dose of the investigator's choice.
4. At 48 h, azacytidine is again added to a final concentration of 5 μM or dose of choice in fresh medium.
5. At 72 h, the treated cells are counted with a hemocytometer and, if necessary, given further treatment with azacytidine, until the cells have undergone at least two doublings to 4×10^6. This is important because azacytidine is incorporated only into the new DNA strand during replication. The

cytotoxic effects of azacytidine mean that the doubling time will likely be longer than for untreated cells. The investigator should monitor the cells and adjust the dose and treatment time accordingly (*see* **Note 4**). At low concentrations (~1 μ*M*) of azacytidine, there may be benefit in a single treatment of TSA (*see* **Note 2**) at a final concentration of 500 nM at the time point of 24 h prior to harvest.

6. Untreated (mock) cells were cultured over an identical period of time with an equivalent volume of PBS and, if TSA is used, for the final 24 h, with an equivalent volume of EtOH (*see* **Note 5**).
7. If Trizol is not used, add trypsin to a final concentration of 0.04% to flask/plate, place in 37°C incubator for 5 min. Pipette media with cells into a 15-mL conical centrifuge tube and centrifuge for 5–10 min at 320*g* in a benchtop centrifuge before nucleic acid isolation.

3.3. Nucleic Acid Isolation

1. Total RNA used for the microarray analysis is isolated from cultured cells using TRIZOL reagent (Invitrogen) and purified using an RNeasy Mini Kit (Qiagen), combined with DNase treatment following the manufacturer's instructions.
2. Run 1–2 μg of total RNA in a 1.5% agarose gel. The RNA quality is confirmed by observation of an approximately 2:1 ratio between the 28S and 18S ribosomal RNA bands which should also appear sharp by eye.
3. For isolation of DNA, the untreated cells are centrifuged in a 15-mL conical polypropylene tube and the media decanted. 4.5 mL of proteinase K buffer, 20 μL of proteinase K, and 500 μL of 10% SDS are added to the cell pellet and incubated at 37°C overnight. An equal volume of 5 mL phenol/chloroform is added, mixed by inversion for 5 min, and centrifuged at 3500*g* for 10 min. The top layer is removed by pipette with care to avoid disturbance of the interface. An equal volume of 5 mL phenol/chloroform is added and the step is repeated. An equal volume of 5 mL chloroform/isoamyl alcohol is added, mixed by inversion for 5 min and centrifuged at 3500*g* for 10 min. The top layer is removed and two volumes of ice-cold 100% ethanol added. The precipitated DNA is spooled out with a pipette tip and air-dried for 2 min, then dissolved in sterile distilled water.

3.4. Labeling and Hybridization to Array

1. The procedure to be followed will depend upon the choice of expression microarray. In our laboratory, total RNA (20 μg) was labeled and simultaneously reverse transcribed into cDNA, using a Fluorescent Direct Label Kit (Agilent Technologies).

The labeled samples were cleaned with a QIAquick PCR purification kit (Qiagen), and then hybridized to the Human Whole Genome 44 K Oligo Microarray for 17 h at 65°C as recommended by the manufacturer (Agilent Technologies). Agilent Feature Extraction software (G25677AA, Agilent Technologies, 2004) was used to analyze the microarray data.

3.5. Intuitive Selection of Genes for Validation

1. In general, global microarray studies result in large amounts of information that needs to be interrogated to prioritize further experimental studies. A SAM (significance analysis of microarrays) statistical test *(8)* is performed on all arrays. A subset of genes can now be selected.
2. We selected a subset of genes that showed at least threefold upregulation of expression after azacytidine treatment in at least three of the four cancer cell lines on the basis that such genes might be expected to be more frequently methylated in primary tumors *(5)*. Some previous 5Aza-dC reactivation studies have reported that selection and analysis of genes with no basal expression in the untreated cells yields a higher proportion of methylated genes after validation *(4, 7)* (*see* **Note 6**). In addition, one can select for genes upregulated after 5Aza-dC treatment by different histological cell type, earlier pathological grade or stage versus later, drug resistant versus drug sensitive or other criteria.
3. The list of genes can be interrogated to prioritize genes for validation. The most useful criteria are an examination of expression status in the normal cell counterpart and the presence and location of a CpG island in the promoter region. Genes that showed no expression in normal cells, according to the Cancer Genome Anatomy Project (CGAP) Serial Analysis Gene Expression (SAGE) database (http:www.cgap.nci.nih.gov) can be excluded from immediate study. The genomic sequence of the upregulated genes is obtained from the GeneCard website (http:www.genecards.org/index.shtml) in order to analyze the promoter region. The presence of CpG islands nearest the transcriptional start site is searched for using both the CpG island revealing programs on the WebGene Website (http:www.itb.cnr.it/sun/webgene) and the CpG Island Searcher (http:cpgislands.usc.edu/). We chose to analyze genes that had a promoter CpG island that met the criteria outlined by Takai and Jones which are GC content "higher than 55%"; a ratio of the observed versus the expected frequency of CpG dinucleotides greater than 65% and a length larger than 200 bp *(9)*. The RepeatMasker Web Server can be used to identify repetitive elements in the promoter region (http:www.repeatmasker.org/) (*see* **Note 7**).

3.6. Bisulfite Modification of DNA and Sequencing of Gene Promoter CpG Islands

1. Genomic DNA (~1 μg) from untreated tumor cell lines, normal cells, and primary tumor specimens are modified by sodium bisulfite as described in **Chapter 14**.
2. Approximately 50 ng of modified DNA is PCR amplified with primers designed to the CpG island of interest over 35 cycles. PCR amplification products are run on a 2% agarose gel with size markers. The gel slice containing the amplified PCR product is cut out and purified using the MiniElute Gel Extraction kit (Qiagen, Valencia, CA). PCR product DNA is sequenced using the ABI 3100A capillary genetic analyzer. Sequences are analyzed using the Sequencher Software, Version 4.2.2. The observation of hypermethylation in the untreated cell lines compared to the normal cell DNA is evidence that the methylation may be a tumor cell-specific alteration.
3. DNAs from a set of primary tumors representative of histological cell type, grade, and stage are analyzed for the methylation status of a gene by bisulfite sequencing or quantitative real-time MSP exactly as described in **Chapters 22** and **23**. This is important because there is evidence that tumor cell lines have more gene methylation than the primary tumor counterparts *(10)*.

4. Notes

1. The stability of azacytidine in aqueous solution stored at −80°C is a potential variable. It is best to aliquot before freezing and to discard an aliquot after thawing. The effect of batch variability or degradation over time on efficacy could be tested by quantitative RT-PCR assay of reactivation of expression of a chosen gene in a cell line.
2. The 5Aza-dC treatment can be combined with the HDAC inhibitor, trichostatin acid (TSA) since there is good evidence that the processes of methylation and deacetylation interact to silence transcription, although it is believed that TSA has less synergistic effect at the relatively higher 5 μM 5Aza-dC dose we used, than when combined with lower doses of 5Aza-dC *(11)*. However, the addition of TSA can result in the upregulation of unmethylated genes and can complicate the subsequent analysis of methylated genes. Our preference would be to examine TSA separately. TSA (Wako, Richmond, VA) is dissolved in absolute ethanol as a 330-μM stock solution, and stored in aliquots at −20°C.
3. Six T75 flasks typically yield 20–100 μg of total RNA by Trizol isolation. Fewer T75 flasks/plates are needed if the investigator chooses to use an RNA amplification step used in some array protocols.

4. If toxicity is too high at 5 μM it is useful to culture the cells with 2.5 μM. For slow-growing cell lines, we have found incorporating a recovery period of >24 h before the third dose of 5Aza-dC to be helpful.
5. Although mock is the better control it seems unlikely that the small amount of PBS would affect the cells so untreated cells are a suitable control.
6. To identify genes with no basal expression in the untreated cells, it is necessary to perform qRT-PCR. However, there are conceptual issues that argue against such an approach. These include that in gender-specific cancers, such as ovarian or prostate, sex differentiation genes can be reactivated; in tumor cells where a classical tumor suppressor gene, for example, *VHL* is functionally inactivated by point mutation, a transcript can still be present, and cell lines can contain subclones, for example, the T24 tumor cell line has a major clone with inactivated $p16^{INK4a}$ but contains a minor clone with functional $p16^{INK4a}$ so again a transcript would be present. Further, an oligonucleotide probe on an array may not discriminate between alternative splice forms of the same gene where expression of not all isoforms is lost with promoter hypermethylation *(12)*, for example, *RASSF1A (13)*.
7. Potential caveats are that SAGE databases are only as good as original studies, that borderline CpG island can still be methylated with functional relevance, and that promoter CpG islands containing repetitive elements need not be eliminated from study since, for example, the *VHL* and $p16^{INK4a}$ TSGs are known to contain repetitive elements in, or near, the promoter CpG island but still to have promoter hypermethylation of functional significance.

Acknowledgements

The author would like to thank Dr Jim Herman at Johns Hopkins University, Baltimore, MD and Dr Dominique Broccoli at Anderson Cancer Center, Savannah, GA for helpful discussion.

References

1. Jones, P. A., Baylin, S. B. (2007) The epigenomics of cancer. *Cell* 128, 683–692.
2. Costello, J. F., Frühwald, M. C., Smiraglia, D. J., et al. (2000) Aberrant CpG-island methylation has non-random and tumour-type-specific patterns. *Nat Genet* 25, 132–138.
3. Sjoblom, T., Jones, S., Wood, L. D., et al. (2006) The consensus coding sequences of human breast and colorectal cancers. *Science* 314, 268–274.
4. Suzuki, H., Gabrielson, E., Chen, W., et al. (2002) A genomic screen for genes upregulated by demethylation and histone deacetylase inhibition in human colorectal cancer. *Nat Genet* 31, 141–149.
5. Ibanez de Caceres, I., Dulaimi, E., Hoffman, A. M., et al. (2006) Identification

of novel target genes by an epigenetic reactivation screen of renal cancer. *Cancer Res* 66, 5021–5028.
6. Liang, G., Gonzales, F. A., Jones, P. A., et al. (2002) Analysis of gene induction in human fibroblasts and bladder cancer cells exposed to the methylation inhibitor 5-aza-2′-deoxycytidine. *Cancer Res* 62, 961–966.
7. Yamashita, K., Upadhyay, S., Osada, M., et al. (2002) Pharmacologic unmasking of epigenetically silenced tumor suppressor genes in esophageal squamous cell carcinoma. *Cancer Cell* 2, 485–495.
8. Tusher, V. G., Tibshirani, R., Chu, G. (2001) Significance analysis of microarrays applied to the ionizing radiation response. *Proc Natl Acad Sci USA* 98, 5116–5121.
9. Takai, D., Jones, P. A. (2002) Comprehensive analysis of CpG islands in human chromosomes 21 and 22. *Proc Natl Acad Sci USA* 99, 3740–3745.
10. Smiraglia, D. J., Rush, L. J., Frühwald, M. C., et al. (2001) Excessive CpG island hypermethylation in cancer cell lines versus primary human malignancies. *Hum Mol Genet* 10, 1413–1419.
11. Cameron, E. E., Bachman, K. E., Myohanen, S., et al. (1999) Synergy of demethylation and histone deacetylase inhibition in the re-expression of genes silenced in cancer. *Nat Genet* 21, 1037.
12. Shames, D. S., Girard, L., Gao, B., et al. (2006) A genome-wide screen for promoter methylation in lung cancer identifies novel methylation markers for multiple malignancies. *PLoS Med* 3, e486.
13. Dammann, R., Li, C., Yoon, J.-H., et al. (2000) Epigenetic inactivation of a RAS association domain family protein from the lung tumour suppressor locus 3p21.3. *Nat Genet* 25, 315–319.

Part IV

Gene-Specific Methylation Analysis

Chapter 14

DNA Methylation Analysis by Bisulfite Conversion, Cloning, and Sequencing of Individual Clones

Yingying Zhang, Christian Rohde, Sascha Tierling, Heinrich Stamerjohanns, Richard Reinhardt, Jörn Walter, and Albert Jeltsch

Abstract

DNA methylation is an essential epigenetic modification in the human genome. For the investigation of DNA methylation patterns, bisulfite conversion and DNA sequencing is a method of choice, because it provides detailed information on the methylation pattern of individual DNA molecules at single CG site resolution. The method is based on the deamination of cytosine residues to uracils in the presence of NaOH and sodium bisulfite. Since methylcytosine is not converted under these conditions, the original methylation state of the DNA can be analyzed by sequencing of the converted DNA. After the conversion reaction, the DNA sequence under investigation is amplified by polymerase chain reaction (PCR) with primers specific for one strand of the bisulfite-converted DNA. The PCR product is cloned and individual clones are sequenced. Here, we describe an advanced protocol for bisulfite conversion, protocols for cloning, and tools for primer design (Methprimer, Bisearch). In addition, we present tools for the web display of primary data and data analysis (BiQ Analyzer, BDPC) and describe the setup of a sequencing and analysis pipeline for medium to high throughput.

Key words: DNA methylation, bisulfite, sequencing, cloning.

1. Introduction

DNA methylation is an essential epigenetic signal involved in development, gene regulation, imprinting, and preserving genome integrity. In mammals, it mainly occurs at cytosines in the context of the CpG-dinucleotide. The bisulfite genomic-sequencing method is based on the selective deamination of cytosine residues but not 5-methylcytosines by treatment with

Yingying Zhang and Christian Rohde contributed equally to the work

Jörg Tost (ed.), *DNA Methylation: Methods and Protocols, Second Edition, vol. 507*

DOI 10.1007/978-1-59745-522-0_14 Springerprotocols.com

sodium bisulfite and the sequencing results from the subsequently generated polymerase chain reaction (PCR) products *(1, 2)*. Currently, bisulfite sequencing of single clones is regarded as the gold standard of DNA methylation analysis, because sequencing of subcloned individual DNA molecules provides the most reliable and detailed information on the methylation pattern for every single CpG site in a relatively long stretch of sequence (300–500 bp). Furthermore, it provides unambiguous methylation information for haplotypes of DNA molecules in a qualitative and quantitative manner and is able to show correlations between the DNA methylation states of different CpG sites. In this chapter, we describe the application of an advanced protocol for bisulfite conversion, optimized protocols for cloning, tools for primer design, the web-based compilation and display of primary data and data analysis, as well as the setup of a sequencing and analysis pipeline for medium- to high-throughput analysis. **Figure 14.1** shows the outline of the method. Briefly, first the genomic DNA is fragmented by digestion with restriction enzymes. Then, in the presence of NaOH and sodium bisulfite, the genomic DNA fragments are denatured to single-stranded DNA and all unmethylated cytosines will be chemically converted to uracils. After bisulfite conversion, the upper and lower strands of the DNA are no longer complementary. In the next step, the DNA sequence under investigation is amplified by PCR with primers specific for one strand of the bisulfite-converted DNA. During this process, uracil will be amplified as thymine. In contrast, if the

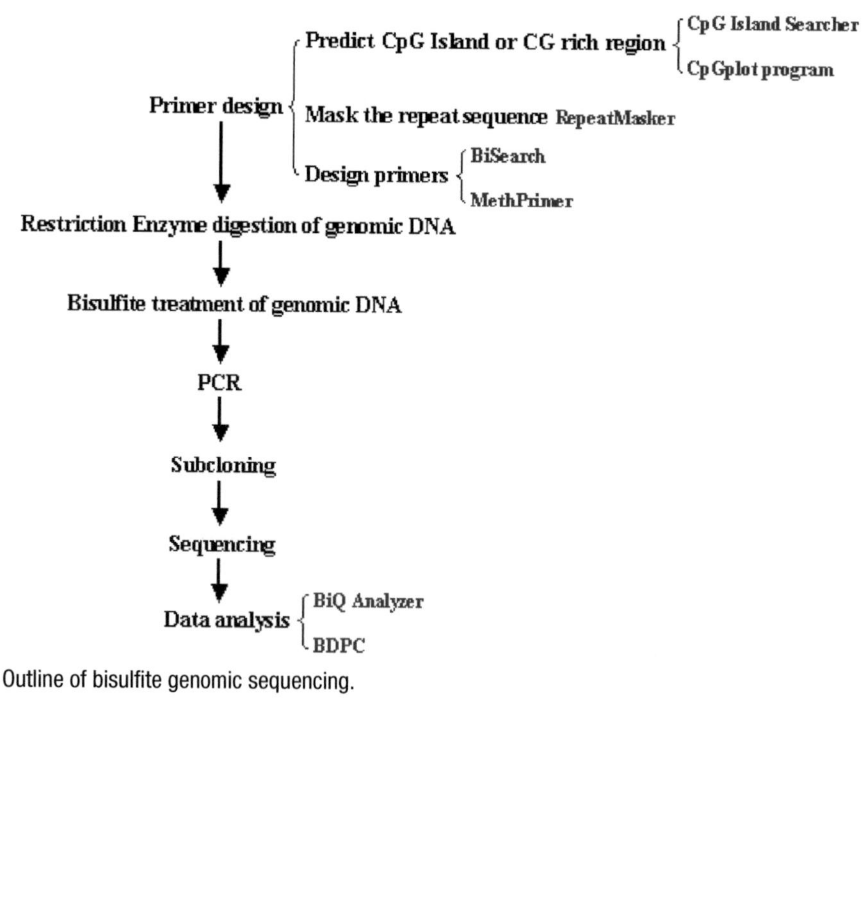

Fig. 14.1. Outline of bisulfite genomic sequencing.

cytosine in the CpG dinucleotide context is methylated, it will not react with bisulfite, such the 5-methylcytosines are remain unchanged and are amplified as cytosine in the PCR.

The amplified PCR product is a pool of DNA molecules. Each of them could have a unique methylation pattern. Therefore, the PCR product is subsequently subcloned and randomly selected clones are sequenced. The cloning results give the methylation information of individual DNA molecules. According to the sequencing results, it can be distinguished if the cytosines in the CpG sites are methylated. The methylation percentage of each CpG site can be calculated according to the number of methylated and unmethylated cytosines in different clones. Finally, sequencing data is analyzed, compiled, and prepared for web presentation and integrated data analysis.

2. Materials

2.1. Programs for Primer Design

1. CpG Island Searcher program: http://www.cpgislands.com.
2. CpGPlot program: http://www.ebi.ac.uk/emboss/cpgplot.
3. RepeatMasker: http://www.repeatmasker.org.
4. BiSearch: http://bisearch.enzim.hu.
5. MethPrimer: http://www.urogene.org/methprimer/index1.html.

2.2. Bisulfite Conversion of Genomic DNA

1. Freshly prepare 2 M NaOH and 0.3 M NaOH in sterile distilled H_2O.
2. Solution I: 1.9 g $NaHSO_3$ (Sigma No. 13438) is dissolved in the mixture of 2.5 mL sterile water and 750 μL of 2 M NaOH (*see* **Note 1**). This solution needs to be freshly prepared.
3. Solution II: 98.6 mg of 6-hydroxy-2,5,7,8-tetramethylchroman-2-carboxylic acid (Sigma No. 238813) is dissolved in 2 mL of dioxane. This solution needs to be freshly prepared.
4. 1 × TE buffer: 10 mM Tris–HCl, 1 mM EDTA, pH 8.
5. Microcon Ultracel YM-50 columns (Millipore).

2.3. PCR and Gel Electrophoresis (Agarose and Polyacrylamide)

1. HotStarTaq DNA Polymerase (Qiagen).
2. HotFirePol DNA Polymerase (Solis BioDyne).
3. Jetsorb Gel Extraction Kit (Genomed).
4. ChargeSwitch® PCR Clean-Up Kit (Invitrogen).
5. 1.2% agarose gels in 0.5 × TBE.
6. PAGE running buffer (10 × TPE buffer): 0.9 M Tris–HCl, 20 mM EDTA, pH 8.2 adjusted with H_3PO_4.
7. For PAGE: Rotiphorese gel 40 (40% solution of acrylamide/bis-acrylamide 29:1, Roth), *N,N,N′,N′*-tetra methylethylenediamine (TEMED) should be stored at 4°C. 10% ammonium persulfate (APS): Prepare 10% solution in water and store the aliquots at −20°C.

2.4. Cloning

1. Luria-Bertani (LB) medium: 1.0% NaCl, 1.0% tryptone, and 0.5% yeast extract, pH 7.0.
2. LB ampicillin (100 mg/mL) agar (1.5%) plates.
3. StrataClone™ PCR Cloning Kit (Stratagene).
4. pGemT T/A Cloning Kit (Promega).

2.5. Sequencing

1. 3730 × 1 ABI 96-capillary sequencer systems (Applied Biosystems).
2. ABI BigDye Terminator kit version 3.1 (Applied Biosystems).

2.6. Result Analysis

1. Chromas: http://www.technelysium.com.au
2. FinchTV (Geospiza): http://www.geospiza.com/finchtv.
3. BiQ Analyzer: http://biq-analyzer.bioinf.mpi-sb.mpg.de *(3)*.

2.7. Result Presentation

1. Data compilation: http://biochem.jacobs-university.de/BDPC/.

3. Methods

3.1. Primer Design

1. Check if there is a high density of CpG-dinucleotides in the region of interest. The presence of a CpG island (CGI) can be determined with the CpG Island Searcher program or CpG-Plot program (*see* **Note 2**).
2. To avoid designing primers in the repetitive DNA interspersed in the genome, use the RepeatMasker software to identify the presence of repeat in the region of interest (*see* **Note 3**).
3. Design primers in the region of interest. The length of the PCR products to be amplified is suggested not to exceed 500 bp (*see* **Note 4**). BiSearch *(4)* and MethPrimer *(5)* are two online primer-designing programs with unique properties of designing primers for bisulfite-converted DNA. Additionally, BiSearch can be used for computational search for mispriming sites of the primers on the bisulfite-treated genome and to predict potential nonspecific amplification products *(4)* (*see* **Note 5**).

3.2. Bisulfite Conversion of Genomic DNA

1. Digest 200–300 ng of genomic DNA with an appropriate restriction enzyme (40 U) at 37°C overnight (*see* **Note 6**). The selected enzyme should not digest DNA within the region of interest (*see* **Note 7**). Set up the reaction in a 20 μL volume.

2. Add 187 μL of solution I to the 20 μL of digested genomic DNA to denature the DNA and convert unmethylated cytosine into uracils. Mix by pipetting up and down.
3. Add 73 μL of solution II to the mixture and gently mix by pipetting up and down.
4. Incubate the mixture in a thermocycler using the following conditions: 15 min at 99°C, 30 min at 50°C, 5 min at 99°C, 1.5 h at 50°C, 5 min at 99°C, and 1.5 h at 50°C (*see* **Note 8**).
5. Add 150 μL of sterile distilled H_2O to the mixture. After mixing by pipetting up and down, transfer the reaction mixture to the Microcon Ultracel YM-50 columns. Place the column in a collection tube (provided in the kit) and centrifuge at 14,000*g* for 15 min (*see* **Note 9**).
6. Carefully separate the column from the collection tube. Discard the filtrate. Place the YM-50 column back into the same collection tube.
7. Add 500 μL of 1 × TE buffer and centrifuge at 14,000*g* for 10 min. Separate the column from the collection tube. Discard the filtrate. Place the YM-50 column back into the same collection tube.
8. Add 500 μL of 0.3 M NaOH to the column and incubate at room temperature for 10 min to desulfonate the DNA. Afterward, centrifuge at 14,000*g* for 10 min. Discard the filtrate.
9. Add 500 μL of TE buffer, centrifuge at 14,000*g* for 10 min and discard the filtrate.
10. Place the column upside-down in a new collection tube provided. Add 50 μL 1 × TE (50°C) in the middle of the sample reservoir. Incubate at room temperature for 1 min and centrifuge at 1,000*g* for 10 min (*see* **Note 10**).
11. Collect the DNA in the collection tube and measure the concentration of the single-stranded bisulfite-treated genomic DNA by UV spectrometry. Store it at −20°C (*see* **Note 11**).

3.3. PCR

1. 1–2 μL of the bisulfite-converted DNA is used as template for PCR in a 25 μL reaction mixture (1 × PCR buffer, 1.5 mM $MgCl_2$, 0.2 mM of each dNTP, 0.4 μM of each primer and 2.5 U of HotStarTaq polymerase). Perform PCR with the following program: 15 min at 95°C, 5 × (30 s at 94°C, 30 s at 65°C, 90 s at 72°C), 5 × (30 s at 94°C, 30 s at 60°C, 90 s at 72°C), 35 × (30 s at 94°C, 30 s at 55°C, 90 s at 72°C), 5 min at 72°C *(6)* (*see* **Note 12**).
2. After the PCR amplification, 5 μL of the PCR product is electrophoresed on a 1.2% agarose gel. For better resolution, an 8% PAGE gel can be prepared. Prepare a 1.5-mm thick gel by mixing 8 mL 1 × TPE buffer, 2 mL acrylamide (40%), 25 μL TEMED, and 25 μL APS (10%) (*see* **Note 13**).

3. Stain the gel and visualize the PCR product under UV-light. An example of the result is shown in **Fig. 14.2**.
4. Purify the PCR product following the manufacturer's instructions of Jetsorb Gel Extraction Kit or ChargeSwitch® PCR Clean-Up Kit (*see* **Note 14**).

3.4. Cloning and Shipping Clones for Sequencing

1. Subclone the purified PCR product using the pGemT T/A Cloning Kit or StrataClone™ Kit (*see* **Note 15**).
2. In our project, biochemical and sequencing facilities were remote. For shipping the clones to the sequencing facility, we picked colonies and transferred to the wells of 96 or 384 well plates containing LB ampicillin agar (1.5%) in each well. After overnight incubation at 37°C, the plates were sealed and shipped by overnight mail.
   A second possibility is to perform PCR on selected clones with primers enclosing the insert. For this parts of a bacterial colony are picked, suspended in 20 μL of water, heated to 95°C for 5 min, spun down, and 1 μL of the supernatant is used for a subsequent PCR reaction. Colony PCR products positive for the insert are shipped by overnight mail for sequencing.

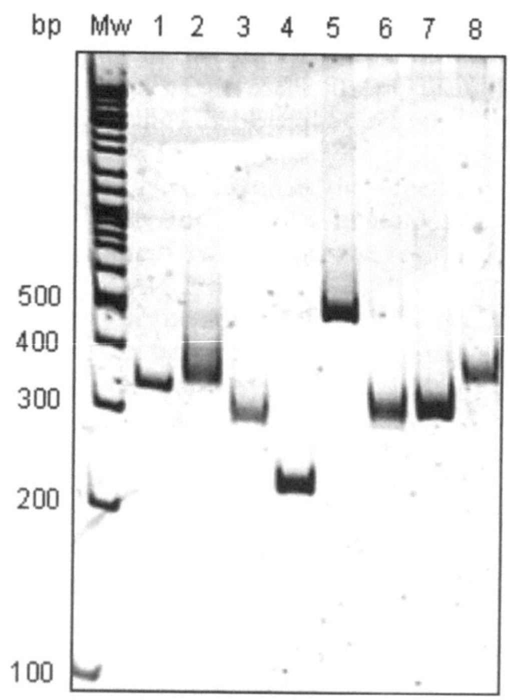

Fig. 14.2. Example of the electrophoretic analysis of eight different PCR products amplified from bisulfite-treated genomic DNA. Mw: DNA Marker (100 bp ladder). Lane 1–8: PCR products with size of 334, 336, 294, 217, 476, 302, 303, 358 bp, respectively.

3.5. Sequencing

1. Plasmid DNA was isolated by a highly automated alkaline lyses procedure, which includes template purification by PEG-precipitation and subsequently adjustment to similar molarity.
2. The primers used were M13-Reverse: 5′-CAG GAA ACA GCT ATG ACC-3′ and T7-Forward: 5′-TAA TAC GAC TCA CTA TAG GG-3′.
3. In 5 μL reactions, 20–30 ng plasmid DNA and 5–10 μM primers were used.
4. Cycling was done in a GenAmp PCR System 9700 using 2.5 min at 96°C, 35 × (20 s at 96°C, 10 s at 50°C, and 4 min at 60°C).
5. DNA sequences were determined using ABI BigDye Terminator v3.1 Kit (Applied Biosystems) using a 3730 × 1 ABI 96-capillary sequencer systems equipped with capillaries of 50-cm separation length.
6. Resulting sequences were processed for quality clipping.

3.6. Result Analysis

1. Extract the sequencing results in FASTA format results using a trace file viewer like Chromas or FinchTV (*see* **Notes 16** and **17**). This is necessary if the subsequent analysis is done with the BiQ Analyzer software. An example of sequencing results of two single clones is shown in **Fig. 14.3**.
2. The BiQ Analyzer software tool *(3)* can be used to import the results of the sequencing of the subcloned PCR products, make the alignment with the original sequence, perform basic statistics of the methylation level, and generate the methylation pattern (*see* **Note 18**). During this process, the data with incomplete bisulfite conversion or low-sequence identity can

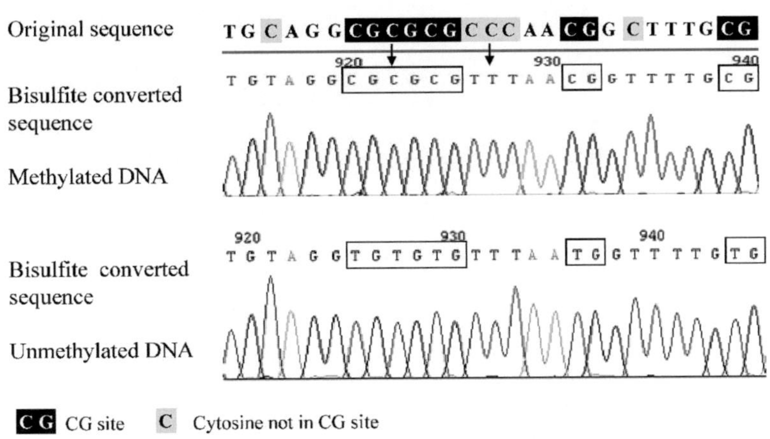

Fig. 14.3. Example of bisulfite sequencing results of single clones. After bisulfite conversion, the unmethylated cytosines are converted to thymines, leaving the methylated cytosine in the CpG site unaltered. According to the sequencing result of single clone, the methylated cytosine and unmethylated cytosine can be distinguished. Original sequence: before bisulfite treatment.

be excluded either with the software or manually. The output file is in HTML format, which can be applied for further analysis as described in the next part.

3.7. Result Presentation

1. To allow the compilation of the BiQ Analyzer data from several PCR products (in our analysis about 1,500 different PCR products in total, comprising about 30,000 clones were investigated), we developed a web interface called bisulfite sequencing data presentation and compilation (BDPC), which reads in the data from the original BiQ Analyzer result files (*see* **Note 19**) *(9)*. BDPC prepares the following output: (1) A collection of Microsoft Excel-compatible tables compiling (i) for each PCR product, the average methylation level, the number of clones analyzed, and the percentage of CpG sites analyzed in each PCR product (which is an indication of the technical quality of the data); (ii) the methylation levels observed at each CpG site; and (iii) the methylation levels of each clone. (2) Publication grade figures in PNG format showing the methylation pattern for each PCR product are embedded in a HMTL file summarizing the methylation data, the genomic sequence, and some basic statistics of the methylation level. (3) A summary file, which presents the methylation pattern of all biological samples analyzed with each amplicon, linked to the individual HTML result files, such that the set can be directly used for presentation of the data in the Internet. (4) A condensed

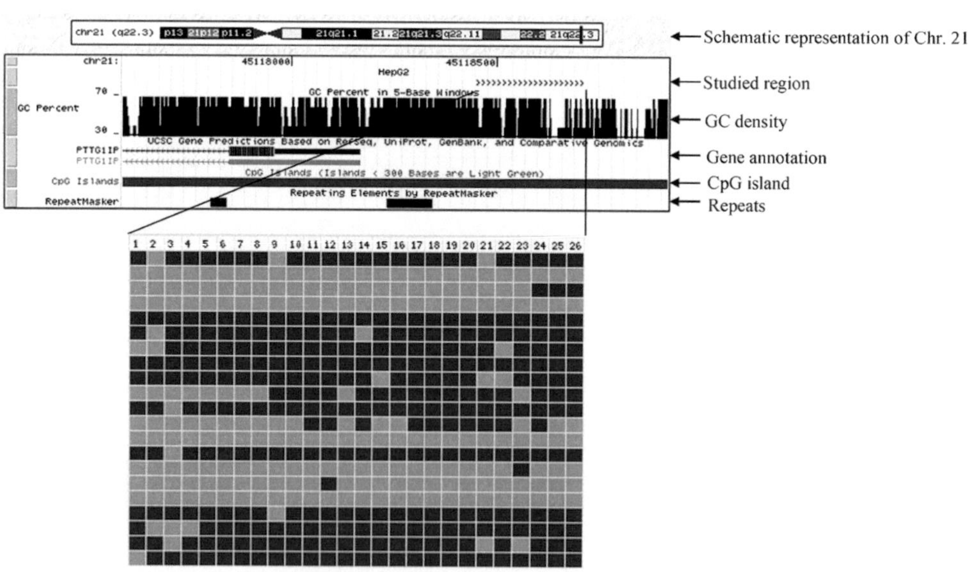

Fig. 14.4. Schematic display of the methylation status of 26 CpGs in a region of Chromosome 21 investigated by bisulfite sequencing of single clones. The region (265 bp) lies within the promoter of the PTTG1IP gene. It is embedded in a CpG island without repeats. Each row corresponds to each clone of bisulfite PCR products. Each column corresponds to one CpG site in the studied region. Grey boxes represent unmethylated CpG sites, black boxes methylated ones.

file containing all primary data in simplified format for further downstream data analysis.

2. An example of the diagrammatic representation of the methylation pattern of a studied region is shown in **Fig. 14.4**.
3. The results can be displayed as a custom track annotation in the UCSC Genome Browser, which allows for easy presentation of data sets. Find an example at http://biochem.jacobs-university.de/BDPC/.

4. Notes

1. Sodium bisulfite dissolves completely after the addition of 2 M NaOH.
2. Both "CpG Island Searcher" and "CpGPlot" can be used to check the presence of a CGI in the target region manually. Currently, the definition of a CGI is not uniform. The criteria for a CGI proposed by Takai and Jones are the most stringent ones *(7)*, which require DNA sequence length of ≥500 bp, GC content of ≥55%, and a ratio of the observed CpG frequency versus the expected one of ≥0.65.
3. The importance of this step is demonstrated in **Fig. 14.4**. The primers should not be designed in the repetitive DNA sequence, which is annotated by the UCSC Genome Browser. Additionally, the UCSC Genome Browser allows DNA sequence download in which repeats are masked by using the RepeatMasker software. One can directly copy the sequence for the next step.
4. Longer fragments are difficult to be amplified due to the fragmentation and degradation of DNA template during bisulfite conversion.
5. As a consequence of the bisulfite treatment, the complexity of the nucleotide sequence is reduced, which promotes mispriming. The two programs described are suggested to be used for primer design. It is also possible to design primers on *in silico* converted DNA manually. In principle, the primer should target a region, which contains several cytosines ideally located in the 3′ part. These cytosines are substituted by thymines, such that converted DNA is specifically amplified. The primers should not contain CpG sites within their sequence to avoid discrimination against methylated or unmethylated DNA. The BiSearch software allows the inclusion of CpG sites in the primer binding site. In this case, Y is used to represent C or T in the sense chain and R is used to represent A or G in the antisense chain *(4)*.
6. The bisulfite-conversion rate is critical to determine the accuracy of the method to define the methylation status. To ensure complete bisulfite conversion, it is not recommended

to start with more than 500 ng genomic DNA in a single reaction.

7. To facilitate strand separation, the genomic DNA is digested with a restriction enzyme into smaller fragments (3–6 kb). For example, BamHI or PstI can be used. This is crucial for the bisulfite-conversion reaction.
8. In this step, the reaction mixture should be split such that the reaction volume fits the size of the wells of the thermocycler.
9. All centrifugation steps are carried out at room temperature.
10. It is important to store the bisulfite-treated genomic DNA in TE buffer, rather than water. In TE buffer, the DNA is more stable and can stored in −20°C for up to 6 months.
11. To avoid the DNA degradation by thawing and freezing, it is recommended to aliquot the bisulfite-treated genomic DNA.
12. It is necessary to use a DNA polymerase, which can use templates containing uracil like Taq polymerase. If difficulties are encountered to obtain the PCR products with the described parameters, an alternative program could be: 15 min at 95°C, 5 × (30 s at 94°C, 30 s at 60°C, 90 s at 72°C), 5 × (30 s at 94°C, 30 s at 55°C, 90 s at 72°C), 35 × (30 s at 94°C, 30 s at 50°C, 90 s at 72°C), 5 min at 72°C. A further alternative: 15 min at 95°C, 5 × (30 s at 94°C, 30 s at 55°C, 90 s at 72°C), 5 × (30 s at 94°C, 30 s at 50°C, 90 s at 72°C), 35 × (30 s at 94°C, 30 s at 45°C, 90 s at 72°0C), 5 min at 72°C. Also nested PCR or semi-nested PCR may help to get specific PCR products. Using a gradient PCR cycler can further optimize the annealing temperature.
13. In acrylamide gels, the separation of DNA usually is better than in agarose gels. But one needs to consider that converted DNA in acrylamide gels migrates differently compared to normal DNA *(8)*.
14. In case that unspecific PCR by-products appear that cannot be removed by the optimization of the PCR protocol, the PCR product may be purified from agarose gels using commercially available kits like QIAquick Gel Extraction Kit.
15. The subcloning kits based on topoisomerase ligation technique are highly efficient even using a small amount of DNA substrate. One can split the material provided in the kit for one reaction to perform three reactions.
16. We suggest using Chromas as a trace file viewer and for DNA sequence extraction. With this software it is possible to edit the annotation if necessary. Typical vector sequence can be defined in Chromas and the DNA insert sequence can be extracted by cutting the vector automatically on export. Furthermore, many sequencing results can be exported in batch.
17. Sometimes misread bases are encountered in the analysis of the sequencing results. For example, there could be

NG-sites, CN-sites, or TN sites at CpG dinucleotide positions. It is recommended to check the chromatogram of the corresponding clones and correct the errors manually wherever possible.

18. In general, the original DNA sequence should be used without the primer sequence to avoid analyzing CpG sites, which are included in the primer sequence. If the amplicon ends on a C, that is followed by a G such that a CpG site is present, the G should be included in the template sequence used for BiQ Analyzer.
19. Sometimes CpG positions are not recognized by the BiQ Analyzer software because of an erroneous alignment of the sequences to the original sequence. Since it is not possible to correct the alignment during analysis, one needs to double check and correct the result manually before uploading the data to the BDPC the software.

Acknowledgements

This work has been supported by the BMBF NGFN2 program. Technical support by S. Becker is gratefully acknowledged.

References

1. Frommer, M., McDonald, L. E., Millar, D. S., et al. (1992) A genomic sequencing protocol that yields a positive display of 5-methylcytosine residues in individual DNA strands. *Proc Natl Acad Sci USA* 89, 1827–1831.
2. Grunau, C., Clark, S. J., Rosenthal, A. (2001) Bisulfite genomic sequencing: systematic investigation of critical experimental parameters. *Nucleic Acids Res* 29, e65.
3. Bock, C., Reither, S., Mikeska T., et al. (2005) BiQ Analyzer: visualization and quality control for DNA methylation data from bisulfite sequencing. *Bioinformatics* 21, 4067–4068.
4. Tusnády, G. E., Simon, I., Váradi, A., et al. (2005) BiSearch: primer-design and search tool for PCR on bisulfite-treated genomes. *Nucleic Acids Res* 33, e9.
5. Li, L. C., Dahiya, R. (2002) MethPrimer: designing primers for methylation PCRs. *Bioinformatics* 18, 1427–1431.
6. Yamada, Y., Watanabe, H., Miura, F., et al. (2004) A comprehensive analysis of allelic methylation status of CpG islands on human Chromosome 21q. *Genome Res* 14, 247–266.
7. Takai, D., Jones, P. A. (2002) Comprehensive analysis of CpG islands in human chromosomes 21 and 22. *Proc Natl Acad Sci USA* 99, 3740–3745.
8. Handas, V., Jeltsch, A. (2004) Anomalous mobility of polymerase chain reaction products after bisulfite treatment of DNA. *Anal Biochem* 333, 196–198.
9. Rohde, C., Zhang, Y., Jurkowski, T. P., Stamerjohanns, H., Reinhardt, R. Jeltsch, A. (2008) Bisulfite sequencing Data Presentation and Compilation (BDPC) web server - a useful tool for DNA methylation analysis. *Nucl. Acids Res.* 36, e34.

Chapter 15

Identification and Quantification of Differentially Methylated Loci by the Pyrosequencing^{TM} Technology

Emelyne Dejeux, Hafida El abdalaoui, Ivo Glynne Gut, and Jörg Tost

Abstract

Most available protocols for gene-specific DNA methylation analysis are either labor intensive, not quantitative, or limited to the measurement of the methylation status of only one or very few CpG positions. Pyrosequencing is a real-time sequencing technology that overcomes these limitations. After bisulfite modification of genomic DNA, a region of interest is amplified by polymerase chain reaction (PCR) with one of the two primers being biotinylated. The PCR-generated template is rendered single stranded and a pyrosequencing primer is annealed to analyze quantitatively CpGs within 120 bases. Advantages of the pyrosequencing technology are the ease of its implementation, the high quality and the quantitative nature of the results, and its ability to identify differentially methylated positions in close proximity. A minimum amount of 10 ng of bisulfite-treated DNA is necessary to obtain high reproducibility and avoid random amplification. The required DNA amount can be provided by an individual sample or a pool of samples to rapidly investigate the presence of variable DNA methylation patterns. The use of pools and serial pyrosequencing, that is, the successive use of several pyrosequencing primers on the same DNA template, significantly reduces cost, labor, and analysis time as well as saving precious DNA samples for the analysis of gene-specific DNA methylation patterns.

Key words: Serial pyrosequencing, pooled DNA, real-time synthesis, bisulfite, epigenotyping, quantification.

1. Introduction

Pyrosequencing is a sequencing-by-synthesis method, in which the incorporation of nucleotides complementary to a template strand is monitored bioluminometrically *(1, 2)* (**Fig. 15.1**). Pyrophosphate (PP_i) is released upon incorporation of the nucleotide(s) by the Klenow fragment of the *Escherichia coli* DNA polymerase I and is used by an ATP sulfurylase to produce ATP

Jörg Tost (ed.), *DNA Methylation: Methods and Protocols, Second Edition, vol. 507*

DOI 10.1007/978-1-59745-522-0_15 Springerprotocols.com

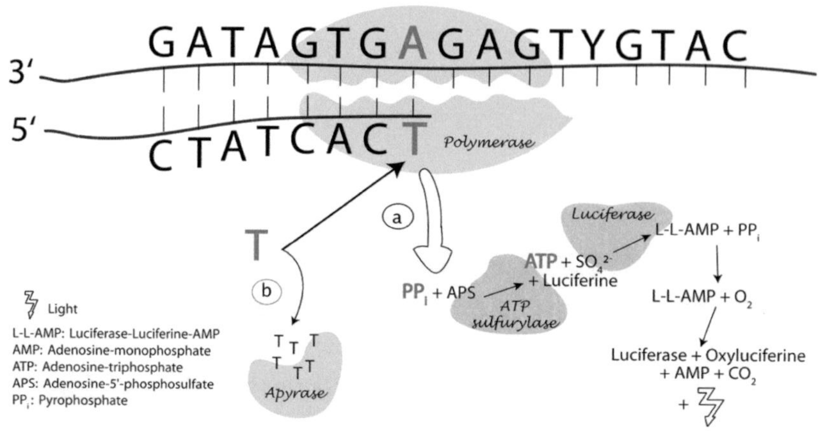

Fig. 15.1. Principle of the pyrosequencing technology. A TTP nucleotide complementary to the adenosine in the template strand is added and incorporated by the polymerase. The pyrophosphate released upon nucleotide incorporation is converted into a light signal through a cascade of enzymatic reactions that are described in detail in the text (**a**). Nonincorporated nucleotides are removed from the reaction mixture by the apyrase (**b**).

from adenosine phosphosulfate *(3)*. This ATP provides the necessary energy for the luciferase to oxidize D-luciferin. The product oxyluciferin is generated in an excited state, which decays to the ground state with the emission of a photon that can be detected by a charge-coupled device (CCD) camera. In contrast to conventional Sanger sequencing that uses a mixture of the four fluorescently labeled terminating ddNTPs and strand elongating dNTPs *(4)*, only one nucleotide is dispensed at a time by an inkjet-type cartridge in pyrosequencing reactions. Unincorporated nucleotides as well as excess ATP are degraded prior to addition of the next nucleotide by an apyrase. The carefully optimized interplay of the kinetics of the four enzymes ensures that at most one nucleotide is present at any time in the reaction mixture permitting clear assignment of the light signal to the incorporation of a specific nucleotide and thereby reconstruction of the sequence synthesized by the iterative addition of nucleotides. The dispensation order of nucleotides can either be predefined for the analysis of known sequences or consist of cyclic ACGT dispensations for *de novo* sequencing. The procedure of the pyrosequencing assay is simple and robust and results are highly reproducible. The instrument provides a great flexibility performing either different assays (up to 96) in parallel or analyzing up to 96 samples with one specific assay. These properties have made pyrosequencing a widely used analysis platform for various biological and diagnostic applications, such as routine (multiplex) genotyping of single nucleotide polymorphisms (SNPs), bacterial typing, or mutation detection *(3, 5)*. The intensity of the bioluminometric response is directly proportional to the amount of incorporated nucleotides, that is, a peak corresponding to the incorporation of two consecutive (and identical) nucleotides will have the double height compared to the signal of a single nucleotide incorporation. The peak heights in the resulting output format, termed pyrogram, thus inform on the extent of homopolymeric sequences and proportions of alleles can be deduced directly from the relative height of the peaks corresponding to variable nucleotide positions. The quantitative nature of the results is the most important characteristic of the pyrosequencing technology – especially with regard to DNA methylation analysis. The bioluminometric response is linear for the sequential addition of up to five identical nucleotides (dCTP, dGTP or TTP, respectively) or three α-S-dATPs. The latter is used instead of dATP which serves as a direct – though less efficient – substrate for the luciferase and would therefore uncouple signal detection from nucleotide incorporation. Pyrosequencing has therefore been used in a variety of applications where quantitative assessment of the relative abundance of two individual nucleotides or short sequences is required such as determination of SNP allele frequencies in pooled samples *(6)*, analysis of copy-number variation *(7)*, karyotyping *(8)* and DNA methylation analysis *(9–11)*.

For DNA methylation analysis, genomic DNA extracted from a tissue sample is treated with sodium bisulfite to "freeze" the methylation status of the cytosines and to translate the epigenetic modification into sequence information *(12)*. A small region of interest is amplified with one of the two amplification primers being biotinylated. This label is subsequently used to generate a single-stranded template to which the pyrosequencing primer is annealed. DNA methylation analysis by pyrosequencing thus permits simultaneous analysis and quantification of the methylation status of several CpG positions in close proximity. This point is of particular interest as successive CpGs might display significantly different levels of methylation as demonstrated in the differentially methylated region of imprinting genes *(13)* as well as at promoters devoid of a CpG island *(14)*. Pyrosequencing combines the advantages of sequence-based approaches, such as inbuilt quality control and resolution of individual CpG sites, with the possibilities of medium to high throughput (for DNA methylation analysis) and the advantages of PCR-based technologies. It features a limit of detection of ~3% for the minor component of a quantitative signal and a quantitative resolution of at least 5% *(10,15)*. The quantitative accuracy of pyrosequencing can be applied to the determination of the global DNA methylation content of a sample *(16, 17)* as well as to gene-specific analyses, such as the identification of genes aberrantly silenced by promoter hypermethylation in cancer *(18,19)*, the distinction between age-related and cancer-associated DNA methylation patterns *(20)* or the analysis of the epigenetic field defect in prostate cancer *(21)*. A diagnostic test for aberrant methylation patterns involved in the imprinting disorders Prader–Willi and Angelman syndromes was proposed using pyrosequencing *(22)* and a recently developed method permits the analysis of allele-specific DNA methylation patterns *(23)*. The broad range of applications combined with the above described advantages has made pyrosequencing a widespread analysis method. However, there are some inconveniences associated with this technology, mainly concerning the size of the amplicon and the sequencing read-length. Due to the thermal instability of the enzymes (especially the luciferase), pyrosequencing has to be carried out at 28°C which limits the size of the amplified amplicon to 300 bp or less as the formation of secondary structures can complicate annealing of the sequencing primer or increase background signals. The limitation in the read-length is mainly due to dilution effects and increasing background due to frame shifts of subpopulations of sequenced molecules. Previously described restrictions in the number of CpGs that can be analyzed simultaneously *(13, 24)* have been overcome by the development of a new operating and analysis software (Q-CpG) that no longer relies on pattern-recognition algorithms. In recent years, we have developed two approaches to improve the throughput, simultaneously reduce cost, labor, and analysis time, and save precious DNA samples. As the template strand used for pyrosequencing is not altered during the pyrosequencing reaction, it can be recovered after the sequencing run for the successive use of several sequencing primers on the same DNA template (serial pyrosequencing) *(25)*. This improvement enables the analysis of an entire region amplified in a single PCR. A novel screening approach identifies differential methylation between two sample groups by creating pools stratified for clinical parameters of interest, for example, cancerous versus matched peritumoral tissue. This method helps to concentrate research efforts and available biological material on genes displaying variable methylation patterns *(15)*. In the protocol described in this chapter we have included all the different steps necessary for the identification of differential methylation of samples and the subsequent analysis in a larger sample cohort. However, some steps such as the construction of pools or the use of serial pyrosequencing are optional and can be omitted from the protocol if the target region is already known or the analysis of a few CpG positions is sufficient.

2. Materials

2.1. Assay Design

1. Design of PCR primers using MethPrimer *(26)* (http://www.ucsf.edu/urogene/methprimer/index1.html).
2. Design of pyrosequencing primers manually or using the commercial PSQ assay design software (Pyrosequencing AB, Uppsala, Sweden).

2.2. Sample Preparation, Calibration Standards and Bisulfite Treatment

1. Proteinase K for DNA extraction.
2. *Sss*I methylase (New England Biolabs; cat. no. M0226L).
3. REPLI-g Mini Kit (Qiagen, Valencia, CA; cat. no. 150025).
4. Quant-iT^{TM} dsDNA broad-range assay kit (Invitrogen, Carlsbad, CA; cat. no. Q33130).
5. SpectraMAX Gemini XPS microplate spectrofluorometer (Molecular Devices, Sunnyvale, CA).
6. MethylEasy bisulphite kit (Human Genetic Signatures, North Ryde, Australia; cat. no. ME001) or the respective high-throughput kits converting 96 samples at a time (cat. no. MEHT002 for the centrifugation protocol or cat. no. METH003 for the vacuum protocol, respectively, Human Genetic Signatures) (*see* **Note 1**).

2.3. Construction of Pools

1. NanoDrop spectrophotometer (NanoDrop Technologies Inc, Wilmington, DE).

2.4. The PCR Amplification

1. Primers for PCR amplification.
2. Biotinylated primers for PCR amplification (Biotez) (*see* **Note 2**).
3. HotStar Taq DNA polymerase (Qiagen; cat. no. 203205).
4. dNTPs.
5. Thermo-Fast® 96–skirted, 96 well plates (ABgene, Epsom, UK; cat. no. AB-0800).
6. Eppendorf 96-Gradient Mastercycler (Eppendorf, Hamburg, Germany).

All reagents should be stored at −20°C.

2.5. Sample Preparation for Pyrosequencing Analysis

1. Streptavidin Sepharose HP beads (GE Healthcare, Uppsala, Sweden; cat. no. 17-5113-01).
2. Vacuum preparation tool *(27)* (Pyrosequencing AB) using the corresponding filter probes (cat. no. 60-0180).
3. Troughs (Pyrosequencing AB, cat. no. 60-0182).
4. Binding buffer: 10 mM Tris–Cl, 2 M NaCl, 1 mM EDTA, 0.1% Tween 20; pH 7.6 (*see* **Note 3**).
5. Denaturing solution: 0.2 M NaOH.
6. Washing buffer: 10 mM Tris–acetate; pH 7.6.
7. Annealing buffer: 20 mM Tris–acetate, 2 mM Mg–acetate; pH 7.6.
8. Thermowell sealing tape.
9. Thermomixer or similar (room temperature).
10. Heating device, for example, heating plate or thermoblock.
11. Thermoplate for sample preparation (Pyrosequencing AB; cat. no. 60-0123).
12. Primers for pyrosequencing.
13. Plate for pyrosequencing analysis, PSQ HS 96 Plate (Pyrosequencing AB; cat. no. 40-0028).

All reagents used for this step should be stored at room temperature except for the streptavidin-coated sepharose beads (+4°C) and the pyrosequencing primer (−20°C).

2.6. Analysis by Pyrosequencing Reaction

1. Pyrosequencer PSQ 96MD System (Pyrosequencing AB; formerly called PSQ 96HS system).
2. Cartridge for reagent dispensation: PSQ HS 96 Dispensing Tip Holder (Pyrosequencing AB; cat. no. 60-0134).
3. PSQ HS 96 Reagent Dispensing Tip (Pyrosequencing AB; cat. no. 40-0030).
4. PSQ HS 96 Nucleotide Dispensing Tip (Pyrosequencing AB; cat. no. 40-0029).
5. Pyrosequencing Kit: PyroGold SQA reagent kit 1 × 96 (Pyrosequencing AB; cat. no. 40-0045).
6. Q-CpG software (Pyrosequencing AB; cat. no. 60-260).

2.7. Serial Pyrosequencing

All required reagents are listed under heading **2.5** and **2.6**.

3. Methods

The protocol for DNA methylation analysis using the pyrosequencing technology can be subdivided into seven steps (**Fig. 15.2**):

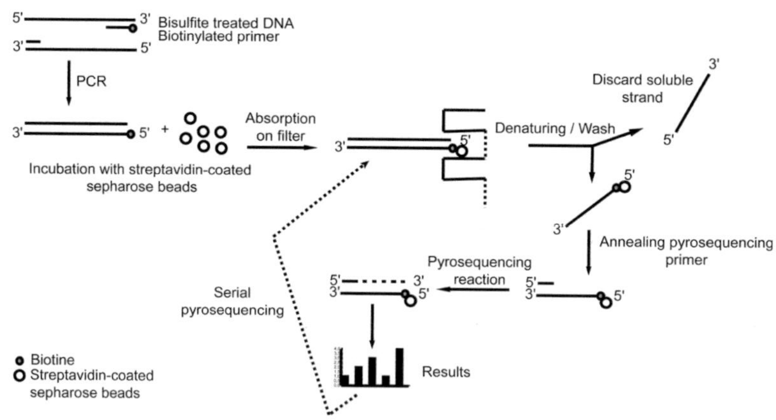

Fig. 15.2. Outline of the different steps of the pyrosequencing procedure. A target region is amplified with one of the two primers being biotinylated. The biotinylated strand is captured on streptavidin-coated sepharose beads which are retained on the filter plate while the non-biotinylated strand is washed off under denaturing conditions. A sequencing primer is annealed to the template strand for the pyrosequencing reaction. After analysis the biotinylated strand can be repurified discarding the *de novo* synthesized complementary strand of the last pyrosequencing reaction. The template is then available for the sequencing of another part of the same target region analyzing additional CpG positions (serial pyrosequencing).

- *The experimental design* for a quantitative gene- or promoter-specific analysis of CpG methylation.
- *The preparation of calibration standards* and*the bisulfite treatment* of samples and standards.
- *Optional: the preparation of pools* to rapidly test for the presence of differential methylation between two groups.
- *The PCR amplification* of the target sequence performed with one of the PCR primers biotinylated.
- *The sample preparation for pyrosequencing analysis*, which requires rendering the PCR product single stranded and subsequent annealing of the pyrosequencing primer to the single-stranded DNA template.
- *The pyrosequencing reaction* synthesizes the complementary strand to the single-stranded DNA template and analyzes quantitatively the CpG (and other polymorphic) positions in the sequence (**Fig. 15.3**).
- *Serial pyrosequencing* enables the analysis of additional CpGs that have not been analyzed in the first pyrosequencing run on the previously used template strand.

3.1. Assay Design

The assay design is probably the most critical step for pyrosequencing-based DNA methylation analysis. Great attention should therefore be paid to the experimental design as this will crucially influence the successful outcome of the assay.

Many standard software tools developed for conventional PCR cannot handle primer design on bisulfite-converted DNA due to the lower complexity of bisulfite-treated DNA. However, some commercial and freely available software have been especially designed for this purpose, such as the CpGWARE primer design software (Chemicon International), MethPrimer *(26)*, Bisearch *(28)* or MethylPrimer Express (Applied Biosys-

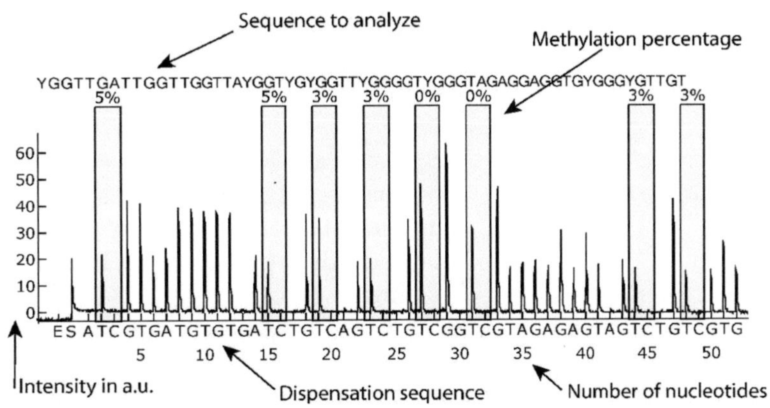

Fig. 15.3. Example of a pyrogram, the output format of the pyrosequencing reaction. Intensity of the signals is given in arbitrary units (a.u.). The pyrogram depicts the analysis of eight CpGs in the CpG islands spanning the transcription start of *CDKN2A*.

tems). The procedure for the design of amplification primers using MethPrimer as performed in our laboratory is described in **Section 3.1.1**. Primers are positioned to amplify the target region irrespective of its methylation status. The subsequent design of the pyrosequencing primers to cover the target region is described in **Section 3.1.2**. Here we describe an approach for the manual design of pyrosequencing primers, which in our hands is as successful as the commercial design software (*see* **Note 4**).

3.1.1. Design of Amplification Primers

1. As the two strands of genomic DNA are no longer complementary after bisulfite treatment, both strands can be analyzed for possible amplification products. MethPrimer takes only the forward strand into account and the reverse strand has to be created manually or using one of a variety of software tools.
2. Primers should be approximately 30 bp in length and the optimum size of an amplification product is ~250 bp, although we successfully analyzed DNA methylation in PCR products up to 350 bp.
3. Place primers in a region containing four or more cytosines that have been converted during bisulfite treatment to ensure that they are only complementary to completely converted DNA as the chemical treatment is rarely complete.
4. Primers should preferentially contain no CpG positions. If it cannot be avoided the maximum number of CpG positions covered should be restricted to one, and this CpG position should not be included in the last five bases from the 3′ terminus to avoid preferential amplification.
5. To ensure specificity, palindromes within primers and complementary sequences between primers as well as degenerated bases and inosine should be avoided.
6. The generated amplicon should be verified for the presence of potential polymorphic positions such as SNPs underlying the annealing sites for the amplification primers. We strongly recommend the redesign of amplification primers annealing to potentially polymorphic sites (even at the 5′ terminus).

3.1.2. Design of Pyrosequencing Primers

1. Sequences are identified where a sequencing primer could be positioned and at least the last five bases from the 3′ terminus do not overlap with any other potentially variable position, including CpGs and SNPs that are retained after bisulfite treatment.
2. The last four or five bases from the 3′ terminus should be verified to be unique in the amplicon by using, for example, the MS Word Find tool. As few as four consecutive nucleotides complementary to a sequence in the amplification product might add to background signal confounding precise quantification.

3. Successfully designed primers should also be checked for primer dimers and possible hairpin structures.
4. As read-lengths of up to 120 bases can be achieved with the PyroGold SQA kit, primers can also be positioned in non-polymorphic regions next to the variable region using part of the nucleotide dispensations to approach the region of interest.
5. At least one cytosine not followed by a guanine should be included in the dispensation order to control for complete bisulfite conversion.
6. The direction of the pyrosequencing primer defines which of the amplification primers needs to be biotinylated. This primer should be checked carefully not to form any hairpin structure.
7. If several pyrosequencing primers are required to cover the region of interest, analysis of a few CpG positions by more than one pyrosequencing primer improves confidence into the acquired results and helps to detect potential technical artifacts.

3.2. Sample Preparation, Calibration Standards and Bisulfite Treatment

A high number of CpG positions in the amplified fragment results in a large sequence difference between completely methylated and nonmethylated molecules after bisulfite treatment. This difference strongly influences the melting and annealing properties during PCR amplification and predisposes for preferential amplification of one allele (PCR bias) *(29)*. As PCR bias is both sequence and strand specific, it is almost impossible to predict. PCR bias can be detected with a known degree of methylation or mixtures of completely methylated and unmethylated DNA prepared as described below.

1. Extract genomic DNA using standard procedures such as phenol/chloroform extraction or commercial kits including a treatment of the DNA with proteinase K as the presence of residual amounts of chromatin proteins in the DNA sample is highly detrimental for complete bisulfite conversion.
2. Prepare completely unmethylated DNA using the Repli-g whole-genome amplification kit following the manufacturer's instructions.
3. Prepare completely (or at least highly) methylated DNA by incubation of genomic DNA with the CpG methylase *Sss*I. Add 7.5 μL of NE-buffer 2, 10 nmol of S-adenosylmethionine (SAM), and 6 U of *Sss*I to 4.5 μg of human genomic DNA in a final volume of 67.5 μL. Incubate the solution at 37°C in a water bath. After 3 h and again after additional 2 h, add 10 nmol of SAM and 6 U of *Sss*I and incubate the reaction overnight at 37°C. Inactivate the enzyme at 95°C for 5 min and store DNA at −20°C until further use.
4. Adjust concentration of DNA samples and calibration standards to 50 ng/μL after measuring the concentration with the

Quant-iT^{TM} dsDNA broad-range assay kit on a spectrofluorometer.

5. Treat samples and standards with sodium bisulfite using the MethylEasy kit following the manufacturer's instructions (*see* **Note 1**).

3.3. Preparation of Pools for Rapid Screening

Pooling of samples provides a rapid means to detect potential variable DNA methylation patterns in candidate genes without prior knowledge about the presence of differential methylation. It permits the analysis of a large number of genes while saving DNA from precious clinical specimens. The methylation patterns of the pooled samples reflect the average of the individual samples contained in the respective pool with high precision and helps prioritize target genes for further evaluation *(15)* (*see* **Note 5**).

1. Measure the concentration of the bisulfite-treated DNA (samples and unmethylated and methylated standards) with a NanoDrop spectrophotometer and adjust the concentration to 20 ng/μL (*see* **Note 6**).
2. Create calibration standards with a known degree of methylation by mixing the respective volumes of the Repli-g amplified and the *Sss*I-treated, bisulfite-converted DNAs in 10% or 25% increments.
3. Constitute pools of samples by mixing equal volumes of samples with the same phenotypic or clinical parameters. There is no restriction on the pool size but methylation events occurring at low frequency might be missed in large pools. Although the quantitative resolution is about 5%, subtle changes attributable to a single or very few differentially methylated samples might be missed. Additional pools of identical etiology can increase the confidence in the results and might also detect less frequent methylation events.

3.4. The PCR Amplification

A strong and specific PCR product obtained by amplification with one of the two primers biotinylated is required for a successful pyrosequencing reaction.

1. An increase in random preferential amplification during PCR is observed with decreasing amounts of template DNA used in the PCR amplification *(13)*. A minimum of 10 ng DNA is necessary to ensure high reproducibility. Usually, we amplify regions of interest using 20 ng of bisulfite-treated genomic DNA. The quantity of DNA can be provided by a single sample or a pool of samples.
2. Reaction conditions are 1 × HotStar *Taq* buffer supplemented with 1.6 mM $MgCl_2$, 200 mM dNTPs, and 2.0 U of HotStar *Taq* polymerase and 5 pmol of forward and reverse primer in a 25 μL volume, one of them being biotinylated (*see* **Note 7**).

3. The PCR program consists of a denaturing step of 15 min at 95°C followed by 50 cycles of 30 s at 95°C, 30 s at the respective annealing temperature determined beforehand by a temperature gradient, and 20 s at 72°C, with a final extension of 5 min at 72°C. Perform 50 cycles of amplification to ensure complete exhaustion of the free biotinylated primer and to yield a strong PCR product (*see* **Note 8**).
4. Deposit 5 μL of the PCR product as well as the positive and negative controls (*see* **Note 8**) on an agarose gel. The entire PCR amplification should be repeated if the slightest signal in the negative controls is detected.
5. PCR products can be stored at +4°C for several days or at −20°C for several months.

3.5. Sample Preparation for Pyrosequencing Analysis

In this step, the previously amplified double-strand DNA is rendered single stranded to enable annealing of the sequencing primer as the pyrosequencing reaction takes place at 28°C due to the thermal instability of the enzyme mix. Template preparation can be applied to a complete 96 well plate as well as to a single well of a 96 well plate.

1. Transfer 5 μL of the PCR product into a new PCR plate, add 40 μL of binding buffer and 2 μL sepharose beads and complete to 80 μL with water. Cover the plate with a sealing tape and incubate the reaction mixture for 10 min at room temperature under constant mixing (1400 rpm). It is crucial that the beads do not sediment (*see* **Note 9**).
2. During this incubation step, prepare the pyrosequencing plate by diluting 4 pmol of the pyrosequencing primer into 12 μL of annealing buffer into the respective wells of the PSQ plate. One or several different pyrosequencing primers can be used on the same plate (*see* **Note 10**).
3. Fill the four troughs of the vacuum preparation tool with 180 mL of 70% ethanol, washing buffer, and water, respectively, and the trough for the denaturing solution with 120 mL 0.2 M NaOH. This different level assures that the NaOH is completely washed off in the following washing step as it might otherwise inhibit the ensuing pyrosequencing reactions.
4. Turn on the workstation, create vacuum in the aspiration device (450 mm Hg), and clean the tips of the filters by immersion in water for several seconds. Remove the PCR plate from the mixer and aspirate the binding mix.
5. Immerse the tips of the filters for 5 s each in the successive baths of ethanol 70%, denaturing solution, and washing solution. Turn over the tool and release the vacuum.
6. Immerse the tip of the filters in the annealing mix of the sequencing plate and shake gently to release the beads into the wells.

7. Check the plate used for incubation with the binding buffer for the presence of remaining beads. The presence of beads indicates incomplete retrieval of the PCR product which might lead to failure of the subsequent pyrosequencing analysis due to insufficient quantity of template material (*see* **Note 11**).
8. Prepared plates can be stored at this point for 2 days at 4°C prior to pyrosequencing analysis.
9. Incubate the sequencing plate for 2 min at 80°C on the thermoplate placed on a heating device. Sealing of the plate is not necessary. Allow the plate to cool down to room temperature.

   For a template-preparation protocol compatible with analysis on the PSQ96MA instrument, please follow the above described procedure and refer to **Note 12** for changes.

3.6. Pyrosequencing Reaction

1. During cooling of the sequencing plate (Step **3.5.9**), program the sequencing run on the Pyrosequencer. We recommend using the Q-CpG software. When using the SNP software, enter new assays well beforehand, since depending on the number of CpGs this might take several hours.
2. The software calculates the quantity of reagents necessary to perform the run. Dispense the reagents and enzyme mix in the appropriate tips placed in the cartridge (*see* **Note 13**). Wear a lab coat and powder-free gloves when handling the dispensing cartridge to prevent dust particles from clogging the tip heads of the Nucleotide Dispensing Tips (NDTs) and be careful not to create air bubbles when filling the tips. For sequencing runs of up to 30 nucleotides the PyroGold SNP reagent kit can be used. The PyroGold SQA reagent kit enables longer read-length of up to 120 nucleotides.
3. Deposit the sequencing plate and the reagents' cartridge in the Pyrosequencer.
4. Perform the dispensation test with a sealed pyrosequencing plate to verify that the dispensing tips are properly working. Droplets should be clearly visible and homogeneous. Change the tips if necessary.
5. Start the pyrosequencing run. The length of a run is proportional to the number of dispensations (one per minute).
6. After the end of the sequencing run, analyze the results with the Q-CpG software (**Fig. 15.4**). In some cases the signals corresponding to part of the sequence might be missing. This is in most cases due to a problem of nucleotide dispensation (tips blocked). Check the point of the tips for large droplets that might have accumulated at a dust particle or due to an adverse electrostatic environment (humidity too low).

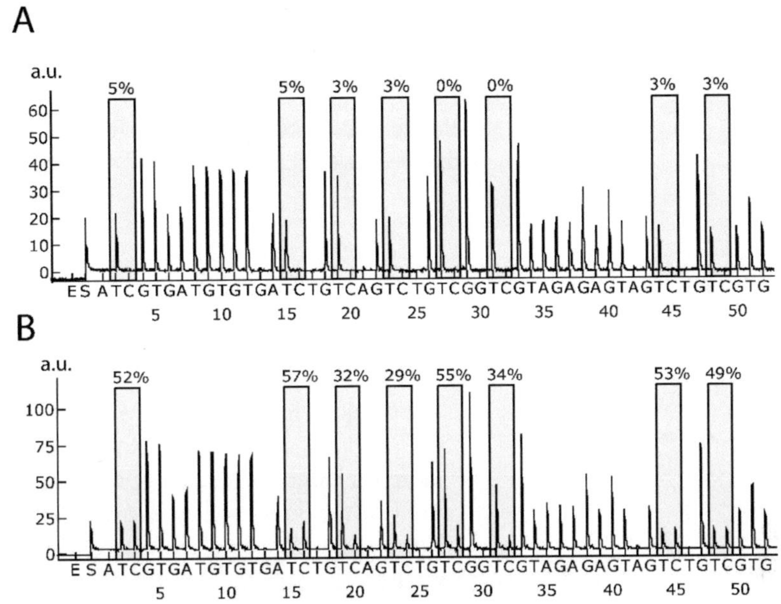

Fig. 15.4. Analysis of DNA methylation patterns in the *CDKN2A* promoter in human hepatocellular carcinoma. Panel A depicts the peritumoral liver tissue corresponding to the tumor shown in panel B. Hardly any methylation is detected in the peritumoral tissue (average 2.7%) while on average 45% methylation is found in the tumor.

7. If the plate should be used for serial pyrosequencing (see below), seal the plate and store the plate at 4°C for not more than 3 days.
8. The Q-CpG software allows exporting the results, which can then be further treated with statistical or graphical software such as Excel®.
9. Remove the cartridge and clean the tips well with pure water.

3.7. Serial Pyrosequencing

If not enough PCR product is available to perform all pyrosequencing reactions on a target region amplified in a single PCR reaction, it is possible to recycle the template strand for additional analyses as the biotinylated strand is not altered during pyrosequencing. In case of dispensation problems during the run, this method also allows to repeat the pyrosequencing run without the need to redo PCR amplification and use more of the sample DNA. Although there is a slight loss in intensity due to incomplete recovery of the biotinylated strand, quantitative results are unaltered for several cycles of pyrosequencing on the same template *(25)*.

1. Prepare a pyrosequencing plate with a new pyrosequencing primer as described in Step **3.5.2** and prepare the workstation as described in **Section 3.5.3**.

2. Add 20 μL binding buffer to the completed sequencing reaction (from **Section 3.6.7**) and resuspend the sepharose beads by vigorous pipetting.

   Purify the mixture without further incubation as described in Steps **3.5.4–3.5.9**.

4. Notes

1. A detailed protocol for the bisulfite conversion of genomic DNA without the use of commercial kits for bisulfite conversion has been previously published by us and others *(20, 24)*. The mentioned bisulfite kit from Human Genetic Signatures is routinely used in our laboratory, but other kits are commercially available from Qiagen, Zymo, or Applied Biosystems probably yielding similar results.
2. Biotinylated primers are more sensitive to storage conditions than unmodified primers and dilutions as well as stock tubes should be kept at −20°C. Dilutions should be aliquoted and not subjected to more than five cycles of freezing and thawing. It should also be noted that the quality might differ substantially from provider to provider. Biotinylated primers should always be ordered HPLC purified to remove free biotin which competes with the PCR product for the binding sites of the streptavidin-coated beads used for template preparation.
3. The different buffers used for template preparation are also commercially available from Pyrosequencing AB.
4. The commercial pyrosequencing design software has been developed for the design of SNP genotyping assays and its quality criteria for successful assay design might be too stringent. Despite a low quality score and the display of a number of error messages, primers designed by the software do often work in practice but should be carefully verified using the appropriate controls and calibration standards. Manual design of the pyrosequencing primers is in our hands as successful.
5. The use of pools requires as detailed knowledge as possible on the available samples. The composition of the pools depends directly on the question to be answered. Samples homogeneous for a distinct clinical or pathological parameter are grouped together, but different pools might be required for different parameters under investigation.
6. If only frequently occurring hyper- or hypomethylation events are of interest, the additional step of adjusting

the concentration can be omitted. However, accuracy is improved if the concentrations are equilibrated *(15)*.

7. Avoid higher amount of primers as excess biotinylated primer may diminish the capture of the amplification product on the streptavidin-coated beads and may give rise to background signals during the pyrosequencing reaction.
8. The high number of amplification cycles could cause amplification of very small amounts of contaminating DNA. Therefore, always include appropriate controls, especially several negative controls. We also strongly recommend a spatial separation of pre- and post-PCR manipulations to reduce the risk of contaminations. Conventional decontamination methods such as dUTP incorporation can only be applied to bisulfite-treated DNA under special conditions *(30,* **Chapter 26***)* as nonmethylated cytosines are converted to dUTPs during bisulfite treatment.
9. In general, 5 μL of PCR product are adequate but depending on the strength of the PCR product on the agarose gel more or less product might be used. However, too little PCR product will lead to insufficient peak height for accurate quantification in the resulting pyrogram while too much PCR product might lead to premature loss of signal due to lack of reagents. In the latter case, it is possible to launch the same plate again modifying the sequences to analyze for each well starting from the first missing nucleotide. A sometimes easier alternative is to repurify the plate following the instructions in the serial pyrosequencing paragraph (**3.7**).
10. Each primer should be checked for background signals in the absence of template by preparing a well of the plate containing annealing buffer and the respective pyrosequencing primer.
11. If no liquid has been aspirated in a well, the respective filter tip should be changed. If the liquid has been aspirated, resuspend beads by adding 20 μL of binding buffer and follow the purification procedure described in Steps **3.5.3–3.5.6** above.
12. The described protocol is also applicable for analysis with a PSQ96MA instrument with the following modifications: Add 40 μL of binding buffer and 4 μL of streptavidin-coated Sepharose beads to 25 μL PCR and complete to 80 μL with H_2O. For pyrosequencing, 15 pmol of sequencing primer are diluted into 40 μL of annealing buffer in a PSQ 96 plate low (Biotage, cat. no. 40-0010). Nucleotides and reagents are dispensed by a PSQ 96 reagent cartridge (Biotage, cat. no. 40-0022).
13. Take the pyrosequencing kit out of the fridge 30 min prior to use. Otherwise, small air bubbles might form in the

cartridge. If repeatedly problems with blocked cartridges are encountered, it might help to centrifuge the nucleotides before use. When several consecutive runs will be performed, the dispensing tips of the cartridge can be filled in advance with a volume sufficient for all runs.

Acknowledgments

This work was supported by the French Ministry of Research and the European Commission under the Integrated Project "Mol-Page" (contract number LSHG-CT-2004-512966).

References

1. Ronaghi, M. (2001) Pyrosequencing sheds light on DNA sequencing. *Genome Res* 11, 3–11.
2. Ronaghi, M., Uhlen, M., Nyren, P. (1998) A sequencing method based on real-time pyrophosphate. *Science* 281, 363–365.
3. Ahmadian, A., Ehn, M., Hober, S. (2006) Pyrosequencing: history, biochemistry and future. *Clin Chim Acta* 363, 83–94.
4. Sanger, F., Nicklen, S., Coulson, A. R. (1977) DNA sequencing with chain-terminating inhibitors. *Proc Natl Acad Sci USA* 74, 5463–5467.
5. Marsh, S. (2007) *Pyrosequencing Protocols. Methods in Molecular Biology* 373, Humana Press, Totowa, NJ.
6. Wasson, J. (2007) Allele quantification and DNA pooling methods. *Methods Mol Biol* 373, 63–74.
7. Pielberg, G., Day, A. E., Plastow, G. S., et al. (2003) A sensitive method for detecting variation in copy numbers of duplicated genes. *Genome Res* 13, 2171–2177.
8. Deutsch, S., Choudhury, U., Merla, G., et al. (2004) Detection of aneuploidies by paralogous sequence quantification. *J Med Genet* 41, 908–915.
9. Colella, S., Shen, L., Baggerly, K. A., et al. (2003) Sensitive and quantitative universal Pyrosequencing methylation analysis of CpG sites. *Biotechniques* 35, 146–150.
10. Tost, J., Dunker, J., Gut, I. G. (2003) Analysis and quantification of multiple methylation variable positions in CpG islands by Pyrosequencing. *Biotechniques* 35, 152–156.
11. Uhlmann, K., Brinckmann, A., Toliat, M. R., et al. (2002) Evaluation of a potential epigenetic biomarker by quantitative methyl-single nucleotide polymorphism analysis. *Electrophoresis* 23, 4072–4079.
12. Frommer, M., McDonald, L. E., Millar, D. S., et al. (1992) A genomic sequencing protocol that yields a positive display of 5-methylcytosine residues in individual DNA strands. *Proc Natl Acad Sci USA* 89, 1827–1831.
13. Dupont, J. M., Tost, J., Jammes, H., et al. (2004) De novo quantitative bisulfite sequencing using the pyrosequencing technology. *Anal Biochem* 333, 119–127.
14. Chelbi, S. T., Mondon, F., Jammes, H., et al. (2007) Expressional and epigenetic alterations of placental serine protease inhibitors: SERPINA3 is a potential marker of preeclampsia. *Hypertension* 49, 76–83.
15. Dejeux, E., Audard, V., Cavard, C., et al. (2007) Rapid identification of promoter hypermethylation in hepatocellular carcinoma by pyrosequencing of etiologically homogeneous sample pools. *J Mol Diagn* 9, 510–520.
16. Issa, J. P., Gharibyan, V., Cortes, J., et al. (2005) Phase II study of low-dose decitabine in patients with chronic myelogenous leukemia resistant to imatinib mesylate. *J Clin Oncol* 23, 3948–3956.
17. Yang, A. S., Estecio, M. R., Doshi, K., et al. (2004) A simple method for estimating global DNA methylation using bisulfite PCR of repetitive DNA elements. *Nucleic Acids Res* 32, e38.
18. Mirmohammadsadegh, A., Marini, A., Nambiar, S., et al. (2006) Epigenetic silencing of the PTEN gene in melanoma. *Cancer Res* 66, 6546–6552.
19. Xinarianos, G., McRonald, F. E., Risk, J. M., et al. (2006) Frequent genetic and epigenetic abnormalities contribute to the deregulation of cytoglobin in non-small cell lung cancer. *Hum Mol Genet* 15, 2038–2044.

20. Kwabi-Addo, B., Chung, W., Shen, L., et al. (2007) Age-related DNA methylation changes in normal human prostate tissues. *Clin Cancer Res* 13, 3796–3802.
21. Rodriguez-Canales, J., Hanson, J., Tangrea, M., et al. (2007) Identification of a unique epigenetic sub-microenvironment in prostate cancer. *J Pathol* 211, 410–419.
22. White, H. E., Durston, V. J., Harvey, J. F., et al. (2006) Quantitative analysis of SNRPN(correction of SRNPN) gene methylation by pyrosequencing as a diagnostic test for Prader-Willi syndrome and Angelman syndrome. *Clin Chem* 52, 1005–1013.
23. Wong, H. L., Byun, H. M., Kwan, J. M., et al. (2006) Rapid and quantitative method of allele-specific DNA methylation analysis. *Biotechniques* 41, 734–739.
24. Tost, J., Gut, I. G. (2007) Analysis of gene-specific DNA methylation patterns by pyrosequencing technology. *Methods Mol Biol* 373, 89–102.
25. Tost, J., El abdalaoui, H., Gut, I.G. (2006) Serial pyrosequencing for quantitative DNA methylation analysis. *Biotechniques* 40, 721–726.
26. Li, L. C., Dahiya, R. (2002) MethPrimer: designing primers for methylation PCRs. *Bioinformatics* 18, 1427–1431.
27. Dunker, J., Larsson, U., Petersson, D., et al. (2003) Parallel DNA template preparation using a vacuum filtration sample transfer device. *Biotechniques* 34, 862–868.
28. Aranyi, T., Varadi, A., Simon, I., et al. (2006) The BiSearch web server. *BMC Bioinformatics* 7, 431.
29. Warnecke, P. M., Stirzaker, C., Melki, J. R., et al. (1997) Detection and measurement of PCR bias in quantitative methylation analysis of bisulphite-treated DNA. *Nucleic Acids Res* 25, 4422–4426.
30. Tetzner, R., Dietrich, D., Distler, J. (2007) Control of carry-over contamination for PCR-based DNA methylation quantification using bisulfite treated DNA. *Nucleic Acids Res* 35, e4.

Chapter 16

Mass Spectrometric Analysis of Cytosine Methylation by Base-Specific Cleavage and Primer Extension Methods

Dirk van den Boom and Mathias Ehrich

Abstract

The analysis of epigenetic changes in genomic DNA has seen an exponentially increasing interest over the last years. Within the field of epigenetics DNA methylation patterns have become of particular interest to the scientific community. The covalent addition of a methyl group to cytosine bases in the CpG dinucleotide sequence holds particular analytical advantages. Working with DNA as an analyte molecule is robust and samples are unproblematic to collect and handle. Also changes in DNA methylation are a dynamic process and the resulting patterns are tightly associated to disease. This combination of robust technical performance and disease-specific methylation patterns might enable DNA methylation as a powerful biomarker in the future. The increased interest has triggered exciting new findings which ultimately show that epigenetic regulation of gene expression is not a binary system. On the contrary, especially the quantitative measure DNA methylation has greatly contributed to the areas of gene regulation, developmental biology, and translational medicine. Performing quantitative methylation measurements in large scale used to be impaired by the limitations of measurement technologies. They either suffered from limited throughput, limited accuracy, high cost, or a combination of those. Here we introduce a new technique that combines candidate gene amplification with base-specific cleavage or primer extension methods and MALDI-TOF mass spectrometric analysis to overcome the described limitations.

Key words: DNA methylation pattern, quantitative analysis, MALDI-TOF MS, base-specific cleavage, primer extension.

1. Introduction

The field of DNA methylation analysis can be categorized in three main applications: whole genome methylation analysis (WGMA, where a snapshot of the methylation status is taken for the majority of CpG islands in the genome), quantitative high-resolution scanning (which provides a quantitative readout for individual

Jörg Tost (ed.), *DNA Methylation: Methods and Protocols, Second Edition, vol. 507*

DOI 10.1007/978-1-59745-522-0_16 Springerprotocols.com

CpG sites in a target region with high accuracy) and ultrasensitive detection (which aims to identify a few methylated molecules in the presence of an excess of unmethylated molecules). Unfortunately, today no single method is able to perform cost-efficient whole genome analysis, highly accurate quantitative methylation analysis, and ultrasensitive methylation detection simultaneously. Although the introduction of new approaches to perform whole genome DNA methylation analysis has greatly advanced our understanding of epigenetic regulation, the results provide a rather basic insight into individual regions, rarely with individual CpG resolution. Consequently, follow-up studies to those whole genome scans are common. Whole genome studies are often performed to identify a set of target genes using a limited number of samples. The results are typically verified in studies with higher spatial resolution. These fine mapping studies typically have two primary goals: (1) the target areas identified by whole genome scans rarely provide methylation data for individual CpG sites. To get more detailed information about a selected target it is necessary to perform a high-resolution scan of the region. Also, when it is important to know the exact boundaries of where methylation differences can be observed, the genomic area of interested can be expanded and the analysis can be performed on the entire CpG-dense region. An example for a CpG island promoter region with strong regional differences is given in **Fig. 16.1**. (2) Because whole genome studies are usually performed in a limited number of samples, the initially obtained results need to be validated in larger sample sets (see also **Fig. 16.2**).

In the following we describe two recently developed methods for the analysis of methylation patterns in individual target regions that can be used for high-resolution mapping in large sample cohorts and for the targeted precise quantitation of individual CpG sites. Both methods use matrix-assisted laser desorption/ionization time-of-flight mass spectrometry (MALDI-TOF MS) for identification of methylated sites and determination of their degree of methylation, but differ in their coverage, and hence use different assay formats. For high-resolution scanning of longer sequence stretches, such as promoter regions, we propose the use of a comparative sequencing method employing base-specific cleavage to identify methylated sites and quantitate the respective degree of methylation *(1–3)*. Once the methylation pattern is fully characterized, primer extension methods can be employed to specifically target only those CpG sites that showed functional relevance in a biological context *(4–6)*. This second method will then also allow multiplexing the analysis of multiple CpG sites from different promoter regions.

Figure 16.3 depicts a scheme of the concept of base specific cleavage in combination with MALDI-TOF MS-based readout of the analytes for quantitative DNA methylation analysis.

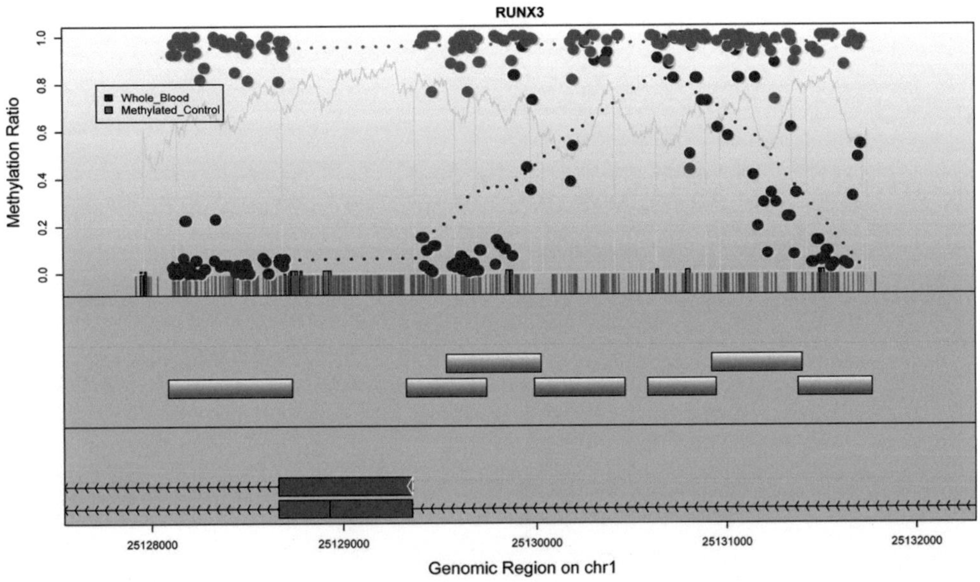

Fig. 16.1. Depicted are regional differences that can be found in CpG island promoter regions. A set of methylated DNAs (*red*) and control DNAs derived from whole blood samples (*blue*) were measured across a total seven amplification targets covering approximately 3.5 kb of genomic sequence. The results are shown as circles with respect to their genomic location in the upper panel. The red bars at the bottom of the upper panel indicate the position of the CpG sites in this area and green rectangles indicate the presence of a conserved transcription factor binding site. The second panel indicates the location of the seven amplification targets and the third panel shows the location of the *RUNX3* exons (blue rectangles) including the transcription start site (TSS, red triangle). The methylation levels around the transcription start site are separated into low and highly methylated. A good example of regional methylation differences can be seen around 700 bp upstream of the TSS, where methylation levels in whole blood control are increasing and drop to low levels again around 2 kb upstream of the TSS. (*see* Color Plate 2)

In brief, the method starts with PCR amplification of the target region from bisulfite-treated DNA, which is followed by in vitro transcription to generate a single-stranded RNA molecule. The RNA strand is then cleaved base specifically in individual reactions either after U or C, determined by the usage of noncleavable nucleotides. The cleavage reaction is driven to completion and the resulting cleavage products represent a well-defined substring of the analyzed target region, which is only dependent on the sequence context and not dependent on the reaction conditions. The cleavage products are then analyzed using MALDI-TOF MS. For analysis of DNA methylation we examine the methylation-dependent C/T sequence changes introduced by bisulfite treatment. Those C/T changes are reflected as G/A changes on the reverse strand and hence result in a mass difference of 16 Da for each CpG site enclosed in the cleavage products generated from the RNA transcript. The mass signals representing nonmethylated DNA and those representing methylated DNA build signal pairs, which are representative for the CpG sites within the analyzed sequence substring. The signal intensities are compared, and the relative amount of methylated DNA can be calculated from this

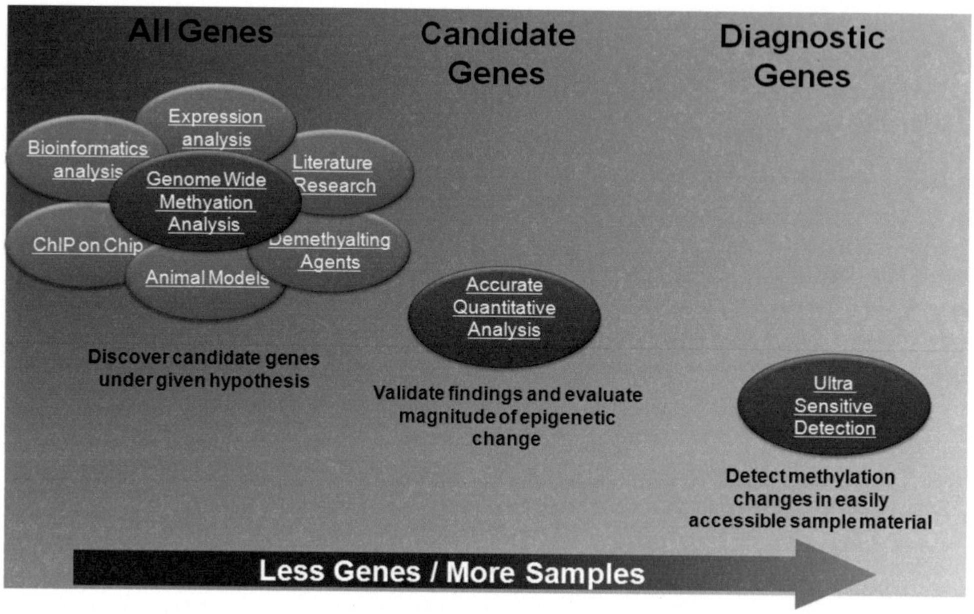

Fig. 16.2. Schematic overview of a typical methylation biomarker discovery experiment. Usually a list of candidate genes is selected using either genome-wide interrogating methods or focused literature search or pathway analysis. These candidate genes are followed up in a larger number of samples using a highly accurate cost efficient method, like sequenom EpiTYPER,® software. Once the clinically relevant genes are identified, a method that is able to detect a small number of methylated molecules, against the background of unmethylated molecules can be employed for clinical testing (e.g., *MGMT* methylation testing in brain cancer). (*see* Color Plate 3)

ratio. The method yields quantitative results for each of these sequence-defined analytical units, which contain either one individual CpG site or an aggregate of subsequent CpG sites. We refer to these analytical units as "CpG units".

A scheme for the targeted analysis of individual CpG sites in a promoter region is shown in **Fig. 16.4**. Single-base and multiple-base primer extension methods in combination with MALDI-TOF MS analysis of primer extension products have been used routinely for the analysis of SNPs and mutations *(7–14)*. The assay format can be adapted fairly simply to the analysis of genomic DNA methylation. As described earlier, the treatment of genomic DNA with bisulfite converts unmethylated cytosines to uracils, whereas methylated cytosines remain unchanged. Any amplification product from the treated DNA will therefore harbor C/T sequence changes, which can be treated as a SNP in terms of the readout by a post-PCR primer extension. In brief, the target region harboring the CpG site of interest is PCR amplified from bisulfite-treated DNA. In a post-PCR primer extension reaction, a primer is annealed immediately adjacent to the CpG site and is extended by a single nucleotide using a DNA polymerase and a mix of four terminators (such as ddNTPs or acycloNTPs). The primer extension reaction will terminate on different nucleotides

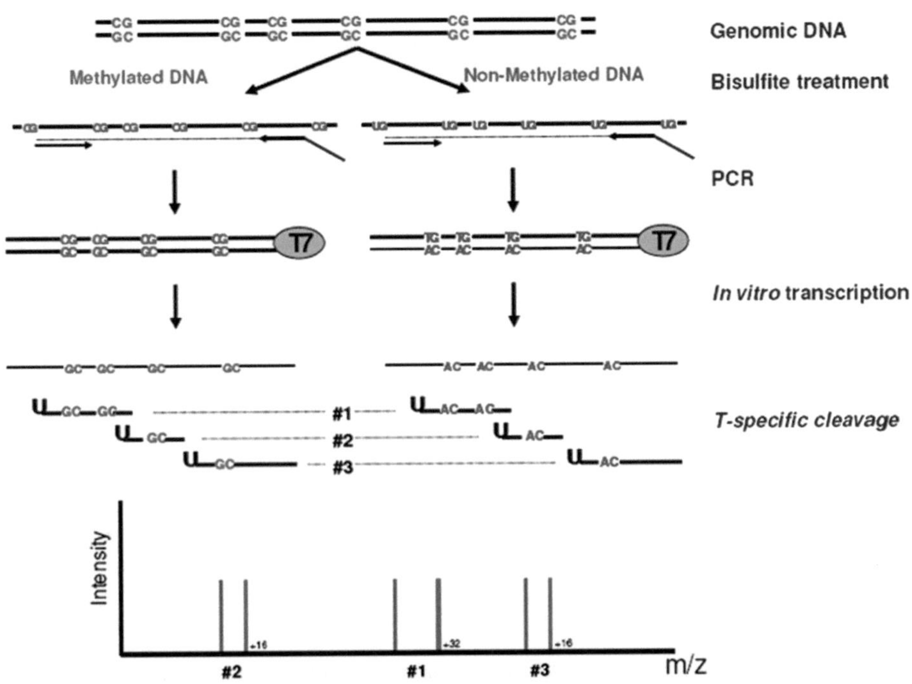

Fig. 16.3. Base-specific cleavage process for quantitative analysis for cytosine methylation by MALDI-TOF MS. As a first step genomic DNA is bisulfite treated to introduce methylation-dependent sequence changes in the genomic DNA, which then can be amplified as a stable signal during PCR amplification. To avoid biased amplification, PCR primers should be designed devoid of CG sites. For base-specific cleavage PCR primers are tagged with a T7 promoter tag (for the reverse primer). Amplified regions can span anywhere from 100 bp to 500 bp, depending on the quality of the genomic DNA used for bisulfite treatment. After PCR amplification, the PCR product is transcribed into a single-stranded RNA from the reverse. The C/T changes introduced during the bisulfite treatment should be represented as G/A changes in this RNA transcript. The RNA transcription can utilize substitution of single rNTP by its dNTP analog (e.g., CTP fully substituted by dCTP). The RNA transcript is then cleaved base specifically by RNase A. This enzyme cleaves at every C and U. If dCTP was used during the transcription, the RNase will cleave base specifically at every uracil. The cleavage process is driven to completion and yields a complex mixture of short oligonucleotides (the base-specific cleavage products), which is then analyzed by MALDI-TOF MS analysis. Detected mass signals can be easily interpreted to identify methylation events in the target sequence. In the depicted case three cleavage products are generated, which each carry one or two CpG sites. A difference in methylation will lead to an A/G sequence change within each of the cleavage products. The corresponding mass signals will therefore shift 16 Da (the mass difference between A and G) if a methylation event occurs. If two or more CpG sites are embedded within a cleavage product mass may shift by multiples of 16 Da. For samples containing a mixture of methylated and nonmethylated target sequence (here indicated in red and green) both mass signals will be observed and the peak area ratio of mass signals representing methylated and nonmethylated sequence can be utilized to estimate the degree of methylation for each CpG site. (*see* Color Plate 4)

(C/T on the forward strand (Δm = 15 Da) or G/A on the reverse strand (Δm = 16 Da)) depending on the methylation status of the target region, and therefore will generate distinct mass signals. The resulting primer extension products are conditioned with ion-exchange resin to remove salt adducts and are then analyzed by MALDI-TOF MS. Each primer and its extension products have a unique molecular mass. The mass signal pattern can therefore be easily interpreted to yield information on the methylation status of the targeted CpG site. Several studies have extensively

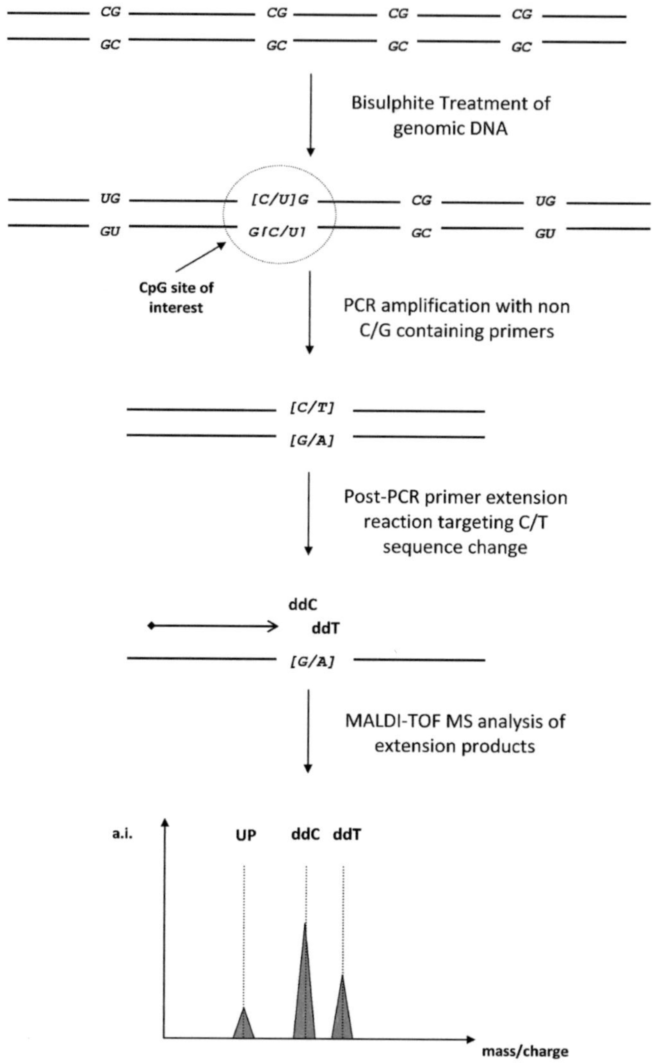

Fig. 16.4. Targeted analysis of the degree of methylation at specific CpG sites using primer extension methods. In comparison to base-specific cleavage that allows the simultaneous analysis of multiple CpGs in a single reaction, this method targets individual CpG sites with proven functional relevance for a precise quantitative methylation analysis. The method is also based on bisulfite treatment of genomic DNA prior to PCR amplification. PCR primers should be designed devoid of CG sites to avoid preferential amplification. Following the amplification reaction, an extension primer is annealed immediately adjacent to the CpG site of interest and is extended by a single base. If the site is methylated, the primer will be extended by a ddC; if the site is unmethylated a ddT will be incorporated. The extension products differ in mass by 15 Da and can therefore be easily distinguished in a subsequent mass spectrometric analysis. The relative amount of methylated and nonmethylated DNA can be estimated with high precision from the peak area ratio of the mass signals representing the two extension products. UP marks unextended primer, that sometimes can be detected in the mass spectrum.

validated the utility of mass signal peak area ratios as a precise estimate of the relative amount of alleles in a sample mixture or pool *(5, 15–17)*. The peak area ratio of mass signals derived from primer extension products representing methylated and unmethylated DNA can, therefore, also be used to estimate the degree of methylation at each of the targeted CpG sites, similar to what is described above for the base-specific cleavage assays.

For methylation analysis, the primer extension assay can create up to three mass signals: a mass signal representing unextended primer, a mass signal representing an unmethylated CpG site, and a mass signal representing a methylated CpG site. The available mass window of a standard linear MALDI-TOF MS is much larger than the three mass positions occupied by a single assay. The technology therefore lends itself to multiplexing several of the primer extension reactions. This can be achieved with careful primer design, for example, by varying the length of the extension primer and ensuring that each unextended primer and extension product of an assay in multiplex creates a unique mass signal. Multiplexing of assays usually starts already at the PCR amplification level and is carried through all subsequent steps as a multiplexed assay. For genotyping multiplexes of up to 40 SNPs have been performed successfully. For quantitative assays we recommend to limit the multiplexing to about 20–25 plexes in order to not compromise precision. Cytosine methylation analysis in particular has the challenge that the four-letter code of A, C, G, and T is reduced to a three-letter code in the bisulfite treatment. Assays should therefore be checked carefully for potential primer interactions and mis-hybridization on the level of PCR and primer extension.

With consideration of the length of this contribution, we focus in the Methods section of this contribution on the details of the base-specific cleavage assay for high-resolution quantitative scanning of promoter regions. Details of primer extension-based methods and their readout by MALDI-TOF MS have been described in detail in earlier issues of methods in molecular biology, which are referenced here *(13, 14)*. We also took care to provide a significant number of references so that the reader can go back to original scientific manuscripts. Finally, we added a Notes section for the primer extension method in order to provide the reader with tips and tricks above and beyond what is described in the scientific literature.

2. Materials

2.1. Bisulfite Treatment

1. EZ-96 DNA Methylation Kit (Zymo Research, Orange County, CA).

2.2. MassCLEAVE™ Protocol

2.2.1. PCR Reagents

1. HotStart Taq DNA Polymerase (5 U/μL, Qiagen, Valencia, CA).
2. HotStart Buffer (contains 15 mM $MgCl_2$, Tris–HCl, KCl, $(NH_4)_2SO_4$, pH 8.7, Qiagen).
3. PCR primers (IDT, San Diego, CA).
4. dNTPs (25 mM each).

2.2.2. Post-PCR Reagents

1. Shrimp Alkaline Phosphatase (Sequenom, San Diego).
2. RNase-free ddH_2O.
3. T7 R and DNA polymerase (Sequenom).
4. T7 R and DNA polymerase buffer (Sequenom).
5. T Cleavage Mix (Sequenom).
6. C Cleavage Mix (Sequenom).
7. DTT (Sequenom).
8. RNaseA (Sequenom).
9. CLEAN Resin (Sequenom).
10. 384-well SpectroChip® (Sequenom).

2.3. Instrumentation

1. MassARRAY Nanodispenser (Sequenom).
2. MassARRAY Compact MALDI-TOF MS (Sequenom).
3. Epityper™ software v1.0 (Sequenom).

3. Methods

3.1. Sodium Bisulfite Treatment

Genomic DNA sodium bisulfite conversion was performed using the EZ-96 DNA Methylation Kit. The manufacturer's protocol was followed using 1 μg of genomic DNA (*see* **Note 1**) and an alternative conversion protocol (a two-temperature DNA denaturation) was used that consisted of 50°C for 15 min and 95°C for 20 s cycled 20 times (*see* **Note 2**).

3.2. Primer Design

Primers are designed using Sequenom's Primer Design software (www.epidesigner.com). The recommended size range for PCR amplicons is 200–550 base pairs (bp) (*see* **Notes 2–4**). Amplification length is limited by the quality of bisulfite-treated DNA (*see* **Note 2**). Highly degraded DNA will only yield stable results for short amplification targets, while high-quality DNA might result in successful amplification of target exceeding 550 bp. During PCR amplification, a T7-promoter tagged reverse primer is incorporated into the amplification product for in vitro transcription. Adding a 10mer-tag sequence to the forward primer balances the PCR primer length. A schematic representation of the primer components is displayed in **Fig. 16.5**. The T7 promoter tag and 10mer tag are added automatically by the software upon export of the designed PCR primers.

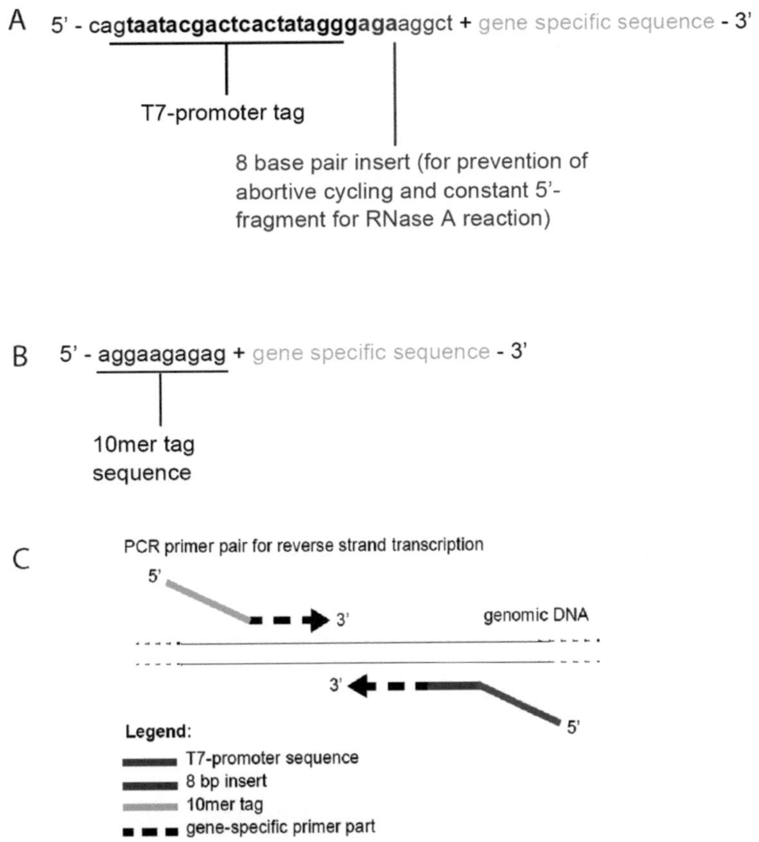

Fig. 16.5. Schematic representation of primer design and tag structure for MassCLEAVE-based methylation analysis. (**A**) Depiction of the T7 promoter tag sequence including an 8 bp tag reducing abortive cycling in the transcription process; (**B**) depiction of the 10mer tag sequence; (**C**) depiction of the PCR primer pair composition and orientation for the generation of reverse transcripts enabling analysis of methylation as G/A sequence changes. (*see* Color Plate 5)

3.3. Negative Control

Run the PCR reaction (3.4) without any bisulfite-treated DNA. The entire homogeneous MassCLEAVE^{TM} (hMC) assay process should be performed from PCR reaction to MALDI-TOF MS analysis. The negative control verifies the occurrence of PCR contamination and indicates excessive formation of primer dimers that lead to specific transcripts and respective cleavage products. Also additional signals can be cross-correlated with those from the negative control.

3.4. Preparing PCR Reactions

The recommended protocol for the pre-PCR portion of the homogeneous MassCLEAVE^{TM} assay is outlined next. Final concentrations are given per 5 μL PCR reaction.

1. Prepare the PCR premix cocktail in a container suitable for holding ≥5.0 mL (e.g., 15-mL tube) by adding the reagents in the order of **Table 16.1**.

Table 16.1
Preparation of a PCR master mix

Order	Reagent	Final concentration for single reaction	Volume for single reaction (μL)	Volume for 384-well microtiter plate (μL)†
1	ddH_2O	N/A	1.42	806.6
2	10 × Hot star buffer	1×	0.50	284.0
3	dNTP mix, 25 mM each	200 μ*M*	0.04	22.7
4	5 U/μL Hot star taq	0.2 U/reaction	0.04	22.7
Total volume			2.00	1,136.0

† Volumes for a 384-well microtiter plate include 48% overhang to account for possible pipetting loss.

2. Add the PCR primers to the PCR premix cocktail as listed in **Table 16.2**.
3. Pipette 4 μL of PCR premix cocktail and primer into a 384-well PCR plate.
4. Pipette 1 μL bisulfite-treated DNA into each well of the 384-well PCR plate. Use 1.00 μL of 10 ng/μL bisulfite-treated DNA to obtain a final concentration of 2 ng/μL DNA per reaction.
5. Seal the microtiter plate with a plate-sealing film. Make sure the edges of the plate sealing film are well sealed.
6. Centrifuge the microtiter plate at 560*g* for 1 min.
7. Cycle the PCR cocktail as follows (*see* **Note 5**): 94°C for 15 min, 45 cycles of 94°C for 20 s, 56°C for 30 s, and 72°C

Table 16.2
Preparation of a primer master mix

Order	Primer	Final concentration for single reaction	Volume for single reaction (μL)	Volume for 384-well microtiter plate†
1	1 μM forward primer	200 nM (1 pmol/reaction)	1.00	variable‡
2	1 μM reverse primer	200 nM (1 pmol/reaction)	1.00	variable‡
Total volume			2.00	variable‡

† Volumes for a 384-well microtiter plate include 48% overhang to account for possible pipetting loss.
‡ The amount of primer will vary due to variations in plate setups.

for 1 min. Terminate the program by an incubation at 72°C for 3 min.

3.5. Removal of Unincorporated dNTPs

PCR reactions need to be treated to reduce the presence of any unincorporated dNTPs, which could interfere with the subsequent RNA transcription step. Shrimp Alkaline Phosphatase (SAP) is used to dephoshorylate deoxynucleotides and render them inactive for the subsequent enzymatic steps (*see* **Note 6**).

1. Prepare the SAP enzyme solution in a 1.5-mL tube by adding the reagents in the order of appearance in **Table 16.3**.
2. Vortex the SAP enzyme solution for 5 s to mix the solution. Be sure to carefully and thoroughly mix the solution.
3. Centrifuge the SAP enzyme solution for 10 s at 2,500*g*.
4. When PCR cycling is complete remove the 384-well PCR microtiter plate from the thermocycler.
5. Add 2.0 μL of SAP solution to each well of the 384-well PCR microtiter plate. To achieve more consistent results it is recommended to use a multichannel pipette or a liquid handling robot.
6. Seal the 384-well PCR microtiter plate with plate-sealing film. Make sure the edges of the plate-sealing film are well sealed.
7. Centrifuge the 384-well sample microtiter plate at 3,000*g* for 1 min.
8. Incubate the 384-well sample microtiter plate at 37°C for 20 min. Inactivate the enzyme at 85°C for 5 min. Cool down and hold at 4°C.
9. While the sample microtiter plate is incubating, begin preparing the hMC reaction cocktail.

3.6. Transcription

3.6.1. Preparing and Adding the Transcription / RNase A Cocktails

After adding SAP, the next step consists of the preparation of the T Cleavage transcription/RNase A and C Cleavage transcription/RNase A cocktails (*see* **Note 7**). In this step the 7-μL PCR/SAP reaction is split up into two new plates. Each plate receives a 2-μL aliquot of the PCR/SAP reaction. 5 μL of

Table 16.3
Preparation of a SAP master mix

Order	Reagent	Volume for single reaction (μL)	Volume for 384-well microtiter plate (μL)†
1	RNase-free ddH_2O	1.70	901
2	Shrimp alkaline phosphatase (SAP)	0.30	159
	Total volume	2.00	1,060

† Volumes for a 384-well microtiter plate include 38% overhang to account for possible pipetting loss.

transcription mix is then added to each well. The first of the two plates will receive the T cleavage mix, while the second plate gets the C cleavage mix. The leftover PCR/SAP reaction can be used to perform a gel check on a 1.5% agarose gel.

1. Prepare the T Cleavage transcription/RNase A cocktail by adding the reagents in the order of appearance in Table 16.4.
2. Prepare the C Cleavage transcription/RNase A cocktail by adding the reagents in the order of appearance in **Table 16.5**.
3. Remove PCR/SAP mix sample microtiter plate from the thermal cycler.
4. Centrifuge the sample microtiter plate and both C/RNase A and T/Rnase A plates at 540*g* for 1 min.
5. When complete, remove the plate-sealing film as each one is utilized.
6. Transfer 2 μL from each well of the PCR/SAP sample into each well of a new 384 microtiter plate. The plate layout shall be preserved (A1 to A1, H14 to H14, etc.).
7. Add 5.0 μL T cleavage mix solution to each well of the new 384 microtiter plate. To achieve more consistent results, it is recommended to use a multichannel pipette or a liquid-handling robot.
8. Seal the sample microtiter plate with plate-sealing film. Make sure the sealing plate film has a good seal along the edge of the plate.
9. Centrifuge the sample microtiter plate at 540*g* for 1 min.

Table 16.4
Preparation of the T cleavage master mix

Order	T cleavage transcription/ RNase A cocktail	Final concentration for single reaction (in 7 μL)	Volume for single reaction (μL)	Volume for one sample microtiter plate (μL)†
1	RNase-free ddH_2O	N/A	3.21	1,602
2	5 × T7 polymerase buffer	0.64 ×	0.89	444
3	T cleavage mix	N/A	0.22	110
4	100 mM DTT	3.14 mM	0.22	110
5	T7 RNA & DNA polymerase	0.09 mg/mL	0.40	200
6	RNase A		0.06	30
	Total volume		5.00	2,496

† Volumes are for a 384-well microtiter plate and include approximately 33% overhang to account for possible pipetting loss.

Table 16.5
Preparation of the C cleavage master mix

Order	T cleavage transcription/ RNase A cocktail	Final concentration for single reaction (in 7 μL)	Volume for single reaction (μL)	Volume for one sample microtiter plate (μL)†
1	RNase-free ddH_2O	N/A	3.21	1,602
2	5 × T7 polymerase buffer	0.64 ×	0.89	444
3	C cleavage mix	N/A	0.22	110
4	100 mM DTT	3.14 mM	0.22	110
5	T7 RNA & DNA polymerase	0.09 mg/mL	0.40	200
6	RNase A		0.06	30
	Total volume		5.00	2,496

† Volumes are for a 384-well microtiter plate and include approximately 33% overhang to account for possible pipetting loss.

10. Transfer 2 μL from each well of the PCR/SAP sample into each well of a second new 384 microtiter plate. The plate layout shall be preserved (A1 to A1, H14 to H14 etc.).
11. Add 5.0 μL of C cleavage mix solution to each well of the second new 384 microtiter plate. To achieve more consistent results it is recommended to use a multichannel pipette or a liquid-handling robot.
12. Seal the sample microtiter plate with plate-sealing film. Make sure the sealing-plate film has a good seal along the edge of the plate.
13. Centrifuge the sample microtiter plate at 540*g* for 1 min.
14. When the transfer of PCR/SAP mix into two new plates is done, the PCR/SAP microtiter plate can be stored or discarded (*see* **Note 8**).

3.6.2. Incubating the Transcription/RNase A Cocktails

1. After adding the transcription/RNase A cocktails (T Cleavage and C Cleavage) to the sample microtiter plates, the next step is to incubate the transcription/RNase A cocktails.
2. Incubate each sample microtiter plate at 37°C for 3 h.
3. Immediately process the plates after incubation. If unable to do so, then freeze them (at −20°C) and store them overnight. Do not store the plates at 4°C or room temperature.

3.7. Conditioning the hMC Reaction Products

After incubating the hMC reaction, the next step is conditioning of the hMC reaction products with Clean Resin. This conditioning step is important to optimize mass spectrometric analysis of the hMC reaction products. This process occurs in four steps.

3.7.1. Prepare a Plate of Clean Resin

1. Transfer Clean Resin from its container onto the 384-well Clean Resin plate using the spoon.
2. Spread into the wells of the Clean Resin plate using the Clean Resin scraper.
3. Scrape excess Clean Resin off the Clean Resin plate using the scraper. Return the excess Clean Resin to its container.
4. Let the Clean Resin stand in the Clean Resin plate for at least 20 min. While letting the Clean Resin stand in the plate, add water to the sample microtiter plate by performing the next procedure.

3.7.2. Add Water to the 384-Well Sample Microtiter Plates

1. Remove the sample microtiter plates from the thermal cycler when the incubation is finished.
2. Centrifuge the sample microtiter plates at 540*g* for 1 min.
3. Prepare a reservoir with at least 21 mL of nanopure water.
4. Remove the plate-sealing film of the T cleavage sample plate.
5. Add 20 μL of nanopure water from the reservoir to each well of the T cleavage sample plate.
6. Seal the sample microtiter plate with plate-sealing film. Make sure the sealing plate film has a good seal along the edge of the plate.
7. Centrifuge the sample microtiter plate at 540*g* for 1 min.
8. Remove the plate-sealing film of the C cleavage sample plate.
9. Add 20 μL of nanopure water from the reservoir to each well of the C cleavage sample plate.
10. Seal the sample microtiter plate with plate-sealing film. Make sure the sealing plate film has a good seal along the edge of the plate.
11. Centrifuge the sample microtiter plate at 540*g* for 1 min.
12. Proceed directly to the next step.

3.7.3. Add Clean Resin to the Sample Microtiter Plate

1. Place the sample microtiter plate, upside-down, onto the Clean Resin plate.
2. Holding the sample microtiter plate and the Clean Resin plate together, flip them over so the Clean Resin falls from the Clean Resin plate into the wells of the microtiter plate.
3. Tap the Clean Resin plate to help the Clean Resin fall into the microtiter plate.
4. Afterwards, examine the Clean Resin plate to make sure all the Clean Resin fell into the microtiter plate wells.
5. Seal the sample microtiter plate with plate-sealing film. Make sure the sealing plate film has a good seal along the edge of the plate.

3.7.4. Rotate and Centrifuge the hMC Reaction Products

1. Rotate the sample microtiter plates using a rotator for 10 min at room temperature. The rotator must rotate the microtiter plate 360° perpendicular to its long axis.
2. Centrifuge the sample microtiter plates at 3,200*g* for 5 min.
3. The hMC reaction products are now ready for transfer to a SpectroCHIP using the MassARRAY® Nanodispenser (*see* **Note 9**).

3.8. Sample Transfer on SpectroCHIP Array

For automated, high-performance analysis of nucleic acid mixtures by MALDI-TOF MS, the sample needs to be transferred from the microtiter plate format to a chip array. The chip array (SpectroCHIP®, SEQUENOM) provides miniaturized prefabricated arrays of matrix sample spots on a silicon chip. The miniaturization improves homogeneity of the sample and leads to increased performance in qualitative and quantitative analysis. While ample volume of sample is generated during the MassCLEAVE process, only 15 nL are required for the subsequent analysis. The corresponding volume of analyte is transferred onto the chip array. This is best performed using a piezoelectric-pump-based dispensing system able to transfer low sample volumes or a pintool system as provided by SEQUENOM (Nanodispensing device). This robotic system transfers analyte from 384-well microtiter plates onto 384-element chip arrays in 9 min. The chip array can be used immediately for MALDI-TOF MS analysis.

3.9. MALDI-TOF MS Analysis

Analysis of chip-transferred samples proceeds in a linear, delayed extraction time-of-flight (TOF) mass spectrometer. Mass spectra are acquired in positive ion mode (all positively charged molecular ions are accelerated). The chips are introduced into the ion source and high-vacuum conditions are applied. Image processing aligns the laser position automatically to the chip element raster for fully automated scanning of each chip position. Each matrix crystal is addressed individually and irradiated with a 337-nm laser pulse of 1 ns duration. The time-resolved mass spectrum is then translated into the mass spectrum by comparison with known calibration standards. Usually 15 single-laser shots are accumulated and averaged into a single spectrum. This average spectrum is then further processed and analyzed using dedicated software (EpiTYPER®, SEQUENOM) that performs baseline correction, peak identification, and quality assessments. Spectrum data quality is analyzed in real time during data acquisition. In case of insufficient mass-spectrum quality, the software will automatically reacquire new data points from the same chip position before it finally moves to the next chip position (*see* **Note 10**). Once all data is acquired, it is saved to an Oracle database. The data can be assessed using client analysis software.

3.10. Data Analysis

The resulting data can be viewed in the EpiTYPER® desktop software. The software will automatically calculate the amount of methylated DNA for each CpG position and displays the results in a table. For better visualization, a graphical representation of the results is also provided. The so-called Epigram figures present the methylation results as colored circles (**Fig. 16.6**). For each sample, a line is drawn that reflects the amplification length and colored circles are positioned to reflect the position of the CpG site within the amplification target. The color with the circle is representative for the methylation status at this position. Different color schemes can be chosen (e.g., yellow indicating nonmethylated DNA and blue indicating methylated DNA). Typically, the results from a large-scale methylation experiment are analyzed using third-party statistical packages (i.e., the R environment for statistical computing, an example is given in **Fig. 16.7**). The build in data export functionality allows formatting the results in a format that can easily be accessed by the aforementioned software packages.

3.11. Validation of Selected CpG Postions by Primer Extension Methods

Details of primer extension based methods and their readout by MALDI-TOF MS have been described in detail in earlier issues of Methods in Molecular Biology™ *(13, 14)*. The Notes section

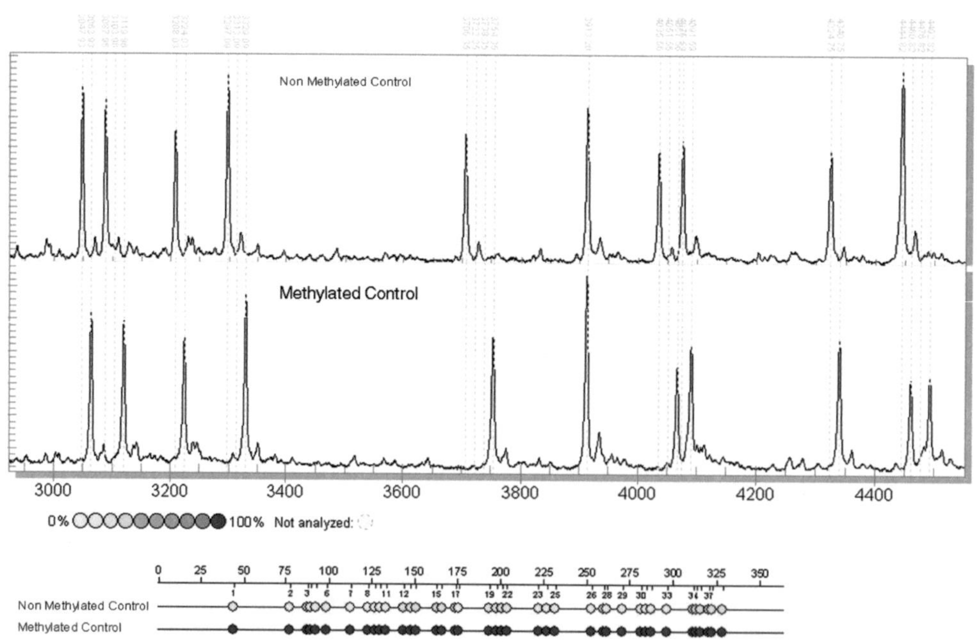

Fig. 16.6. The upper panel shows mass spectra derived from nonmethylated DNA (*top*) and fully methylated DNA (*bottom*). While the start tag, which does not contain CpG residues, remains unchanged at 3,913 Da, all other signal are shifted by 16, 32, or 48 Da indicating the presences of one, two, or three methylated CpG sites, respectively. The *lower panel* shows how this data is visualized in an Epigram (see text). (*see* Color Plate 6)

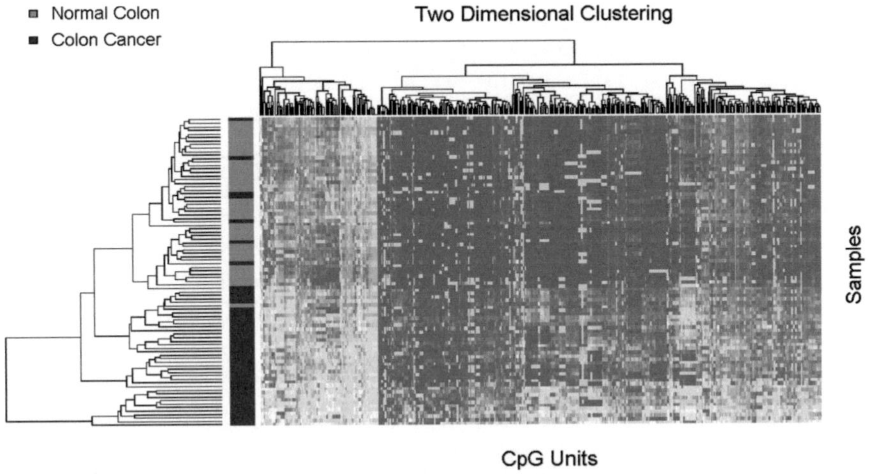

Fig. 16.7. We analyzed CpG island promoter methylation in 48 pairs matched colon cancer and adjacent normal tissue samples. The methylation data was filtered to exclude poor quality data (less than 90% of all data available) and then used in a two-dimensional hierarchical clustering algorithm. This figure presents the results in a pseudo color image (heatmap), where red indicates no methylation and highly methylated CpG sites are shown in yellow. The type of tissue samples is marked as a colored sidebar on the left side of the figure (*orange* = normal colon tissue, *blue* = colon cancer tissue). The results indicate generally higher methylation levels in the group of colon cancer tissues, which leads to a separation of the two groups in the unsupervised clustering. (*see* Color Plate 7)

provides a number of tips and tricks above and beyond what is described in the scientific literature (*see* **Notes 11–14**).

4. Notes

1. Getting enough sample material to have 1 μg genomic DNA available for bisulfite treatment is often challenging. Although smaller DNA amounts can be used for bisulfite treatment, the recovery of DNA is depending on large-enough elution volumes during the final cleanup step. In our hands, elution volumes around 100 μL work best. During PCR amplification we recommend using 10 ng bisulfite-treated DNA (~3,000 genomic copies, see also **Note 4**). In the standard 5-μL PCR the maximum volume of DNA solution that can be used is 2.44 μL. Consequently we recommend using more than 400 ng during bisulfite treatment.
2. The reproducibility of replicated measurements is mainly a function of DNA molecules that are available for PCR amplification. How many molecules are available depends on two main factors: (1) quality of the bisulfite-treated DNA and (2) amplification length. Because the chemical conversion of DNA during bisulfite treatment introduces random strand breaks, not all input DNA molecules are available for amplification. The random nature of this degradation also

implies that it is less likely to find intact DNA molecules for longer amplification lengths. When in doubt about the DNA quality we recommend to perform a quality check of the bisulfite-converted DNA. A set of amplicons of increasing length covering the same genomic area (preferably hemimethylated regions) is used in replicate measurements. At a certain amplification length, the variance of replicated measurements will show a drastic increase, indicating that at this length only few molecules are available for amplification. Target amplicons should be designed to match this upper limit *(18)*.

3. When analyzing methylation patterns in the promoter region of target genes we find high regional variability in about 20% of the cases. For biomarker discovery studies, where the methylation behavior of a target region is unknown, we recommend to design overlapping amplicons that cover the entire promoter region ± 200 bp.
4. To allow unbiased amplification from bisulfite-treated DNA, the amplification primers have to be void of any CpG residues. Consequently, the PCR primers contain only three different nucleotide bases, which greatly reduce the hybridization specificity compared to genomic DNA. Thus, co-amplification of undesired targets and/or unconverted genomic DNA is a commonly observed problem. Co-amplification of a second target manifests with additional signals in the mass spectrum, which are unaccounted for by the software. To avoid amplification of unconverted genomic DNA, it is recommended to include multiple cytosine residues, which are not located in CpG dinucleotides (ideally more than four). Also when possible it is preferable to chose primer pairs where the left and right primers are of the same length or where the left primer exceeds the right primer by one or two bases.
5. Before incubating the cocktail, review the SEQUENOM Primer Design software for methylation analysis (www.sequenom.com) to determine the correct annealing temperature to heat the cocktail in the 45-cycle segment of this incubation procedure. Adjust annealing temperature according to melting temperature of the primers. It is recommended that PCR incubation conditions be optimized to obtain a defined, single band on the agarose gel (without any additional nonspecific bands). A PCR product free of by-products is important to obtain clean MassCLEAVE™ spectra.
6. When working with the SAP enzyme please note that it should not be left in water for more than 8 h. Therefore, the SAP enzyme solution should be prepared immediately before adding it to the reactions.

7. The C-specific cleavage reaction provides highly accurate results for selected CpG sites. C-specific cleavage on the reverse strand is equivalent to G-specific cleavage on the forward strand, which means that each CpG sites gets cleaved in the middle. Hence, a cleavage fragment can only contain one cytosine residue (A or G on the reverse strand). Although this cleavage pattern provides very specific results for individual CpG sites, its usability is limited by low CpG coverage. Because CpG islands tend to be rich not only in C's but also in G's the C-specific cleavage reaction produces many typically short fragments. These short fragments might be too short to be analyzed or their base composition is identical to another short fragment. In both of those instances, the quantitative analysis of CpG methylation is impaired and should be avoided. Therefore, it is recommended to use the C-specific cleavage reaction after *in silico* modeling of the cleavage pattern and checking whether the desired answers can be obtained.
8. Leftover PCR can be used for further analysis or for gel check. If desired, apply 0.5 μL of PCR product on 1.5% agarose gel to confirm successful PCR amplification and amplification specificity.
9. If transfer of the hMC reaction products to a SpectroCHIP® array is not possible, then store the microtiter plate at −20°C until ready to do so. Before storing each microtiter plate, use adhesive sealing foil on its edges to make sure that there is a good tight seal. Do not store the microtiter plate for more than 2 weeks. After storing a microtiter plate of hMC reaction products, thaw and centrifuge the microtiter plate for 3 min at 540*g* before transferring the reaction products to a SpectroCHIP array.
10. The automated quality control of the analysis software does not work well when the PCR reaction creates a lot of primer dimer product. The amplification product that results from such cross-hybridization of primers contains the T7 promoter sequence, and consequently will result in transcription and cleavage products. The expected cleavage products for the beginning and the end of the amplicon sequence will be generated and therefore recognized as matching. These matching signals might be enough to suggest to the software that the spectra quality is sufficient for analysis. We recommend establishing the assay in well-known control DNAs and excluding assays that yield strong primer dimer products (**Fig. 16.8**).

4.1. Additional Notes: Primer Extension

11. PCR primer design: Design of PCR primers should follow the same rules described above for base-specific cleavage. The specific targeting of only individual CpG sites will allow

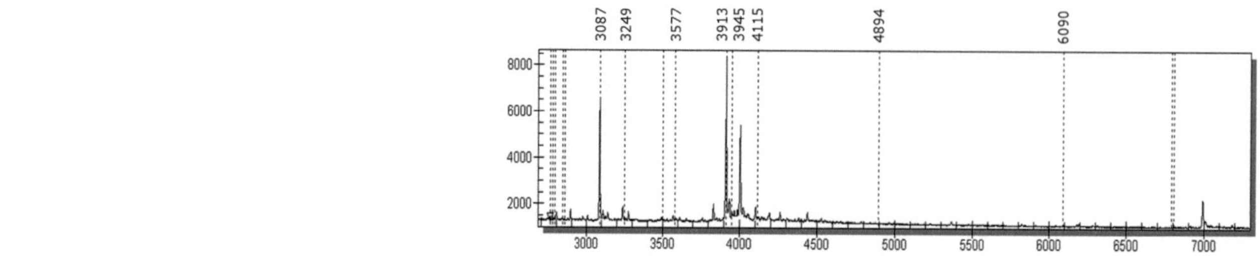

Fig. 16.8. Shown is mass spectrum that results from the formation of primer dimers during PCR amplification. The *dotted lines* indicate where mass signals are expected based on the input sequence. The only expected signals can be found at 3,087 and 3,913 Da. Both of these signals are derived from the primer component of the amplified target sequence. Signals that would be a result from amplification of bisulfite-treated genomic DNA (3,249, 3,577, 3,945, 4,115, 4,894, 6,090, and others) cannot be observed. Furthermore, two unexpected mass signals (around 4,000 and 7,000 Da) are found, which are likely generated by unexpected amplification products resulting from primer dimer formation.

fordesign of shorter amplicons, which could be of additional benefit for methylation analysis of genomic DNA derived from paraffin-embedded tissue and for methylation analysis of circulating nucleic acids in blood/plasma. A respective website for available design software is listed in 3.2.

12. Design of extension primers: It is preferred to design the extension primers toward CpG sites in regions with few CpG sites in close proximity. In very CpG-rich regions, however, it may not be possible to place an extension primer adjacent to a CpG site without the primer overlapping with neighboring CpG sites. To avoid introducing a quantitation bias by selective hybridization, these extension primers should utilize degenerate nucleotides in the respective CpG positions not targeted by the extension reaction.
13. Post-PCR primer extension cycling profile: Due to the conversion of unmethylated cytosines to uracil sequences preceding CpG sites are usually very A/T rich. The primer/template complexes are therefore more labile and high annealing temperatures as well as high temperatures in the extension step should be avoided. The post-PCR primer extension step can usually be run as a two-step profile consisting of a denaturation and combined annealing/extension step.
14. Data analysis: As described for the analysis of base-specific cleavage, the peak area ratio of mass signals representing methylated and nonmethylated CpG sites should be used as a quantitative measure of the degree of methylation. Mass signal-related information such as peak area and peak height as well as calculated frequencies can be exported from the data acquisition/analysis tools in standard formats on an assay and sample-specific basis and thus are available

for further statistical analysis using public domain software. Sample and assay-based quality filters should be determined based on each full data set, for example, one would carefully screen results at the limit of detection and limit of quantitation of the method (5% methylated or nonmethylated DNA, respectively) to reduce over interpretation of spurious results. Since the biggest variance in the overall methylation analysis process is introduced during bisulfite treatment and PCR amplification, we recommend replication of experiments in the PCR amplification step to validate results. Replication at the bisulfite treatment step could significantly deplete the genomic DNA stock.

References

1. Hartmer, R., Storm, N., Boecker, S., et al. (2003) RNase T1 mediated base-specific cleavage and MALDI-TOF MS for high-throughput comparative sequence analysis. *Nucleic Acids Res* 31, e47.
2. Ehrich, M., Nelson, M. R., Stanssens, P., et al. (2005) Quantitative high-throughput analysis of DNA methylation patterns by base-specific cleavage and mass spectrometry. *Proc Natl Acad Sci USA* 102, 15785–15790.
3. Schatz, P., Dietrich, D., Schuster, M. (2004) Rapid analysis of CpG methylation patterns using RNase T1 cleavage and MALDI-TOF. *Nucleic Acids Res* 32, e167.
4. Jurinke, C., Denissenko, M. F., Oeth, P., et al. (2005) A single nucleotide polymorphism based approach for the identification and characterization of gene expression modulation using MassARRAY. *Mutat Res* 573, 83–95.
5. Jurinke, C., Oeth, P., Van Den Boom, D. (2004) Maldi-tof mass spectrometry: a versatile tool for high-performance DNA analysis. *Mol Biotechnol* 26, 147–164.
6. Tost, J., Schatz, P., Schuster, M., et al. (2003) Analysis and accurate quantification of CpG methylation by MALDI mass spectrometry. *Nucleic Acids Res* 31, e50.
7. Gut, I. G. (2001) Automation in genotyping of single nucleotide polymorphisms. *Hum Mutat* 17, 475–492.
8. Tost, J., Gut, I. G. (2005) Genotyping single nucleotide polymorphisms by MALDI mass spectrometry in clinical applications. *Clin Biochem* 38, 335–350.
9. Tost, J., Gut, I. G. (2002) Genotyping single nucleotide polymorphisms by mass spectrometry. *Mass Spectrom Rev* 21, 388–418.
10. Jurinke, C., van den Boom, D., Cantor, C. R., et al. (2001) Automated genotyping using the DNA MassArray technology. *Methods Mol Biol* 170, 103–116.
11. Jurinke, C., van den Boom, D., Cantor, C. R., et al. (2002) The use of MassARRAY technology for high throughput genotyping. *Adv Biochem Eng Biotechnol* 77, 57–74.
12. Rodi, C. P., Darnhofer-Patel, B., Stanssens, P., et al. (2002) A strategy for the rapid discovery of disease markers using the MassARRAY system. *Biotechniques* Suppl, 62–69.
13. Storm, N., Darnhofer-Patel, B., van den Boom, D., et al. (2003) MALDI-TOF mass spectrometry-based SNP genotyping. *Methods Mol Biol* 212, 241–262.
14. Wjst, M., van den Boom, D. (2005) Matrix-assisted laser desorption/ionization time-of-flight mass spectrometry. *Methods Mol Biol* 311, 125–138.
15. Bansal, A., van den Boom, D., Kammerer, S., et al. (2002) Association testing by DNA pooling: an effective initial screen. *Proc Natl Acad Sci USA* 99, 16871–16874.
16. Mohlke, K. L., Erdos, M. R., Scott, L. J., et al. (2002) High-throughput screening for evidence of association by using mass spectrometry genotyping on DNA pools. *Proc Natl Acad Sci USA* 99, 16928–16933.
17. Werner, M., Sych, M., Herbon, N., et al. (2002) Large-scale determination of SNP allele frequencies in DNA pools using MALDI-TOF mass spectrometry. *Hum Mutat* 20, 57–64.
18. Ehrich, M., Zoll, S., Sur, S., et al. (2007) A new method for accurate assessment of DNA quality after bisulfite treatment. *Nucleic Acids Res* 35, e29.

Chapter 17

Melting Curve Assays for DNA Methylation Analysis

Tomasz K. Wojdacz and Alexander Dobrovic

Abstract

The ability of sodium bisulfite to modify cytosines in a methylation-dependent manner allows the conservation of DNA methylation information during PCR amplification. PCR products amplified from bisulfite-modified DNA have significantly different base compositions according to whether they originate from methylated or unmethylated variants of the target template. Different base compositions give rise to different thermal properties of the PCR products. Hence, melting analysis of amplification products in methylation studies allows the determination of whether the PCR products originate from methylated or unmethylated templates. Here, we briefly review recent advances in methodologies based on melting analyses of PCR products derived from bisulfite-modified templates and provide a methodology for methylation-sensitive high-resolution melting.

Key words: Methylation, melting curve, sodium bisulfite, high-resolution melting, PCR bias, Methylation-sensitive high-resolution melting (MS-HRM).

1. Introduction

The introduction of bisulfite modification of genomic DNA enabled the general use of PCR amplification in methylation studies *(1)*. Sodium bisulfite deaminates unmethylated cytosines to uracils leaving 5-methylcytosines intact. As a consequence, methylated cytosines are amplified during the subsequent polymerase chain reaction (PCR) as cytosines whereas unmethylated cytosines are amplified as thymines. Hence, the base composition of the PCR product depends on the 5-methylcytosine content of the template.

The two complementary strands of DNA are held together by hydrogen bonds and stacking interactions. Dissociation of double-stranded DNA is known as DNA melting or denaturation

Jörg Tost (ed.), *DNA Methylation: Methods and Protocols, Second Edition, vol. 507*

DOI 10.1007/978-1-59745-522-0_17 Springerprotocols.com

and can be induced either by increased temperature or denaturing chemicals. The dissociation of the triple hydrogen bond between C and G requires more energy than the dissociation of the double hydrogen bond between T and A, therefore GC-rich sequences melt at relatively higher temperatures compared to AT-rich sequences.

The melting of a DNA amplicon often consists of a series of progressive dissociations of shorter domains within the sequence, which due to their local GC content have different melting temperatures. The sequence-dependent melting of an amplicon across a denaturing gradient is described as a "melting profile".

The melting profile of an amplicon can be determined by subjecting it to a gradually increasing temperature in the presence of an intercalating fluorescent dye, which emits fluorescence when bound (intercalated) to double-stranded DNA. At temperatures below the initiation of the melting process, the intercalating dye saturates the PCR product resulting in high levels of fluorescence. As the temperature rises, the fluorescence levels are stable until the point at which the double-stranded amplicon begins to separate into single strands and a marked drop in fluorescence is observed as the dye is released from the double-stranded DNA. Thus, by monitoring the fluorescence during the increase of the temperature, it is possible to determine the melting profile of a PCR amplicon.

The melting profiles of PCR products originating from methylated and unmethylated variants of the same template are significantly different due to their different GC content. Therefore, the methylation status of an unknown sample can be determined by comparing the melting profile of the sample PCR product with the melting profiles of PCR products obtained from the amplification of methylated and unmethylated control templates.

The first application of melting curve analysis to the profiling of DNA methylation was reported in 2001 *(2)*. This methodology was not generally adopted, as SYBR Green I, the principal fluorescent dye used at that time, could not be used at concentrations that fully saturated the PCR products, thereby blunting the resolution of the methodology.

Current advances in fluorescence detection technology, new algorithms for data calculation, and the use of novel dyes have allowed the development of high-resolution melting analysis (HRM) *(3)*. We applied HRM to methylation analysis in the methodology that we called methylation-sensitive high-resolution melting (MS-HRM). Together with a novel approach to primer design (*see* **Section 3.1**), which allows the control of PCR bias and hence highly sensitive detection of low-level methylation, MS-HRM is proving to be a powerful new methodology for methylation analysis.

2. Materials

2.1. Bisulfite Modification of Genomic DNA

As an alternative to the procedure described in this chapter, some users may prefer to use one of the kits currently available from commercial suppliers.

1. 3 M sodium bisulfite (Sigma S9000): Dissolve 0.57 g of sodium bisulfite in 1 mL of water (this step may take several minutes as sodium bisulfite does not dissolve easily).
2. 10 mM hydroquinone (Merck 8.22.333.0250): Prepare a 40 mM solution of hydroquinone: by dissolving 0.132 g of hydroquinone in 30 mL of water, and dilute it to 10 mM solution (10 mL of 40 mM hydroquinone solution plus 40 mL of water). Note that hydroquinone should be handled with care.
3. 3 M NaOH and 0.1 M NaOH: The 0.1 M NaOH should be made from the 3 M NaOH stock on the day of the procedure. Discard the 3 M NaOH stock when it becomes cloudy.
4. Microcon YM-100 centrifugal filter unit (Millipore).
5. Eppendorf microcentrifuge 5417R.
6. TE buffer: 10 mM Tris–HCl, 1 mM EDTA, pH 8.0.

2.2. Instrumentation

A platform with a combined thermal cycler and a fluorescence detector is ideal to perform in-tube melting analyses. PCR amplification of bisulfite-modified template could be performed prior to melting analyses on any thermocycler, but real-time monitoring of PCR amplification constitutes an important quality control step in the experiments, allowing the elimination of samples where amplification fails, without the need to run gel electrophoresis.

The first experiments using melting profiles to differentiate methylated and unmethylated PCR products were performed on the Lightcycler® 2.0 (Roche, Penzberg, Germany) *(2)*. Recently, a new generation of instruments capable of HRM analyses has been developed with superior data capture abilities and improved software. In our laboratory, we have used two HRM systems: the Roche Lightcycler® 480 and the Corbett RG600 (Corbett Life Science, Sydney, Australia) as their real-time PCR capacity allows the monitoring of the amplification which is invaluable for quality control.

2.3. DNA Saturating Dyes

The very first intercalating dye reported to be suitable for melting-based analyses of methylation was SYBR Green I *(2)*. However, SYBR Green I at the concentration allowing saturation of PCR product was toxic to all the polymerases we have tested (unpublished data). PCR amplifications in the presence of SYBR Green I were not robust and the very low yield of PCR products for post-PCR melting analyses compromised our experiments. The new generation of saturating dyes includes the LC Green family (Idaho Technology Inc., Salt Lake City, UT), Syto®9 (Invitrogen, Carlsbad, CA), ResoLight (Roche), and Eva Green (Biotium Inc., Hayward, CA). These dyes fully saturate the PCR product at concentrations that do not inhibit PCR amplification and allow for both real-time monitoring of PCR amplification and subsequent in-tube melting analyses. These dyes can be used with most of the commercially available polymerases as an additive to the PCR reaction mix. The rapid development of HRM has resulted in the introduction of high-resolution melting master mixes e.g. LightCycler® 480 High Resolution Melting Master (Roche) and SensiMixHRM™ (Quantace Ltd., London, UK).

3. Methods

3.1. Sodium Bisulfite Treatment of Genomic DNA

Bisulfite conversion of genomic DNA consists of three steps: sulfonation of cytosines, hydrolytic deamination of the cytosine sulfonates to uracil sulfonates, and alkaline desulfonation of the uracil sulfonates. Many parameters have to be taken into account during the performance of the above steps and therefore many protocols have been published with different conditions for each step of conversion. In our experience, the most important parameters of the conversion are initial (sufficient) denaturation of the DNA template, incubation time with sodium bisulfite mixture (we have observed few problems with incomplete conversion when incubation times of 16 h were used) and the method used to recover bisulfite-modified DNA that gives the highest yield possible. A protocol utilizing a column-based recovery procedure *(4)* showed the highest recovery rates. Our modified version of that protocol was published in *(5)* and is described below.

This operation should preferably be carried out in a safety cabinet or fume hood.

1. Mix 0.1–1 μg of genomic DNA with water to a final volume of 16 μL.
2. Add 1.1 μL of 3 M NaOH to denature the DNA, and incubate at 37°C for at least 15 min. Then place directly on ice and proceed promptly with the remainder of the protocol.
3. Add 173 μL of freshly prepared 3 M sodium bisulfite and 10 μL of the hydroquinone solution, mix with the pipette, and incubate in the dark for 16 h (overnight) at 50°C.
4. Dilute the bisulfite reaction with water to a total volume of 350–400 μL.
5. Transfer this solution to an assembled Microcon YM-100 centrifugal filter unit (Millipore).
6. Centrifuge at 700*g* for 10 min.

7. Discard the filtrate and add 300 μL of water to the upper chamber and centrifuge for 10 min at 700*g*.
8. Repeat step 7.
9. Discard the filtrate, add 350 μL of 0.1 M NaOH to the upper chamber, and centrifuge for 6 min at 700*g*.
10. Discard the filtrate, add 350 μL of water to the upper chamber, and centrifuge at 700*g* for 8 min.
11. Elute the sample by adding 50 μL of TE buffer; use the pipette for mixing of TE with the sample on the column membrane, and let stand for 15 min.
12. Invert the device and collect the bisulfite-converted DNA in a clean tube (*see* **Note 1**).

3.2. Primer Design for PCR Amplification

The primers for amplification of bisulfite-modified DNA for MS-HRM studies have to amplify the sequence of interest regardless of its methylation status (MIP – methylation independent primers) *(6)*.

It is important to note that when using one primer set for the amplification of two templates with different GC content, the PCR will be biased toward the template with lower GC content (**Fig. 17.1**) *(5, 7)*. The most commonly followed rules in primer design advise not to include any CG nucleotides into the primer sequence and, if this is not possible, to mismatch the C from CG with T *(8)*. In our experience, following the above rules frequently led to amplifications which showed strong bias toward the unmethylated sequence *(5)*.

We proposed new guidelines for methylation-independent primer (MIP) design where inclusion of a limited number of CGs toward the 5′ end of the sequence allowed us to manipulate PCR bias (*5, 6,* **Fig. 17.1**). When primers contain a limited number of CpGs are used, their relative binding affinity for methylated and unmethylated templates is dependent on the annealing temperature. Hence, by manipulating the annealing temperature of the PCR amplification with MIP primers, one can shift the PCR bias from the unmethylated to the methylated sequence and make the assay highly sensitive for methylation detection. If there is no methylated template available, the primers with limited CpGs will amplify unmethylated sequence even at

GENOMIC SEQUENCE (forward strand)
CGTTTGCGACTTGGTGAGTGTCTGGGTCGCCTCGCTCCCGGAAGAGTGCGGAGCTCTCCCTCGGGACGGTGGCAGCCTCGAGTGGTCCTGCAGG
METHYLATED STRAND
CGTTTGCGATTTGGTGAGTGTTTGGGTCGTTTCGTTTTCGGAAGAGTGCGGAGTTTTTTTCGGGACGG**TGGTAGTTTCGAGTGGTTTTGTAGG**
UNMETHYLATED STRAND
CGTTTGTGATTTGGTGAGTGTTTGGGTTGTTTTGTTTTTGGAAGAGTGTGGAGTTTTTTTTTGGGATGG**TGGTAGTTTTGAGTGGTTTTGTAGG**

Fig. 17.1. The example of primer design for MS-HRM assay targeting the promoter of the *MGMT* gene (chr10:131, 155, 538-131, 155, and 631 by UCSC Genome Browser, March 2006). The primers were designed to include a limited number of CpGs that allowed the control of PCR bias during PCR amplification. Bold print – primer binding sites, underlined – CpGs.

high annealing temperature. This allows not only the determination of the methylation status of the sample but also, at the same time, the confirmation of the unmethylated status of the template. Apart from this point, good practice for primer design, such as matching primer melting temperatures, selection against unconverted sequences (inclusion of a T derived from a non-CpG at or near the 3′ end of the primer) and prediction of a low likelihood of primer dimer formation should be followed for the design of MIP primers. The melting temperature of MIP primers should be around 65°C (e.g., using Oligonucleotide properties calculator http://www.basic.northwestern.edu/biotools/oligocalc.html for calculation of the melting temperature of the primers) allowing an annealing temperature of 60°C, which is an empirical value at which our primer design gives good results in our hands. Each primer set should be extensively tested on the mixes of methylated and unmethylated controls for the performance and the extent of PCR bias (*see* **Note 2**).

3.3. Predicting the Melting Behavior of the Sequence of Interest

Long sequences can have very complex melting profiles due to the fact that they consist of multiple small melting domains. Ideally, the melting profile of the sequence of interest should consist of a single melting domain that gives only single drop in the fluorescence and makes the melting results easy to interpret. Amplicons of 100 bp or less often comprise only one major melting domain and therefore the recommended length of the PCR amplicon for melting analyses is around this size. A rough estimate of the melting profile of the sequence of interest can be made on freely available tools like POLAND: http://www.biophys.uni-duesseldorf.de/local/POLAND//poland.html, or the MELT94 algorithm (available from http://web.mit.edu/osp/www/melt.html).

The complex melting profile of a PCR product containing more than one melting domain may complicate melting analyses as drops in fluorescence observed for short melting domains within the sequence can mask the fluorescence changes specific for the methylated/unmethylated alleles.

3.4. Predicting the Melting Temperature of Methylated/Unmethylated PCR Amplicons

The tools described in **Section 3.3** can also be used to estimate the melting temperature of methylated and unmethylated variants of the template. Sequences originating from methylated alleles, where the C within CpG dinucleotides has not been changed, will have a higher melting temperature in comparison to sequences originating from unmethylated allele where all the C were changed into Ts. The difference in melting temperature of unmethylated and methylated templates depends on the number and density of CpGs in the amplicon. It is important to make sure that the difference in melting temperature between the methylated and unmethylated versions of the sequence is large enough to unambiguously distinguish both alleles during melting. When low-quality/resolution fluorescence detection systems are used the melting temperature differences between methylated and unmethylated PCR product has to be significant. With HRM systems, very small differences in melting temperature can be unambiguously resolved as the specification of the systems allow for highly sensitive fluorescence acquisition and precise control of temperature ramp rates (*see* **Note 3**).

3.5. PCR Amplification

PCR reagents from different suppliers differ in their ability to amplify bisulfite-modified DNA (unpublished data). The differences tend to be polymerase dependent and good-quality polymerases are necessary for successful PCR amplification of bisulfite-modified template. Hot Start protocols give superior results to regular amplification techniques when bisulfite-modified DNA is used as a template (*see* **Note 4**), (unpublished data). 0.5–1 U polymerase with standard dNTP concentration (200 μM each) is sufficient for most applications. The role of Mg^{+2} concentration and input template amount will be discussed separately

3.5.1. Mg^{+2} Concentration

The amplification of bisulfite-modified DNA is rarely as robust as the amplification of genomic DNA, presumably due to the degradation of DNA during incubation with bisulfite. As a consequence, the use of high concentrations of Mg^{+2} to enhance PCR amplification is essential in many cases. The Mg^{+2}concentration in polymerase buffers is generally insufficient to give a high yield of PCR product from bisulfite-modified DNA. In our experience, most amplifications for melting analyses require a Mg^{+2} concentration of 2.5–3.5 mM. In general, the concentration of Mg^{+2} in amplifications of bisulfite-modified template has to be empirically adjusted for each assay and is dependent on the bisulfite-modification protocol, the PCR reagents used and input of bisulfite-modified template.

3.5.2. Bisulfite Template Input for PCR Amplification

The main problem in methylation studies is the degradation of DNA during bisulfite modification, and as high as 90% degradation of the template has been reported *(9)*. Our experience shows that the greater the quantity of DNA that is subjected to bisulfite modification the less the extent of degradation and more template available for PCR amplifications. In cancer research, especially, the amount of available sample DNA is generally a limiting factor in methylation studies (*see* **Note 5**). The use of carrier DNA (e.g., herring or salmon sperm DNA) can partly solve the problem allowing for higher recovery rates when small amounts of genomic DNA are subjected to bisulfite modification. The sensitivity of the melting assay is directly correlated to the input DNA for bisulfite modification. With MS-HRM, we were able to reproducibly detect 1–0.1% methylated sequence in unmethylated background when 1 μg genomic DNA was used for bisulfite modification. Post-modification DNA recovery rates also depend on the system used to purify DNA. In our experience, procedures using purification columns are superior to precipitation protocols, especially when an inexperienced person performs the bisulfite modification.

3.5.3. PCR Cycling Parameters

The PCR amplification of bisulfite-modified DNA may require up to 50–60 cycles of amplification to obtain a sufficient yield of PCR product for melting analyses. The number of cycles depends on the assay and real-time monitoring of the amplification allows for the adjustment of the cycles number. The PCR should be stopped where possible just before amplification reaches the plateau phase. When amplification is carried on too long, the by-products of PCR may disturb the melting profile of the sequence of interest (*see* **Note 6**).

For an example, *see* **Note 7**.

3.6. Re-Annealing of PCR Product and the Design of Temperature Gradient for Melting Analyses

Before subjecting PCR products to temperature gradients, the PCR product should be re-annealed. The protocol for re-annealing of the PCR product should consist of a denaturation step at 95°C for 1 min and fast cool down time and hold for 1 min (to allow re-annealing of all the DNA strands) at the temperature from which acquisition of the fluorescence for melting procedure starts. The range of temperatures used in the temperature gradient has to be investigated empirically and it depends on the melting temperatures of unmethylated and methylated PCR product. The gradient has to fully cover the melting temperatures of methylated and unmethylated PCR products (*see* **Notes 3, 6 and 7**).

3.7. Acquisition of the Fluorescence

The precise and accurate acquisition of fluorescence, and small temperature transition rates are the foundations of high-resolution melting analysis. On the first generation of LightCyclers, the temperature ramp rates could be as low as 0.05°C/s. This allowed, in combination with continuous fluorescence acquisition, the attainment of detailed melting curves. Fluorescence acquisition systems in the new generation of the instruments have even higher specifications allowing collection of higher quality data.

We have tested two of the HRM instruments available on the market, the LC480 (Roche) and the Rotor-Gene RG-6000 (Corbett). The settings for data collection in melting experiments giving us satisfactory results for the LC480 system were 50 fluorescence acquisition points per degree centigrade. The corresponding temperature ramp rate for 50 acquisitions per degree was calculated automatically by the instrument by taking into account the time needed at each degree for sufficient acquisition of fluorescence. For the RG-6000, the HRM default settings were used consisting of the continuous acquisition of the fluorescence with a temperature ramp rate of 0.1°C and a 2 s hold on each step. The parameters of the melting gradient can be adjusted individually for each melting assay on both the LC480 and the RG-6000 (*see* **Note 7**).

3.8. Analysis of the Results

3.8.1. Derivative Peak

Melting curves are generated by continuous acquisition of fluorescence from the samples subjected to the linear temperature gradient. For basic analyses, the melting curves can be converted to peaks by plotting the negative derivative of fluorescence over the temperature (–d*F*/d*T*) versus temperature (**Fig. 17.2**). The top of the peak represents the highest drop of fluorescence during melting and can be interpreted as the melting temperature of the PCR product (Tm). In methylation studies, two peaks, one for unmethylated (lower Tm) and one for methylated (higher Tm), are obtained from the control samples. Comparing the peaks of an unknown sample with controls scores the methylation status of the unknown sample (*see* **Note 1**). Heterogeneously methylated templates give a broader peak due to the formation of heteroduplexes (*see* **Note 8**).

3.8.2. Direct Visualization of Melting

The high-resolution data collected on HRM instruments allowed the development of new algorithms for melting curve analyses. After PCR amplification, even replicates of the same sample can differ in the amount of PCR product amplified and therefore display different fluorescence levels. The differences between samples are especially pronounced when HRM detection is used, which does not allow for direct comparison of the curves as the

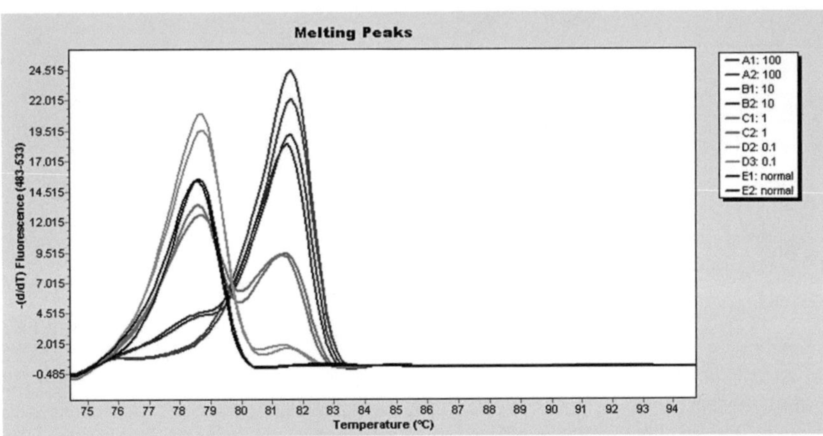

Fig. 17.2. Melting peak analyses (first derivative of the melting curves) of the *MGMT* gene. Details of the MS-HRM assay are given in *(6)* and **Note 7**. The melting curves were derived from samples with known methylated to unmethylated template ratios (ranging from 100% to 0.1%) of fully methylated template diluted in unmethylated template. The experiments were performed on the LC480. (*see* Color Plate 8)

HRM curves have to be normalized for starting and ending levels of fluorescence to make them comparable (**Fig. 17.3a**). The new algorithms supplied with HRM instruments allow for normalization of the starting levels of fluorescence. After normalization, the similarly shaped curves, which were visually not readily comparable before, group together. If unknown samples are run with standards of known methylated to unmethylated template ratios, the level of methylation in unknown samples can be estimated by comparing their melting curves to the melting curves of standards (**Fig. 17.3b**, *see* **Notes 2 and 6**).

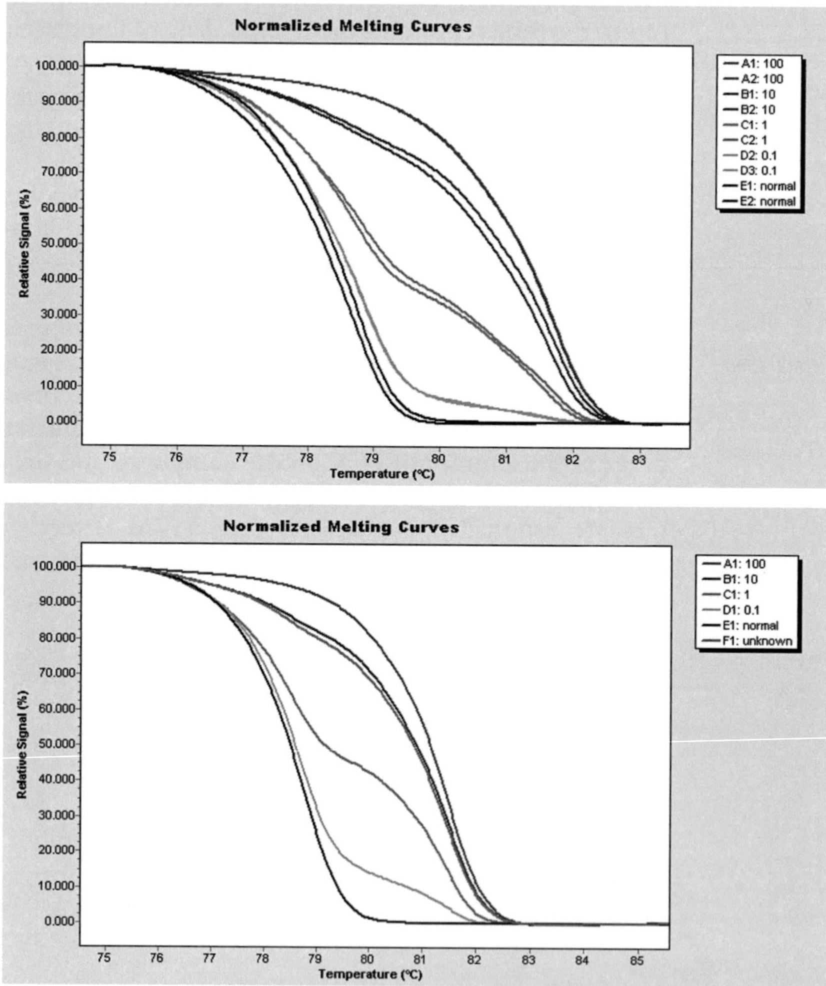

Fig. 17.3. (**a**) An example of "Gene Scan" analyses (*see* **Section 3.8.1**) of melting curves of the *MGMT* MS-HRM assay (*see* **Note 7**), in which melting curves derived from mixtures of methylated and unmethylated template were normalized for input fluorescence. The use of normalization allows similarly shaped curves to be grouped together. (**b**) An example of the estimation of the methylation levels of an unknown sample (*red*) on the basis of the similarity of its normalized melting profile to normalized melting profiles of standards of known methylated to unmethylated template ratios. The unknown sample shows methylation level at around 10% as its normalized melting profile is similar to the melting profile of the standard with 10% methylated template. (*see* Color Plate 9)

4. Notes

1. Bisulfite-modified DNA should be used up as soon as possible after modification. For longer storage: aliquot the stock DNA into small amounts that are thawed for each run. Repetitive freezing and thawing of the bisulfite DNA hastens its degradation.
2. Standards for MS-HRM analyses/primer optimization can be obtained by mixing of bisulfite-modified methylated and unmethylated controls. An equal amount of DNA has to be used for bisulfite modification of controls prior to mixing. Unmethylated control can be DNA from any tissue where the locus of interest is not methylated. The methylation status of the locus of interest in control DNA has to be investigated prior to analyses as some loci have different methylation status in different tissues. We routinely use CpGenome^{TM} Universal Methylated DNA (Millipore Co.) for the methylated control. *Sss*I-treated DNA is an alternative to CpGenome^{TM} Universal Methylated DNA, but complete methylation of the genomic DNA by *Sss*I enzyme can be difficult to achieve.
3. The numerical calculations of the melting temperatures of the methylated and unmethylated PCR products are a good estimate; nevertheless, the range of the temperatures for melting experiments has to be adjusted/corrected after the first run of the experiments.
4. A three-step PCR amplification protocol including an extension step was superior to a two-step protocol. When a two-step setup is used, some polymerases may generate by-products and/or incompletely elongated amplicons that interfered with melting analyses.
5. MS-HRM analysis can also be performed on formalin-fixed tissues. Since many of these will be extensively degraded, the success rate is increased by using relatively small (less than 100 base pairs) amplicons.
6. In MS-HRM, the sensitivity of detection of lower levels of methylation depends on the extent of PCR bias during PCR amplification. As the PCR bias can be adjusted by annealing temperature, the range of the standards can be designed to each experiment individually, and depends on the goals of the user.
7. The MS-HRM protocol used in the methylation-screening experiments of the *MGMT* gene performed on the LC 480. Primers: F-CGTTTGCGATTTGGTGAGTGTT and R-ACCCCGCCCTACCCTATAAATTC. PCR cycling and HRM analysis conditions were as follows: initial activation 10 min at 95°C and 50 cycles of 5 s 95°C, 5 s at 63°C and 5 s at 72°C. Subsequently the product was denatured for 1 min

at 95°C, re-annealed by fast cooling and held for 1 min at 75°C. The HRM analyses were performed in the temperature interval 70–95°C with 50 acquisitions/°C and the default fluorescence temperature gradient parameters selected by the instrument (*see* **Section 3.7**). **Figures 17.2** and **17.3** show an example of the analysis of the results.

8. Heterogeneously methylated templates can be observed in many amplifications. On derivative curves, these are characterized by a broader melting peak typically starting before the unmethylated peak and extending into the methylated peak area. This is due to the formation of heteroduplexes, between heterogeneously methylated templates. An important advantage of MS-HRM is that, unlike many other methods, it allows the detection of heterogeneous methylation.

Acknowledgements

Tomasz K. Wojdacz would like to thank Dr Lise Lotte Hansen for the greatly appreciated support and always good advice. The Danish Medical Research Council and Harboefonden supported financially work of T K Wojdacz. Alexander Dobrovic was supported by grants from the National Health and Medical Research Council of Australia, the Cancer Council of Victoria and the US Department of Defense Breast Cancer Research Program. We also would like to thank Dr. Michael Hoffmann and Roche Diagnostics, Penzberg, Germany for help with the publication of this chapter.

References

1. Wang, R. Y., Gehrke, C. W., Ehrlich, M. (1980) Comparison of bisulfite modification of 5-methyldeoxycytidine and deoxycytidine residues *Nucleic Acids Res* 8, 4777–4790.
2. Worm, J., Aggerholm, A., Guldberg, P. (2001) In-tube DNA methylation profiling by fluorescence melting curve analysis. *Clin Chem* 47, 1183–1189.
3. Wittwer, C. T., Reed, G. H., Gundry, C. N., et al. (2003) High-resolution genotyping by amplicon melting analysis using LCGreen. *Clin Chem* 49, 853–860.
4. Boyd, V. L., Zon, G. (2004) Bisulfite conversion of genomic DNA for methylation analysis: protocol simplification with higher recovery applicable to limited samples and increased throughput. *Anal Biochem* 326, 278–280.
5. Wojdacz, T. K., Hansen, L. L. (2006) Reversal of PCR bias for improved sensitivity of the DNA methylation melting curve assay. *Biotechniques* 41, 274–278.
6. Wojdacz, T. K., Dobrovic, A. (2007) Methylation-sensitive high resolution melting (MS-HRM): a new approach for sensitive and high-throughput assessment of methylation. *Nucleic Acids Res* 35, e41.
7. Warnecke, P. M., Stirzaker, C., Melki, J. R., et al. (1997) Detection and measurement of PCR bias in quantitative methylation analysis of bisulphite-treated DNA. *Nucleic Acids Res* 25, 4422–4426.
8. Clark, S. J., Harrison, J., Paul, C. L., et al. (1994) High sensitivity mapping of methylated cytosines. *Nucleic Acids Res* 22, 2990–2997.
9. Warnecke, P. M., Stirzaker, C., Song, J., et al. (2002) Identification and resolution of artifacts in bisulfite sequencing. *Methods* 27, 101–107.

Chapter 18

Methylation SNaPshot: A Method for the Quantification of Site-Specific DNA Methylation Levels

Zachary Kaminsky and Arturas Petronis

Abstract

As the role for epigenetic signals in genome regulation becomes increasingly understood, the ability to accurately measure levels of DNA methylation at individual cytosines throughout the genome is becoming increasingly important. In contrast to traditional methods for the quantification of cytosine methylation, such as cloning and sequencing of PCR fragments amplified from sodium bisulfite-modified DNA, recent developments have created a fast and effective alternative called methylation-sensitive single nucleotide primer extension (Ms-SNuPE). The following protocol outlines the steps necessary to design and perform Ms-SNuPE experiments using the SNaPshot® chemistry and associated capillary electrophoresis platforms available through Applied Biosystems.

Key words: Epigenetics, DNA methylation, Ms-SNuPE, SNaPshot, sodium bisulfite.

1. Introduction

Methylation of DNA at CpG dinucleotides has been recognized as a key regulator of critical genomic functions, such as transcriptional regulation, silencing of repetitive DNA, and formation and stabilization of heterochromatic regions of the genome (e.g., centromeric and pericentromeric regions). DNA methylation plays a critical role in normal human development, is a key regulator of genomic imprinting, and has been shown to modulate some gene expression responses to environmental stimuli, in some cases by changing the methylation status of a single CpG *(1)*. DNA methylation aberrations have been implicated in numerous rare developmental disorders such as Fragile X syndrome *(2)*, Beckwith–Wiedemann syndrome *(3)*, and Prader–Willi and

Jörg Tost (ed.), *DNA Methylation: Methods and Protocols, Second Edition, vol. 507*

DOI 10.1007/978-1-59745-522-0_18 Springerprotocols.com

Angelman syndrome, among others *(4)*, as well as in various forms of cancer *(5–7)*. DNA methylation is therefore a prime target for study and represents a promising target for diagnostic and therapeutic advances in the future.

The "gold standard" method for the quantification of DNA methylation is based on the sodium bisulfite-based fine mapping of individual CpG dinucleotides, the preferred target of DNA methylation in the human genome. Sodium bisulfite modification deaminates nonmethylated cytosines (C) to uracils, which are subsequently amplified as thymines (T) by polymerase chain reaction (PCR). The end result is a C/T polymorphism in the sequence whose ratio is relative to the original levels of methylated to nonmethylated cytosines. Traditional methods to interrogate these sequence differences involve cloning and sequencing of a large number of sequences per individual to get an accurate quantification of the methylation polymorphism *(8, 9*, **Chapter 14***)*. However, such an approach is very labor intensive and time consuming. In recent years, efficient methods based on single nucleotide primer extension reactions have been developed to quickly and accurately quantify levels of DNA methylation. As indicated by the name, the reaction is based on repeated annealing of a primer exactly one base pair upstream of a target CpG and extending the primer by the incorporation of a single fluorescent dideoxynucleotide. Subsequently, the proportion of the incorporated fluorescent signals that represent C or T can be measured by electrophoresis. As this principle of allele differentiation has been widely applied to single nucleotide polymorphisms (SNPs), some researchers have coined the term methylation-sensitive single nucleotide primer extension (Ms-SNuPE). The method is not limited to one analysis platform, with some groups using radioactively labeled dNTPs or mass spectrometry for quantification *(10, 11*, **Chapter 16***)*, our laboratory uses the ABI SNaPshot reaction platform to perform Ms-SNuPE. In the ABI SNaPshot reaction, multiplexed primers of variable size, targeting different CpG dinucleotides, are cycled with fluorescent dideoxynucleotide terminators that halt the reaction after incorporation. Upon capillary electrophoresis, the different-sized primers migrate at different rates through the polymer matrix, allowing the proportions of incorporated fluorescent signal to be correlated with each target region (**Fig. 18.1**). As the proportion of fluorescent signal is representative of the original proportions of sodium bisulfite-converted C and T, a quantitative measure of DNA methylation is obtained at various positions.

Regions of interest to epigenetic research often have a high GC content, such as the CpG island regulatory elements that are often associated with housekeeping gene promoters. Therefore, strategies for Ms-SNuPE primer design are of particular importance, especially avoiding the incorporation of potentially

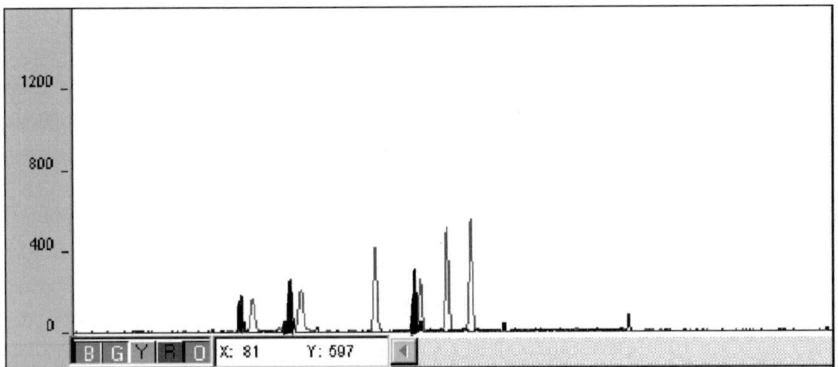

Dye/Sample Peak	Minutes	Size	Peak Height	Peak Area	Data Point
Y, 1	5.71	24.01	190	1974	2141
Y, 2	5.96	31.01	259	2362	2236
Y, 3	6.61	48.96	315	2972	2480
Y, 4	7.74	79.92	92	317	2901
Y, 5	8.79	109.92	63	247	3296
Y, 6	9.13	119.92	66	260	3422
R, 1	5.78	25.66	170	1768	2166
R, 2	6.02	32.61	217	2560	2256
R, 3	6.40	43.27	426	3912	2399
R, 4	6.65	49.86	263	2300	2493
R, 5	6.80	53.71	519	4512	2549
R, 6	6.93	57.10	567	4795	2598

Fig. 18.1. Example output of a multiplexed SNaPshot reaction interrogating six separate CpG dinucleotide positions. Shaded peaks represent the proportion of C while unshaded ones represent T. Positions 1, 2, and 4 display approximately 50% methylation, while the remaining CpG positions are entirely unmethylated. Positional coordinates for each peak as well as the peak height information necessary to calculate the methylation percentage for each CpG are displayed below the peak readout.

polymorphic positions into the Ms-SNuPE primer annealing region. Mismatches in the primer annealing region can introduce large biasing effects to the measured levels of DNA methylation *(12)*. Primer design is, therefore, one of the key aspects of this protocol, in addition to performing sodium bisulfite modification, followed by the Ms-SNuPE reaction and quantification. These four steps will be covered in detail in the sections to follow.

2. Materials

2.1. Post Sodium Bisulfite Modification PCR Primer Design

1. Computer with Internet connection.
2. Account with an oligonucleotide synthesis company for ordering primers.

2.2. Ms-SNuPE Primer Design

1. Same as **Section 2.1**.

2.3. Sodium Bisulfite Modification

1. Sterile water, preferably freshly degassed under a vacuum (*see* **Note 1**).

2. Fresh 3 *M* NaOH solution. Dissolve 3 g NaOH pellets in 25 mL of the degassed water.
3. Fresh 0.1 *M* NaOH solution made from a dilution of above.
4. Fresh hydroquinone solution. Dissolve 0.22 g hydroquinone (Sigma) in 10 mL degassed water. Keep this solution shielded from light.
5. Saturated sodium bisulfite solution. Bring 10.8 g sodium bisulfite (Sigma) to 16 mL final volume in preheated degassed water (55°C). Invert to mix until solution is fully saturated. Add 2.6 mL 3 *M* NaOH solution and 1.0 mL hydroquinone solution. Mix well (*see* **Note 2**).
6. Microcon YM-50 columns (Millipore) or (for high-throughput) Montage PCR96 96-well filtration plates (Millipore) (*see* **Note 3**).
7. TE buffer: 10 mM Tris–HCl, 1 mM EDTA, pH 7.5.

2.4. PCR

1. HotStar Taq Master Mix (Qiagen).
2. Forward and reverse PCR primers, 20 μM in TE buffer.

2.5. Gel Electrophoresis

1. 1 × TBE Buffer: 89 mM Tris–boric acid, 2 mM EDTA.
2. 5 mg/mL Ethidiumbromide.
3. 100 bp DNA ladder (50 ng/μL).

2.6. PCR Clean-Up

1. Qiagen Minelute gel extraction kit (Qiagen).

2.6.1. Option 1: Qiagen Gel Extraction

2.6.2. Option 2: Exo I/SAP Digestion

1. Exonuclease I (20 U/μL) and corresponding 10 × reaction buffer.
2. Shrimp Alkaline Phosphatase (10 U/μL) and corresponding 10 × reaction buffer.

2.7. The SNaPshot Reaction

1. SNaPshot Multiplex Reaction Kit (Applied Biosystems).
2. 5 × Sequencing buffer (Applied Biosystems).
3. Ms-SNuPE primers, resuspended in TE Buffer to a final concentration of 20 μM.

2.8. Removal of Unincorporated ddNTPs

1. Calf intestinal phosphatase (CIP) (10 U/μL) and corresponding 10 × reaction buffer.

2.9. Capillary Electrophoresis

1. ABI 3100 Avante Genetic Analyzer (Applied Biosystems).
2. Genescan 3.1 software (Applied Biosystems).
3. Pop4 polymer (Applied Biosystems).
4. 36-cm array (Applied Biosystems).
5. 10 × EDTA buffer (Applied Biosystems).
6. Optical plate with denaturation cover and septum (Applied Biosystems).

7. GS120 LIZ size standard (Applied Biosystems).
8. DS-02 Matrix standard kit (Applied Biosystems).
9. Hi Di formamide (Applied Biosystems).

3. Methods

3.1. Post Sodium Bisulfite Modification PCR Primer Design

One of the most critical steps for successfully performing Ms-SNuPE reactions is the assay design stage concerning the design of both bisulfite-PCR and Ms-SNuPE primers. More specifically, it is critical that the primers designed to produce the post sodium bisulfite-modified PCR amplicon do amplify efficiently and specifically. Obtaining a good yield from a post-bisulfite-treatment PCR can be a challenge due to the high (~90%) degradation of genomic DNA that occurs during bisulfite treatment *(13)*. Bisulfite treatment can also affect the specificity of some PCR primers because the degenerated genetic code (all unmethylated cytosines converted to thymines) reduces the sequence complexity of some regions and thus produces a risk for nonspecific PCR amplification. Finally, because the methylation status of any CpG dinucleotide is not known a priori, PCR primers should avoid internal CpG incorporation where possible to avoid mispriming. The following steps should be taken to improve the chances of successful post-bisulfite amplification.

1. Select a region of interest within the genome around which to design primers. Be sure to avoid designing primers in repetitive regions of the genome, which can be identified using Repeat Masker in the UCSC genome browser (http://genome.ucsc.edu). The amplification efficiency of the reaction is often dependent on amplicon size, such that a smaller amplicon tends to amplify better. Amplicons less than 300 bp in length can often be amplified with a single PCR. For larger PCR amplicons (>400 bp), sometimes a nested or semi-nested PCR strategy is required. We have not exceeded 600 bp. The optimal amplicon size is, of course, a factor of the region of interest. Once a PCR amplicon has been produced, Ms-SNuPE can be performed on any CpG site within the amplicon and so, depending on the target CpGs, the ideal amplicon should be designed on a case-by-case basis.
2. If the region of interest is a repetitive region, design at least one of the primers in a unique region of the genome to target that specific repetitive element. If the repetitive region is large and if placing a primer in a unique genomic region would cause the amplicon of interest to exceed ~600 bp, a useful strategy is to perform a nested or semi-nested PCR, ensuring that one of the first set of primers is in a unique region of the genome

for the first amplification. The subsequent PCR in the second step can then target your amplicon sequence of interest.

3. Once the general target region has been determined, begin designing PCR primers for the region. Online PCR design software such as MethPrimer (http://www. urogene.org/methprimer/index1.html) performs well and allows the user to specify various parameters such as amplicon size, excluded regions, and primer TM, among others, and returns converted (specific to bisulfite-modified sequence) primer sequences with the specified parameters *(14)*. MethPrimer does not require the electronic conversion of the sequence of interest to a post-bisulfite sequence. However, if the software fails to produce primers for a specific region, it is possible to design primers by hand or use alternative primer design software. In such cases, it will be necessary to manually convert your sequence of interest into a post-bisulfite-modified sequence by converting all "C" to "T" prior to designing primers. When designing primers manually, it is best to select a region that contains some positions that contain cytosines in a non-CpG context to specifically amplify completely converted sequences and avoid homopolymers of any nucleotide that might bind nonspecifically or form hairpin structures (*see* **Note 4**).

3.2. Ms-SNuPE Primer Design

All Ms-SNuPE primers should be designed so that the 3′ end of the primer terminates exactly 1 bp upstream (5′) of the target C of the CpG dinucleotide. Because Ms-SNuPE primers are distinguished by size, it is important to have primers synthesized and purified by HPLC to limit primer size variants. Primers can be designed in either the forward orientation (complementary to the antisense strand) or the reverse orientation (complementary to the sense strand) (*see* **Note 5**). In this way, the design of Ms-SNuPE primers is relatively easy; however, designing the most efficient assay will be dependent on the number of target CpGs and the surrounding CpG density. These factors will affect both the multiplexing capability of the assay and the subsequent clarity of the final electrophoretic profile (**Fig. 18.1**). At this point, it is useful to distinguish between two primer design scenarios that will affect the ability to multiplex the reaction, these being whether there are CpG dinucleotides within the primer binding sequence or not.

3.2.1. Design of Multiplex Assays (When there are "no" CpG dinucleotides in the Ms-SNuPE primer annealing region)

1. **Primer Size:** If there are no CpG dinucleotides in the primer binding sequence, multiplexing (performing more than one single nucleotide primer extension reaction in the same tube) is possible (*see* **Note 6**). When considering the design of a multiplexed assay, the cost of primer synthesis should be weighed against the cost of running reactions separately as for the

electrophoretic separation of the various target CpG dinucleotides, the primers to be multiplexed must differ in length by at least 8 bp and 4 bp below and above a total primer length of 40 bp, respectively. For example, if a second Ms-SNuPE primer is multiplexed with an existing 20 bp primer, the second primer should be at least 28 bp while if the original was 40 bp, the second could be either 32 bp or 44 bp. In our experience, 15–17 bp is the lower limit for primer size, and in an ideal situation of 10 multiplexed reactions, the longest primer would therefore be around 65 bp. Of course, the sequence features of your target oligonucleotide of interest may limit the total number of possible primers to multiplex at once.

2. **Primer TM:** All primers to be multiplexed together should be designed with similar TM. The suggested SNaPshot reaction cycling conditions have an annealing temperature of 50°C and so primers should be designed with a TM of 50°C. The annealing temperature in the cycling protocol can be manipulated to better suit the primers. However, to ensure equal performance in the cycling reaction, all primers should be designed with similar TM (± 3°C). Because of the necessity to vary primer length while keeping the same range of primer TM, primer lengths should be varied by adding noncomplementary tails to the 5′ end of each primer, with primer TM being calculated only for that section of the primer that is complementary to the target amplicon. The sequence 5′-GACT-3′ × N can be used as a nonbinding tail as it does not form hairpin loops (*see* **Note 7**).
3. **Primer Orientation:** Ms-SNuPE Primers complementary to the T-rich sense and A-rich antisense strands, respectively, must not be multiplexed together to avoid primer dimer formation.

3.2.2. *Design of Single Assays (*When there "are" CpG dinucleotides in the Ms-SNuPE primer annealing region*)*

In general, the presence of one or more CpG dinucleotides within the Ms-SNuPE primer binding region requires an additional step in the primer design (*see* **Note 8**). This step is simply to design degenerative or "wobbling" bases at the C position of any upstream CpG dinucleotides within the primer binding region. If designing primers complementary to the antisense strand (Forward Primers), primers should be designed with Y's in place of C's, while reverse strand primers should be designed with R's at the potentially polymorphic positions. The incorporation of wobbling bases in the primer makes multiplexing primers infeasible in terms of downstream data interpretation (*see* **Note 9**).

3.3. *Sodium Bisulfite Modification*

1. Adjust volume of the DNA sample (50 ng–2 μg) to 10 μL (*see* **Note 10**).
2. Transfer DNA sample(s) to PCR tubes (or 96-well plate for high-throughput processing). Add 1.1 μL of freshly prepared 3 M NaOH solution. Centrifuge and seal the tubes (or plate).

3. Place in a thermocycler for 20 min at 42°C.
4. Spin down tubes/plate to catch condensation and carefully open seal. Add 120 μL of fresh sodium bisulfite solution, seal plate/tube with a fresh lid, invert a few times to mix, and then spin-down.
5. Place in a thermocycler for 4–5 h at 55°C (*see* **Note 11**).
6. Remove from thermocycler, spin down, and carefully remove lid.
7. Add 100 μL of sterile water.
8. Transfer each sample to a Microcon YM-50 column or (for high-throughput) the corresponding well in a Montage PCR96 96-well filtration plate.
9. Draw solution through filtration matrix by either centrifugation at maximum speed (Microcon YM-50 column) or vacuum (Montage PCR96 96-well filtration plate), until wells are visibly empty of solution (~4–5 min). DNA remains on the matrix.
10. Desalt DNA by adding 175 μL of sterile water to each well and drawing the solution through the matrix (via centrifugation or vacuum) as before. Repeat this step twice more.
11. Desulfonate by adding 175 μL of fresh 0.1 *M* NaOH. Draw solution through the matrix (via centrifugation or vacuum).
12. Perform a final washing step by drawing 175 μL of sterile water through the matrix.
13. Recover DNA by adding 50 μL of TE Buffer and incubation for 2 min.
14. If using Microcon YM-50 columns, carefully vortex for additional 30 s. Uncap Microcon unit, separate sample reservoir from filtrate cup, and place sample reservoir upside down into a new vial. Spin for 3 min at 1780*g* in invert spin mode to elute DNA. If using Montage PCR96 96-well filtration plate, use a plate shaker to release DNA from filtration matrix for 10 min at 500 rpm. Remove eluted DNA solution from individual wells using a pipette and filter tips.
15. Remove an aliquot for whole genome amplification if needed (*15*, **Chapter 27**). Otherwise, store bisulfite-treated DNA at −20°C (or −80°C for long-term storage).

3.4. Post Sodium Bisulfite Modification PCR

1. In PCR strip tubes, add approximately 4 μL of the eluted sodium bisulfite-treated sample as template for PCR.
2. Add forward and reverse PCR primers to a final concentration of 0.5 μM in the final reaction (0.5 μL of each forward and reverse 20 μM PCR primer stock).
3. To this, add 10 μL of Qiagen HotStar Taq Master Mix, which contains all reagents necessary for PCR, and 5 μL of ddH_2O, bringing the final reaction volume to 20 μL.
4. Cycle in a thermocycler according to the following specifications (*see* **Note 12**):

95°C – 15 min
{95°C – 30 s, 50°C – 45 s, 72°C – 30 s} 40 cycles
72°C – 5 min
Cool to 4°C

3.5. Amplicon Evaluation

To insure that the target amplicon has been amplified, the PCR needs to be evaluated by agarose gel electrophoresis.

1. Make a 2% TBE agarose gel. First, add the volume of 1 × TBE buffer required to fill the electrophoresis tray to an Erlenmeyer flask followed by 2 × weight per volume of biotechnology grade agarose. For example, to make 100 mL of 2% TBE agarose gel, add 2 g of agarose to 100 mL of 1 × TBE.
2. Heat the mixture by microwaving for approximately 3 min until solid agarose is no longer visible.
3. Allow the gel to cool for a few minutes and stir in 1 μL per 100 mL of ethidiumbromide (5 mg/mL) to achieve a final concentration of 0.5 μg/mL in the solution (*see* **Note 13**). When the gel is cool, it can be poured into the gel tray equipped with loading combs and allowed to polymerize (*see* **Note 14**).
4. Run approximately 5 μL of the PCR product on a 2% TBE agarose gel against 3 μL of 50-ng/μL 100-bp ladder to determine if the band of interest was specifically amplified. If there is nonspecific amplification, but the band is present and clearly distinguishable at the correct size, the rest of the PCR product can be run on a second gel and a scalpel or razor used to excise the band. Subsequently extract the amplicon using Qiagen's Minelute Gel Extraction kit, after which the SNaPshot reaction can be performed (*see* **Note 15**) (PCR Clean Up Option 1). If a single PCR amplicon was produced for the desired fragment, use the more rapid treatment involving digestion with exonuclease 1 and shrimp alkaline phosphatase (Exo I/SAP, PCR Clean Up Option 2).

3.6. PCR Clean Up

3.6.1. Option 1: Qiagen Gel Extraction

1. Weigh the excised gel fragment within an Eppendorf tube first by zeroing an analytical balance on an empty Eppendorf and then weighing the tube containing gel.
2. To that tube, add 3 volumes per weight of buffer QG. For example, to 100 mg of gel, add 300 μL of buffer QG. Incubate at 50°C for approximately 10 min, vortexing every 2 min, until no visible agarose remains.
3. Add the contents to a provided microtube column and spin at high speed (18,890*g*) in a microcentrifuge for 1 min.
4. Discard the flow through, add 500 μL of QG to the column, and repeat the spinning procedure including removal of flow through.

5. Add 750 μL of buffer PE, spin at high speed for 1 min, remove the flow-through, and then repeat the spin, without adding any additional buffers, to dry the column in preparation for elution.
6. Remove the column and place it in a newly labeled Eppendorf tube. Add 10 μL of elution buffer EB to the center of the column matrix and let stand for 1 min. Spin for 1 min at high speed (*see* **Note 16**).
7. Repeat the elution procedure to increase the purification yield, ending with a final volume of ~20 μL.

3.6.2. PCR Clean Up Option 2: Exo I/SAP Digestion

Digestion with ExoI/SAP can be more cost-effective than column-purification techniques and faster when large numbers of amplicons are prepared for the SNaPshot reactions.

1. Make a master mix containing 2 μL of ExoI and 5 μL of SAP per 15 μL of PCR sample and one-tenth of the final volume of 10 × NEB buffer 4.
2. Incubate samples at 37°C for 1 h followed by 15 min at 75°C.

3.7. Performing SNaPshot Reaction Cycling

1. The ABI protocol guidelines suggest using between 0.01 pmol and 0.40 pmol of amplicon template per reaction (*see* **Note 17**).
2. In PCR strip tubes, add a master mix comprised of 2 μL of SNaPshot Multiplex Ready Reaction Mix, 3 μL of 5 × Sequencing Buffer, 1 μL of 2 μM Ms-SNuPE primer (0.2 μM in the final reaction), and template and adjust the volume to 10 μL with ddH_2O (*see* **Note 18**).
3. Seal the strip tubes and place them in a thermocycler (preferably, but not necessarily an ABI 9700 thermocycler (*see* **Note 19**)), perform 25 cycles of 95°C for 10 s, 50°C for 5 s, and 60°C for 30 s, followed by a final cool down to 4°C (*see* **Note 20**).

3.8. Removal of Unincorporated ddNTPs

In order to remove downstream fluorescent noise in the data analysis stage created by residual ddNTP terminators, calf intestinal phosphatase CIP digestion is required.

1. Make a master mix by adding 1 μL of CIP to 9 μL of NEB buffer 3. Add 3 μL of the master mix to each SNaPshot reaction well.
2. In a thermocycler, incubate at 37°C for 1 h followed by 15 min at 75°C and then allow cooling to 4°C (*see* **Note 21**).

3.9. ABI Electrophoresis and Data Quantification

The following protocol assumes the use of an ABI 3100 Avante Genetic Analyzer using POP4 polymer, a 36-cm Array that has been loaded with the DS-02 Matrix Standards. It should be noted that the capillary electrophoresis for SNaPshot can be performed on other ABI platforms.

1. Make a master mix containing 9 μL of Hi Di Formamide and 0.5 μL of GS 120 LIZ size standard per sample. Pipette 9.5 μL of this master mix into an ABI optical plate.
2. To each of these wells, add 0.5 μL of the CIP digested SNaPshot reaction mixture (*see* **Note 22**). Load samples in the required order, that is, in columns from top to bottom (not in rows from left to right) for maximal efficiency of the ABI 3100 Avante genetic analyzer (*see* **Note 23**).
3. Place a clean rubber denaturation lid on the plate and denature the samples in a thermocycler for 5 min at 95°C followed by immediate placement into an ice bath for 2 min. Replace the denaturation lid with a clean septum prior to running on the ABI 3100 Avante genetic analyzer.
4. In the ABI data collection software, open a new plate editor, name the project, and select Genescan. Create a valid sample name in the plate-coordinate rows that correspond only to those wells in the optical plate that contain your sample. Select orange for the analysis color, 3100 Avante for the project selection, Dye Set E5 for the dye set, SNP36_POP4DefaultModule as the Run Module, and GS120Analysis.gps for the Analysis Module.
5. Place your optical plate in the correct orientation in the machine, link your plate, and run. The SNaPshot electrophoresis takes about 30 min to complete four samples, assuming the use of an array with four capillaries.
6. After the run is complete, open the Genescan 3.1 Analysis software and select "New Project".
7. In the project window, go to "Add Samples" and select the plate name in the data extractor folder where the data is stored.
8. Set the marker to the color orange, which corresponds to the fluorescent signal of the GS 120 LIZ size standard. Clicking on a sample name will open a window that will show the various traces as well as quantitative estimates of the peak heights and positions (*see* **Note 24**). Optimal peak heights by default are above 50 intensity units on the Y axis (*see* **Note 25**). Incorporation of different ddNTP terminators will result in polymorphic positions having two peaks at a given location; however, due to slight differences in elecrophoretic migration, these peaks (such as a *C* and *T* peak) should appear to be separated by a distance of approximately 1 bp.
9. In the cases where Ms-SNuPE primers were designed with no "wobbling" positions, determine the peak position that corresponds to this primer (*see* **Note 22**) in the data display and determine the percentage of cytosine methylation (C^{met}) according to the following formula:

$$\%C^{\mathrm{met}} = 100x\frac{(\text{Peak Height}\,C)}{(\text{Peak Height}\,C + \text{Peak Height}\,T)}$$

10. In cases where Ms-SNuPE primers were designed with "wobbling" positions, the $\%C^{met}$ can be determined by the following formula:

$$\%C^{met} = 100x \frac{{}_n\Sigma_1(\text{Peak Height}\, C_1 \ldots C_n)}{({}_n\Sigma_1\, \text{Peak Height}\, C_1 \ldots C_n + {}_n\Sigma_1\, \text{Peak Height}\, T_1 \ldots T_n)}$$

where, C_1 is the peak height of the first C peak and C_n is the last (*see* **Note 26**). For Ms-SNuPE primers designed in the reverse orientation, look for G (Blue) and A (Green) peaks in place of C (Black) and T (Red) peaks, respectively.

4. Notes

1. Oxygen in water can reduce the efficiency of sodium bisulfite conversion.
2. It is possible that not all the sodium bisulfite will dissolve. If any substrate remains, centrifuge the solution and use the supernatant.
3. Using the Montage PCR96 96-well filtration plate (Millipore) for high-throughput sample processing requires a suitable vacuum manifold.
4. If either of the PCR primers is designed directly adjacent to the C position in a CpG dinucleotide, this PCR primer can later be used as a Ms-SNuPE primer.
5. Keep in mind that depending on what strand was targeted by the post-bisulfite-modification PCR, only one strand C of a palindromic CG will have a potentially polymorphic position. Therefore, whether using forward and reverse Ms-SNuPE primers, the target C of a given CpG should be the same.
6. Our laboratory uses an ABI 3100 Avante electrophoresis platform using the recommended Pop 4 polymer matrix. We have multiplexed up to six reactions successfully. However, the ABI SNaPshot manual suggests an upward multiplexing capacity of 10 primers in a reaction.
7. Investigate visually the 5′ upstream region of the primer region and avoid complementarity. For example, if the "T" of the GACT does not match, but "GAC" does, the repeating pattern should be shifted to 5′-TGAC-3′, for example.
8. Because of the unknown methylation status of these CpGs a priori, the target amplicon could contain polymorphic sequence at these positions. Experiments in our laboratory demonstrated a strong bias of the measured methylation percentage in cases where there is a polymorphic position in the primer annealing region and that this bias increases as the polymorphic position approaches the 3′ end of the Ms-SNuPE primer *(12)*.

9. The perfectly complementary primer in the degenerative mixture will bind the target and accurately measure the methylation percentage at the target CpG. However, this polymorphic mixture of primers is likely to have different electrophoretic potential and thus will separate, sometimes causing multiple peaks to be observed *(12)*. Quantification is covered in the data interpretation section but the presence of multiple peaks for a single primer would make the distinction of multiplexed primer peaks difficult. Therefore, it is advised to run Ms-SNuPE primers containing degenerate positions separately.
10. In our experience, the optimal starting amount of genomic DNA is 250 ng–1 μg. Small amounts of DNA (e.g., 50 ng or below) may not provide enough useable template for direct use following sodium bisulfite treatment, but can be used for DNA methylation profiling following WGA of the sodium bisulfite-treated DNA (**Chapter 27**).
11. It can be beneficial to ramp the reaction up to 95°C for 1 min each hour to ensure that the DNA remains single stranded.
12. The following cycling conditions generally work well as a universal program for amplification. Further optimization of primer annealing temperature may be required if the amplification does not work. If possible, run a positive control of sodium bisulfite-modified template that has been known to work in the past to rule out a failed bisulfite reaction in the event of PCR failure.
13. Ethidiumbromide is light sensitive, store in a dark place. Take care when handling, it is highly toxic.
14. To save time, a larger volume of gel can be made and stored in a 50°C incubator so that it is ready for the pouring stage. This solution is stable for about a month before the Ethidiumbromide decomposes.
15. It is necessary to purify the post-sodium bisulfite amplicon that serves as template for the SNaPshot reaction. Residual dNTPs could incorporate in place of the fluorescent ddNTP terminators provided in the SNaPshot reaction kit and other residual reaction components such as remaining PCR primers might bind the amplicon and incorporate fluorescent signals.
16. To avoid the Eppendorf tube lids snapping off, place tubes to be eluted side by side in pairs, with the lids crossed and facing inward toward the center of the centrifuge rotor.
17. Simply adding 1–2 μL of purified PCR product to the reaction will generally work without further quantification and calculation.
18. The addition of 5 × Sequencing Buffer is not necessary, as the original protocol calls for 5 μL of SNaPshot Multiplex Ready Master Mix. However, this buffer allows the more

expensive reaction component to be diluted, and thus one can achieve significantly more reactions at lower cost.

19. When performing SNaPshot reaction cycling, a fast ramping speed is desirable for optimal results.
20. If low signals are observed for a given primer or sample, the amount of starting template can be increased or the annealing temperature for the cycling reaction can be adjusted. Ms-SNuPE primer amounts can also be adjusted up or down to modulate signal intensity. However, the amount of template can be a limiting factor in cases of low signal, especially in cases where the PCR amplicon was observed to be weak during gel electrophoresis.
21. SNaPshot reactions can be stored at this point overnight at 4°C or for longer periods at −20°C prior to performing capillary electrophoresis on the ABI 3100 genetic analyzer.
22. Up to 2 μL each of SNaPshot mix and GS 120 LIZ size marker can be added to increase signal intensity as necessary, adding Hi Di Formamide to a final volume of 10 μL.
23. Because of the orientation of the ABI optical plate in the genetic analyzer, the capillary is oriented from top to bottom and thus will draw samples most efficiently if samples are loaded accordingly.
24. Peak positions are relative to the sizes of the peaks of the GS 120 LIZ size marker. The sizes are best thought of as relative sizes in that, for two Ms-SNuPE primers with expected sizes of 20 bp and 28 bp, these may appear as peaks at 25 bp and 37 bp. However, their relative sizes will identify which peaks correspond to which primer and thus target CpG.
25. There is no specified upper limit to the peak height. However, peaks should have a clear defined tip, not a flattening which signifies saturation of the fluorescent signal. When there is a flattened peak top, the data is no longer quantitative.
26. Experiments in our laboratory have demonstrated that this method of quantification is accurate within 5% of the true DNA methylation level as measured by sodium bisulfite-based cloning and sequencing *(12)*.

References

1. Szyf, M., Weaver, I. C., Champagne, F. A., et al. (2005) Maternal programming of steroid receptor expression and phenotype through DNA methylation in the rat. *Front Neuroendocrinol* 26, 139–162.
2. Van Esch, H. (2006) The Fragile X premutation: new insights and clinical consequences. *Eur J Med Genet* 49, 1–8.
3. Weksberg, R., Shuman, C., Smith, A. C. (2005) Beckwith-Wiedemann syndrome. *Am J Med Genet C Semin Med Genet* 137, 12–23.
4. Butler, M. G. (2002) Imprinting disorders: non-mendelian mechanisms affecting growth. *J Pediatr Endocrinol Metab* 15, Suppl 5, 1279–1288.

5. Ramachandran, K., Soloway, M. S., Singal, R., et al. (2007) The emerging role of epigenetics in urological cancers. *Can J Urol* 14, 3535–3541.
6. Arai, T., Takubo, K. (2007) Clinicopathological and molecular characteristics of gastric and colorectal carcinomas in the elderly. *Pathol Int* 57, 303–314.
7. Nelson, W. G., Yegnasubramanian, S., Agoston, A. T., et al. (2007) Abnormal DNA methylation, epigenetics, and prostate cancer. *Front Biosci* 12, 4254–4266.
8. Petronis, A., Gottesman, I. I., Kan, P., et al. (2003) Monozygotic twins exhibit numerous epigenetic differences: clues to twin discordance? *Schizophr Bull* 29, 169–178.
9. Dahl, C., Guldberg, P. (2003) DNA methylation analysis techniques. *Biogerontology* 4, 233–250.
10. Gonzalgo, M. L., Liang, G. (2007) Methylation-sensitive single-nucleotide primer extension (Ms-SNuPE) for quantitative measurement of DNA methylation. *Nat Protoc* 2, 1931–1936.
11. Tost, J., Schatz, P., Schuster, M., et al. (2003) Analysis and accurate quantification of CpG methylation by MALDI mass spectrometry. *Nucleic Acids Res* 31, e50.
12. Kaminsky, Z. A., Assadzadeh, A., Flanagan, J., et al. (2005) Single nucleotide extension technology for quantitative site-specific evaluation of metC/C in GC-rich regions. *Nucleic Acids Res* 33, e95.
13. Grunau, C., Clark, S. J., Rosenthal, A. (2001) Bisulfite genomic sequencing: systematic investigation of critical experimental parameters. *Nucleic Acids Res* 29, e65.
14. Li, L. C., Dahiya, R. (2002) MethPrimer: designing primers for methylation PCRs. *Bioinformatics* 18, 1427–1431.
15. Mill, J., Yazdanpanah, S., Guckel, E., et al. (2006) Whole genome amplification of sodium bisulfite-treated DNA allows the accurate estimate of methylated cytosine density in limited DNA resources. *Biotechniques* 41, 603–607.

Chapter 19

Bio-COBRA: Absolute Quantification of DNA Methylation in Electrofluidics Chips

Romulo Martin Brena and Christoph Plass

Abstract

DNA methylation is the best-studied epigenetic modification, and in mammals it describes the conversion of cytosine to 5-methylcytosine in the context of CpG dinucleotides. In recent years, it has become evident that epigenetic mechanisms are severely disrupted in human neoplasia, and evidence suggests that alterations of DNA methylation patterns may be an integral mechanism in the etiology of other diseases such as bipolar disorder and schizophrenia. The main effect of altered DNA methylation is the disruption of normal patterns of gene expression through genomic instability and hypermethylation of CpG islands, which together could lead to uncontrolled cell proliferation. DNA methylation can be reversed through pharmacological intervention via the systemic administration of DNA methylation inhibitors. Thus, the ability to accurately quantify DNA methylation levels in genomic sequences is a prerequisite to assess not only treatment efficacy, but also the effect of the DNA methylation inhibitors on bystander tissues. Several methods are currently available for the analysis of DNA methylation. Nonetheless, accurate and reproducible quantification of DNA methylation remains challenging. Here, we describe Bio-COBRA, a modified protocol for combined bisulfite restriction analysis (COBRA) that incorporates an electrophoresis step in microfluidics chips. Microfluidics technology involves the handling of small amounts of liquid in miniaturized systems. Bio-COBRA provides a platform for the rapid and quantitative assessment of DNA methylation patterns in large sample sets. Its sensitivity and reproducibility also make it an excellent tool for the analysis of DNA methylation in clinical samples.

Key words: Bio-COBRA, electrofluidics chips, quantification, DNA methylation, DNA methylation inhibitor, dynamic range, 2100 Bioanalyzer.

1. Introduction

Epigenetic modifications, such as DNA methylation, are defined as heritable changes to the DNA with the potential to alter gene expression without affecting the primary DNA sequence. Over the past decade, it has become apparent that aberrant epigenetic

Jörg Tost (ed.), *DNA Methylation: Methods and Protocols, Second Edition, vol. 507*

DOI 10.1007/978-1-59745-522-0_19 Springerprotocols.com

alterations are a common feature of human cancer, playing an important role in their development and progression *(1, 2)*. In mammalian genomes, DNA methylation occurs mainly in the context of 5′-CpG-3′ dinucleotides *(3–5)*. In the human genome, specifically, almost 90% of all CpG dinucleotides are found in repetitive sequences and are normally methylated. Of the remaining 10%, most stay methylation free and are located in 0.5–4.0 kb sequence stretches termed CpG islands *(6, 7)*. Importantly, most CpG islands are in close proximity of genes or actually span gene promoters. The significance of this observation rests upon the fact that genes are consistently silenced or downregulated when their associated CpG island is methylated *(8)*. Because of its potential to abrogate gene expression, DNA methylation has been proposed as one of the two hits in Knudson's two-hit hypothesis required for oncogenic transformation *(1)*.

Studies have shown that aberrant DNA methylation can be detected in secretions and body fluids of patients years prior to the clinical diagnosis of cancer, suggesting that aberrant DNA methylation may be an early occurrence in the process of malignant transformation *(2, 9)*. In light of this evidence, much effort is being devoted to further characterize aberrant DNA methylation patterns in several tumor types in an attempt to uncover specific DNA methylation patterns that might afford clinical diagnostic or prognostic value *(10–16)*. However, given the fact that normal DNA methylation patterns among individuals can vary, the specificity of one or several aberrant DNA methylation events might rest not only on which particular CpG dinucleotides are methylated, but also on their methylation frequency *(17)*. This scenario presents a challenge for the DNA methylation field, since the search for aberrantly methylated loci useful for early disease detection, prognosis, or assessment of disease risk may entail focusing on subtle changes in DNA methylation levels. Thus, a need exists for a screening technique that will allow for the rapid and reliable evaluation of DNA methylation in large sample sets, while at the same time providing quantitative information on the level of aberrant DNA methylation and spatial information as to which CpG dinucleotides are preferentially methylated in a genomic region of interest. Here, we describe a protocol for the coupling of combined bisulfite restriction analysis (COBRA), followed by quantification of the restriction fragments on the Agilent™ 2100 Bioanalyzer platform for a rapid, accurate, and cost-efficient quantification of methylation patterns in any DNA sample. COBRA makes use of the principle that bisulfite treatment of genomic DNA translates the epigenetic information encoded by cytosine methylation into sequence differences, which result in the presence or absence of restriction enzymes recognition sites in a methylation-dependent manner. Quantitative analysis of the relative amounts of restriction fragments allows for the accurate assessment of DNA methylation levels at CpG sites within the respective recognition sites. The main strength of the Bio-COBRA assay is that the DNA methylation status of all DNA molecules in a PCR product is assessed for each sample, which eliminates the need for sequencing individual clones. Bio-COBRA results are calibrated to a standard curve in order to extract *true* DNA methylation percentages. However, our method can be performed without a standard curve, if the analysis desired is a relative comparison of DNA methylation levels in a sample set. It should be noted, however, that Bio-COBRA analysis is limited to DNA sequences which possess at least one restriction enzyme site with a least one CpG dinucleotide in its recognition sequence. Our method also provides an excellent platform for the screening of DNA methylation in genomic regions where aberrant DNA methylation is known to occur; especially in genomic areas suspected of holding diagnostic or prognostic value. However, Bio-COBRA can also be utilized as a discovery tool to detect novel aberrant DNA methylation sites. Conversely, it is important to note that only the methylation status of the CpG dinucleotide or dinucleotides within the recognition sequence of the restriction enzyme utilized in the assay are interrogates for their methylation status. The assay is easy to adopt into molecular biology labs and could be used as a rapid and affordable prescreening tool that provides quantitative DNA methylation data. The Bio-COBRA assay can be performed on 12 samples in less than 1 h. However, if the protocol is started at the DNA isolation step, approximately 48 h would be required to complete the entire procedure.

2. Materials

2.1. Preparation of a DNA Methylation Standard for PCR Bias Correction

1. Human genomic DNA isolated from peripheral blood lymphocytes (PBL) (Aviva Systems Biology, San Diego, CA). Store at −20°C.
2. *Sss*I methylase (20 U/μL) (New England Biolabs, Beverly, MA).
3. 20 mM *S*-adenosyl methionine (SAM) (New England Biolabs).
4. QiaQuick gel extraction kit (Qiagen, Valencia, CA).
5. ND-1000 spectrophotometer (NanoDrop Technologies, Wilmington, DE).
6. Sonicator.

2.2. DNA Isolation from Samples

1. Liquid nitrogen.
2. Lysis buffer: 10 mM Tris–HCl, pH 8.0, 25 mM EDTA, 100 mM NaCl, 0.5% w/v SDS. Store at room temperature.

3. Proteinase K (10 mg/mL) (New England Biolabs). Aliquot and store at −20°C.
4. TE: 10 mM Tris–HCl, pH 7.4, 1 mM EDTA. Store at room temperature.
5. Phenol ("carcinogen").
6. Chloroform.
7. Dialysis tubing (Membrane Filtration Products, Seguin, TX). Store at 4°C.
8. Mortar and pestle.
9. 100% ethanol. Store at −20°C.
10. Water bath.

2.3. Bisulfite DNA Treatment

1. 3 M $NaHSO_3$ pH 5.0 dissolved in water. Prepare fresh before use. Adjust the pH of the solution by adding 10 M NaOH (~150 μL per 10 mL).
2. 10 mM hydroquinone ("toxic") dissolved in water. Prepare fresh before use, protect from light.
3. 3 M NaOH dissolved in water. Prepare fresh before use.
4. 5 M sodium acetate dissolved in water. Store at room temperature.
5. Hybridization oven.
6. Isopropanol. Store at room temperature.
7. QiaQuick gel extraction kit (Qiagen).

2.4. PCR Amplification

1. Platinum Taq DNA polymerase (Invitrogen, Carlsbad, CA).
2. 10 × PCR buffer: 166 mM $(NH_4)SO_4$, 670 mM Tris–HCl, pH 8.8, 67 mM $MgCl_2$, 100 mM β-mercaptoethanol. Aliquot and store at −20°C.
3. 10 mM dNTP mix. Aliquot and store at −20°C.
4. Oligonucleotide primers (10 pmol/μL). Store at −20°C.
5. DNase–RNase free water.
6. Thermal cycler.

2.5. Restriction Enzyme Digestion of PCR Products

1. Restriction enzyme with a CpG dinucleotide in its recognition sequence (e.g., *Bst*UI, *Hpy*CH4 IV, *Hha*I) (New England Biolabs).
2. Bovine serum albumin (New England Biolabs). Store at −20°C.
3. DNase–RNase free water.
4. Concentrator 5301 (Eppendorf, Hamburg, Germany) or any vacuum dryer.
5. 37°C incubator or water bath.

2.6. Polyacrylamide Gel Electrophoresis

1. 29% (w/v) acrylamide/1% (w/v) *N,N'*-methylene-*bis*-acrylamide solution ("neurotoxin when unpolymerized, avoid contact with skin"). Store at 4°C.
2. 10% (w/v) ammonium persulfate ($(NH_4)_2S_2O_8$) dissolved in water. Store at 4°C.

3. *N,N,N,N′*-Tetramethyl-ethylenediamine (TEMED) ("toxic, flammable")(Invitrogen).
4. 10 × TBE: 0.90 M Tris, 0.90 M boric acid, 10 mM EDTA. Dilute in water to 1 × (v/v) as working solution. Store at room temperature.
5. 0.001% (w/v) ethidium bromide ("carcinogen") dissolved in water.
6. Electrophoresis chamber.
7. Electrophoresis power source.
8. Gel imaging system.

2.7. Electrophoresis in Microfluidics Chips

1. Agilent 2100 Bioanalyzer (Agilent, Santa Clara, CA).
2. DNA 1000 LabChip (Agilent).
3. Chip priming station (Agilent).

3. Methods

3.1. Preparation of a DNA Methylation Standard for PCR Bias Correction

1. Fragment the genomic DNA utilizing a sonicator at 70% power for at least 2 min. DNA fragments should not be larger than 2–3 kb. Electrophoresis in a 1% agarose gel could be performed to check the efficacy of the sonication step, as the power settings of different equipment brands tend to vary.
2. To in vitro methylate genomic DNA, prepare the following reaction mix: 15 μg fragmented DNA, 160 U *Sss*I, 3.2 μL SAM, 40 μL enzyme buffer, water (to a final volume of 400 μL). Incubate the reaction at 37°C for 4 h. Two hours into the incubation, add an extra 2 μL of SAM and incubate for the remaining 2 h.
3. Recover the DNA from the reaction utilizing QiaQuick gel extraction kit columns. Mix the 400 μL reaction with 1.2 mL of QG buffer and 400 μL of isopropanol. Vacuum the mix through a column. Wash twice with 800 μL of PB buffer and spin the column at 16,100*g* for 2 min. Eluate the DNA with 200 μL of EB buffer.
4. Perform the in vitro methylation reaction on the recovered genomic DNA one more time to ensure complete conversion of cytosine to 5-methylcytosine. DNA recovery from the columns can be maximized by performing the elution step 4 times, each with 50 μL of elution buffer. In vitro methylated DNA can be stored at −20°C for up to 6 months.
5. Adjust the concentration of the 100% in vitro methylated DNA to 20 ng/μL. Sonicate and concentration-adjust PBL genomic DNA also to 20 ng/μL. Mix the 100% in vitro methylated DNA with the PBL genomic DNA in ratios to obtain the 12-point DNA methylation gradient shown in **Table 19.1** (*see* **Note 1**).

Table 19.1
Preparation of an in vitro DNA methylation standard

DNA methylation	0%	1.6%	3.2%	6.4%	12.5%	25%	50%	75%	87.5%	93.6%	96.8%	100%
In vitro methylated DNA (μL)	–	0.8	1.6	3.2	6.25	12.5	25	37.5	43.75	46.8	48.4	50
Sonicated PBL DNA (μL)	50	49.2	48.4	46.8	43.75	37.5	25	12.5	6.25	3.2	1.6	–
Total Volume (μL)	50	50	50	50	50	50	50	50	50	50	50	50
DNA amount (μg)	1.0	1.0	1.0	1.0	1.0	1.0	1.0	1.0	1.0	1.0	1.0	1.0

3.2. DNA Isolation from Samples

Since the genomic DNA samples used for Bio-COBRA are bisulfite treated, the isolation of high molecular weight DNA, as detailed below, is not required. Thus, commercially available kits, such as DNeasy (Qiagen) may be utilized to rapidly obtain low molecular weight DNA. However, the protocol provided yields, in general, larger amounts of DNA in comparison to DNA isolation kits.

1. Freeze in liquid nitrogen 50–100 mg of human tissue with 1.0 mL of lysis buffer.
2. Crush the frozen tissue to a fine powder with a mortar and pestle.
3. Add 10 μL of proteinase K to the crushed tissue and incubate at 60°C for at least 1 h. The incubation step is complete when tissue is completely dissolved and no DNA clumps are visible.
4. Extract the DNA using phenol/chloroform.
5. Dialyze the DNA samples against 4 L of 10 mM Tris–HCl, pH 8.0, for 1 h to remove all traces of phenol. Repeat the dialysis step twice.
6. Precipitate the DNA using 2.5 volumes of cold 100% ethanol.
7. Resuspend the DNA in TE to a final concentration of 20–40 ng/μL.

3.3. Bisulfite DNA Treatment (DNA Standard and Samples)

1. Prepare all reagents that are needed for bisulfite treating of DNA fresh, shortly before they are needed. Prevent their exposure to light by wrapping them in foil.
2. Add ~150 μL of 10 M NaOH per 10 mL of 3 M $NaHSO_3$ solution to raise its pH to 5.
3. Adjust the concentrations of all DNA samples to 20 ng/μL.

4. Add 5 μL of 3 M NaOH to 1 μg of DNA (50 μL volume) and incubate at 42°C for 30 min.
5. Following the incubation step, add 515 μL of 3 M $NaHSO_3$, pH 5.0, and 30 μL 10 mM hydroquinone to each sample (600 μL final volume).
6. Mix all reagents, wrap the samples in foil (to exclude light), and incubate at 50°C for 16 h in a hybridization oven.
7. Purify the bisulfite-treated samples using a QiaQuick gel extraction kit. To do so, add 1.8 mL of QG buffer and 600 μL of isopropanol to each sample (3.0 mL final volume).
8. Mix all reagents well and vacuum through a spin column. Wash the column twice with 800 μL of PB buffer and spin the column at 16,100*g* for 2 min. Eluate the DNA with 50 μL of EB buffer.
9. Add 5 μL of 3 M NaOH to each sample (50 μL volume) and incubate at 42°C for 30 min. Following the incubation, add 10 μL of 5 M sodium acetate to each sample.
10. Purify the samples using a QiaQuick gel extraction kit by adding 195 μL of QG buffer and 65 μL of isopropanol (325-μL final volume).
11. Vacuum the mix through a spin column. Wash the column twice with 800 μL of PB buffer and spin the column at 16,100 g for 2 min. Eluate the DNA with 300 μL of EB buffer. The bisulfite-treated DNA can be stored for up to 1 year at −20°C.

3.4. PCR Amplification

Design oligonucleotide primers to amplify the target sequence of interest. In order to reduce the PCR bias toward the preferential amplification of either methylated or unmethylated sequences, the primer binding sites should not contain CpG dinucleotides. Free online software to aid in the design of primers for DNA methylation analysis is available (http://www.urogene.org/methprimer/index1.html *(18)* and http://bisearch.enzim.hu) *(19)* (*see* **Note 2**).

1. Amplify by PCR the bisulfite-treated DNA methylation standard and samples of interest using the following reaction mix: 10 μL bisulfite-treated DNA, 5 μL 10 × PCR buffer, 1 μL 10 mM dNTP mix, 1 U Platinum Taq polymerase, 2 μL oligonucleotide mix (10 pmol each), and DNase–RNase free water (50 μL final volume).
2. Place the PCR reactions in a thermal cycler and activate the polymerase by incubating at 95°C for 10 min. Carry out the amplification reaction for 35 cycles, using 96°C as the denaturing temperature in each cycle: for example, [95°C for 10 min, (96°C for 30 s, 60°C for 30 s, 72°C for 30 s) × 35 cycles, 72°C for 10 min]. Dimethylsulfoxide (DMSO) might be added to the PCR reactions in order to help amplify the target sequence. However, high DMSO concentrations (>5%)

can inhibit the DNA polymerase. Thus, consider adding more units of enzyme if high DMSO concentrations are needed for the desired PCR reaction to work (*see* **Note 3**).

3. Check for PCR amplification by running 5 μL of each reaction in an 8% polyacrylamide gel. Run the gel for at least 1 h at 250 V. Visualize the PCR products by staining the gel with a 0.001% ethidium bromide solution.
4. Clean the remaining PCR product (45 μL) using a QiaQuick gel extraction kit by adding 135 μL of QG buffer and 45 μL of isopropanol (225 μL final volume) to each sample.
5. Vacuum the mix through a spin column. Wash the column twice with 800 μL of PB buffer and spin the column at 16,100 g for 2 min. Eluate the DNA with 50 μL of water.
6. Concentrate the eluated PCR product to a final volume of 7 μL. Place the samples in a vacuum concentrator heated to 60°C (if a heat setting is available). It takes 10–20 min for the samples to reach the desired volume (depending on the number of samples being concentrated). Concentrating the PCR samples is required in order to achieve a DNA concentration within the dynamic range of the Agilent DNA 1000 LabChip chemistry (5–50 ng/μL). This step is absolutely necessary if the expected restriction pattern is comprised of multiple (four or more) restriction fragments.

3.5. Restriction Enzyme Digestion of PCR Products

1. Digest the PCR products with the appropriate restriction enzyme. The choice of enzyme is based on the restriction sites present in the target sequence. Preferably, the restriction digestion should not generate more than eight potential restriction fragments. Mix the following reagents and incubate at the appropriate temperature for at least 2 h: 7 μL PCR product, 10 U restriction enzyme, 1 μL restriction enzyme buffer, 0.1 μL BSA, and DNase–RNase free water (10-μL final volume).
2. Check the restriction digestion by running 5 μL of each reaction in an 8% polyacrylamide gel. Run the gel for at least 1.5 h at 250 V. Consider using a loading dye, such as Yellow-Sub (GENEO BioProducts, Hamburg, Germany), that will not migrate over the expected restriction fragments. Visualize the restriction digestion by staining the gel with a 0.001% ethidium bromide solution. Check to ensure complete digestion was obtained by including appropriate controls, such as 100% in vitro methylated DNA. Incomplete digestion in the 100% methylated sample could be due to either incomplete in vitro methylation or incomplete enzymatic digestion. Consider checking the in vitro methylation efficiency by PCR amplifying and digesting an aliquot of the *Sss*I-treated DNA prior to preparing the DNA methylation standard. Restriction digestion in the 0% sample could be due to either basal DNA

methylation in PBL DNA or incomplete bisulfite conversion (*see* **Note 4**).

3.6. Electrophoresis in Microfluidics Chips

The Agilent DNA 1000 chemistry must be equilibrated to room temperature before mixing and loading the micro chip. Reset the plunger of the Chip Priming Station to 1 mL before removing primed chips so as to avoid spills over the station's rubber seal. A compromised seal could result in insufficient pressure to be delivered to the chip's capillary network

1. Follow exactly the protocol provided by Agilent for the DNA 1000 LabChip. Load 1 μL of the restriction digestion into a DNA 1000 LabChip well. Perform the electrophoresis step by running the Agilent 2100 Bioanalyzer (*see* **Note 5**).
2. Export the raw data (CSV file) for each of the samples and plot the raw data from each sample (fluorescence and migration time) in Excel. The Agilent 2100 Bioanalyzer includes a software package that can be used to determine the peak area of each of the peaks generated during the electrophoretic run for each of the samples. However, we strongly encourage the use of peak height, instead of peak area, for calculating DNA methylation percentages *(20)* (*see* **Note 6**).

3.7. Calculation of DNA Methylation Percentages

1. Measure the peak height for each of the restriction fragments and calculate the total DNA methylation level by using the following formula: fluorescence of methylated products / (fluorescence of methylated products + fluorescence of unmethylated product) as shown in **Fig. 19.1**. To help determine

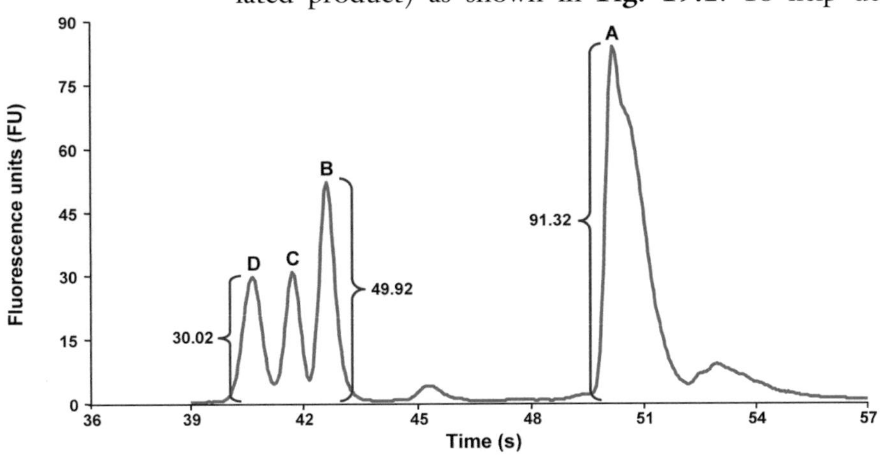

Fig. 19.1. Electropherogram generated by plotting Bioanalyzer raw data into Excel. The Y axis in the plot indicates fluorescence intensity and the X axis indicates migration time in seconds. Each peak in the electropherogram represents a DNA fragment. As in conventional DNA electrophoresis, DNA fragments migrate according to their molecular weight. Since smaller fragments are detected first, their fluorescence intensities are recorded the earliest in the plot (at 39 s in the example provided). Peak A is the result of the signal generated by the undigested portion of the PCR product. Peaks B, C, and D each represent a DNA fragment product of the restriction digestion carried out on the PCR product. The percent methylation of the sample shown is calculated utilizing the formula provided in item **3.7.1**. The fluorescence units for each peak are calculated by measuring the height of each peak (examples A, B, and D).

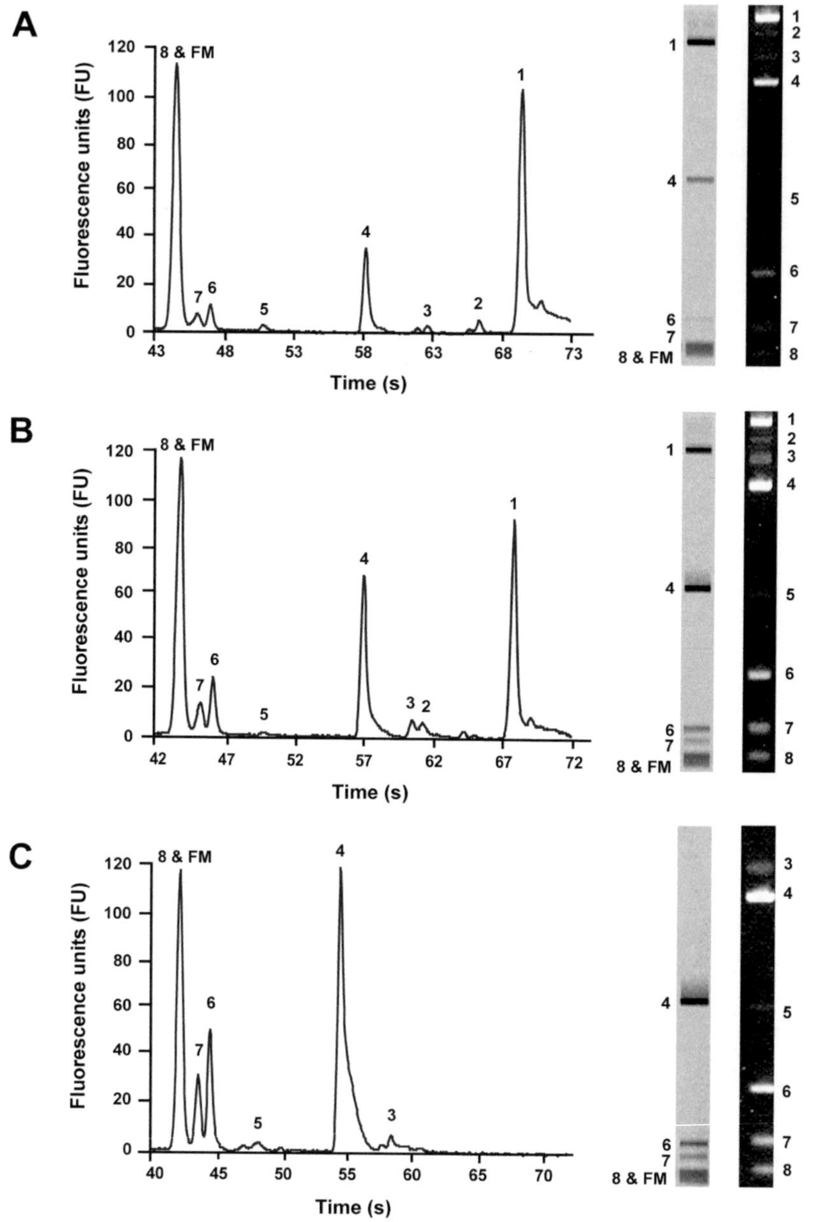

Fig. 19.2. (**A**) Restriction digestion of a 12.5% methylated sample electrophoresed in an 8% polyacrylamide gel (*right lane*). Virtual gel for the same sample automatically generated by the Bioanalyzer (*left lane*) and the corresponding electropherogram (plot). The numbers to the left and the right of the gel lanes correspond to the peaks in the electropherogram. It should be noted that in this example, a 17-bp fragment (indicated with the number 8) co-migrates with the front marker. The front marker (FM) is a 15 bp DNA fragment included in all Bioanalyzer runs and is used to calibrate the system. The overlapping of the 17 bp fragment with the front marker makes the 17 bp fragment unquantifiable. However, the elimination of this fragment from the quantification process would affect all samples equally, thus reducing the bias created by the elimination of one digestion product. Also, the low fluorescence intrinsically emitted from such a short fragment is, in most cases, only a small fraction of the total fluorescence generated by all the restriction fragments in the sample. (**B**) Same set up as in A, but for a 50% methylated sample. (**C**) Same set up as in A and B, but for a 96.5% methylated sample.

which peak corresponds to which restriction fragment, it is useful to generate a plot of the DNA size marker that is added to the chip in each experiment. The fragment size for each of the DNA markers is available in the Agilent DNA 1000 LabChip chemistry product sheet. Truncated peaks or lack of high molecular weight peaks indicate the early titration of the fluorochrome. This is likely due to DNA concentrations well above the dynamic range of the electrophoresis system (tested up to 65 ng/μL). If this phenomenon is observed, perform the restriction reaction in a larger volume (e.g., 20 μL) (*see* **Note 7**).

2. Plot the 12 points of the DNA methylation standard in Excel and generate a function. Utilizing the equation derived from that function allows for the calculation of the true methylation percentages for the experimental samples, since it corrects for any potential amplification bias inherent to differences in the PCR amplification efficiency of methylated versus unmethylated alleles.
3. Electrophoresis of digested PCR products generated from PCR reactions with average efficiency (~10–30 ng/μL) yield fluorescence peaks well above background. The function derived from the 12 point DNA methylation standard is usually linear or logarithmic, depending on the difference in PCR amplification efficiency between the methylated and unmethylated alleles. A comparison among an 8% polyacrylamide gel, a virtual gel generated by the Bioanalyzer and the corresponding electropherogram is shown in **Fig. 19.2**.

4. Notes

1. The accuracy of the gradient rests, first, on the precise determination of the DNA concentration of both, the 100% in vitro methylated DNA and the sheared PBL DNA samples. Using a NanoDrop spectrophotometer can help obtain accurate absorbance readings while minimizing sample loss. Second, careful measurement of the indicated volumes is essential to maintain the appropriate ratio of the two reagents.
2. PCR amplicons, whenever possible, should be less than 400 bp in length and should not contain more than three sites for the chosen restriction enzyme. Given the nature of the quantitative scheme used in Bio-COBRA, over three restriction sites of interest per amplicon could potentially generate a large number of restriction fragments (derived from fully methylated and partially methylated alleles), which would increase the likelihood of producing overlapping peaks on the Bioanalyzer's electropherogram (overlapping peaks result in lower

quantitative accuracy). By the same token, amplicons with more than three restriction sites of interest have a relatively higher chance of generating fragments under 25 bp, which is the current lowest resolution limit for the Agilent DNA 1000 chemistry employed in this assay.

3. Utilizing DMSO at various concentrations (1–5%) with concurrent changes in annealing temperature may help in the optimization of PCR conditions. It might also be useful to reduce the extension time of the PCR cycle to 10–15 s if larger than expected products are generated.
4. Bisulfite convert 100 ng of the *Sss*I-treated DNA, PCR amplify the CpG island of a housekeeping gene, and directly sequence the PCR product. If C and T peaks overlap at any CpG dinucleotides, the in vitro methylation reaction was not complete. If only C peaks are detected at all CpG locations, the enzymatic digestion was not complete.
5. The DNA 1000 Lab-on-a-Chip can accommodate 12 samples per run. Since each chip can only be used once, it is recommended that samples be processed, whenever possible, in multiples of 12 as a way to maximize the cost-efficiency of the assay. If fewer than 12 samples are to be assayed, 1 μL of water should be added to each unused well to maintain constant reaction volumes in the entire chip.
6. For the purpose of the assay, accurate quantification of DNA methylation levels can be best achieved by measuring the peak height of each of the signals in the electropherogram *(21)*. To calculate the peak height of a signal, simply place the mouse arrow at the top of the peak in the electropherogram (graphed in Excel).
7. In performing nucleic acids electrophoresis in a DNA 1000 Lab-on-a-Chip, DNA fragments migrate according to size. Thus, small DNA fragments are detected before larger fragments are. In the electropherogram, the X axis represents time in seconds, increasing from left to right. Thus, peaks toward the left of the electropherogram are generated by the smaller DNA fragments that resulted from the restriction digestion. Conversely, larger fragments migrate further to the right.

Acknowledgement

We would like to thank Herbert Auer and Dr. Karl Kornacker, both of whom were involved in the development and testing of this protocol.

References

1. Jones, P. A., Laird, P. W. (1999) Cancer epigenetics comes of age. *Nat Genet* 21, 163–167.
2. Jones, P. A., Baylin, S. B. (2002) The fundamental role of epigenetic events in cancer. *Nat Rev Genet* 3, 415–428.
3. Clark, S. J., Harrison, J., Frommer, M. (1995) CpNpG methylation in mammalian cells. *Nat Genet* 10, 20–27.
4. Franchina, M., Kay, P. H. (2000) Evidence that cytosine residues within 5′-CCTGG-3′ pentanucleotides can be methylated in human DNA independently of the methylating system that modifies 5′-CG-3′ dinucleotides. *DNA Cell Biol* 19, 521–526.
5. Jabbari, K., Bernardi, G. (2004) Cytosine methylation and CpG, TpG (CpA) and TpA frequencies. *Gene* 333, 143–149.
6. Gardiner-Garden, M., Frommer, M. (1987) CpG islands in vertebrate genomes. *J Mol Biol* 196, 261–282.
7. Larsen, F., Gundersen, G., Lopez, R., et al. (1992) CpG islands as gene markers in the human genome. *Genomics* 13, 1095–1107.
8. Szyf, M., Pakneshan, P., Rabbani, S. A. (2004) DNA demethylation and cancer: therapeutic implications. *Cancer Lett* 211, 133–143.
9. Baylin, S. B., Herman, J. G., Graff, J. R., et al. (1998) Alterations in DNA methylation: a fundamental aspect of neoplasia. *Adv. Cancer Res.* 72, 141–196.
10. Bastian, P. J., Palapattu, G. S., Lin, X., et al. (2005) Preoperative serum DNA GSTP1 CpG island hypermethylation and the risk of early prostate-specific antigen recurrence following radical prostatectomy. *Clin Cancer Res* 11, 4037–4043.
11. Friedrich, M. G., Chandrasoma, S., Siegmund, K. D., et al. (2005) Prognostic relevance of methylation markers in patients with non-muscle invasive bladder carcinoma. *Eur J Cancer* 41, 2769–2778.
12. Levenson, V. V. (2004) DNA methylation biomarkers of cancer: moving toward clinical application. *Pharmacogenomics* 5, 699–707.
13. Brock, M. V., Gou, M., Akiyama, Y., et al. (2003) Prognostic importance of promoter hypermethylation of multiple genes in esophageal adenocarcinoma. *Clin Cancer Res* 9, 2912–2919.
14. Usadel, H., Brabender, J., Danenberg, K. D., et al. (2002) Quantitative adenomatous polyposis coli promoter methylation analysis in tumor tissue, serum, and plasma DNA of patients with lung cancer. *Cancer Res* 62, 371–375.
15. Maruyama, R., Toyooka, S., Toyooka, K. O., et al. (2001) Aberrant promoter methylation profile of bladder cancer and its relationship to clinicopathological features. *Cancer Res* 61, 8659–8663.
16. Maruyama, R., Toyooka, S., Toyooka, K. O., et al. (2002) Aberrant promoter methylation profile of prostate cancers and its relationship to clinicopathological features. *Clin Cancer Res* 8, 514–519.
17. Fraga, M. F., Ballestar, E., Paz, M. F., et al. (2005) Epigenetic differences arise during the lifetime of monozygotic twins. *Proc Natl Acad Sci USA* 102, 10604–10609.
18. Li, L. C., Dahiya, R. (2002) MethPrimer: designing primers for methylation PCRs. *Bioinformatics* 18, 1427–1431.
19. Tusnady, G. E., Simon, I., Varadi, A., et al. (2005) BiSearch: primer-design and search tool for PCR on bisulfite-treated genomes. *Nucleic Acids Res* 33, e9.
20. Brena, R. M., Huang, T. H., Plass, C. (2006) Quantitative assessment of DNA methylation: potential applications for disease diagnosis, classification, and prognosis in clinical settings. *J Mol Med* 84, 1–13.
21. Brena, R. M., Auer, H., Kornacker, K., et al. (2006) Accurate quantification of DNA methylation using combined bisulfite restriction analysis coupled with the Agilent 2100 Bioanalyzer platform. *Nucleic Acids Res* 34, e17.

Chapter 20

Restriction Digestion and Real-Time PCR (qAMP)

Christopher C. Oakes, Sophie La Salle, Jacquetta M. Trasler, and Bernard Robaire

Abstract

DNA methylation in mammals has been shown to play many important roles in diverse biological phenomena. Here we describe a simple and straightforward method that quantitatively measures site-specific levels of DNA methylation in a quick and cost-effective manner. The quantitative analysis of DNA methylation using real-time PCR (qAMP) technique involves the digestion of genomic DNA using methylation-sensitive and methylation-dependent restriction enzymes followed by real-time PCR. This approach generates accurate and reproducible results without the requirement for prior treatment of the DNA with sodium bisulfite.

Key words: DNA methylation analysis, method, restriction enzyme, real-time PCR.

1. Introduction

Epigenetic modifications play a central role in genome control. DNA methylation, one such modification, takes place at approximately 30 million CpG sites throughout the genome and involves the transfer of a methyl group to the fifth carbon of the cytosine ring via the activity of DNA methyltransferases (DNMTs). Methylation of DNA is generally associated with transcriptional repression in addition to highly specialized processes such as genomic imprinting and X chromosome inactivation *(1)*. It is also increasingly apparent that modifications to DNA methylation patterns are involved in cancer development as well as aging *(2, 3)*.

Several methods offering varying degrees of resolution and sensitivity can be employed to survey DNA methylation at an overall genomic level or at specific sequences. The introduction of bisulfite-genomic sequencing was a significant advance in the

Jörg Tost (ed.), *DNA Methylation: Methods and Protocols, Second Edition, vol. 507*

DOI 10.1007/978-1-59745-522-0_20 Springerprotocols.com

ability to establish the methylation status of all CpG dinucleotides within a given region and remains unsurpassed in the degree of resolution and information it provides *(4, 5)*. However, bisulfite sequencing remains time consuming and costly, especially when the methylation state of a large number of loci has to be investigated. Furthermore, sodium bisulfite conversion causes significant sample loss due to DNA degradation *(6)*, and may not be conducive for the generation of quantitative results *(7)*.

Here we describe a simple and straightforward technique that can provide a quantitative evaluation of the methylation status of multiple CpG sites in a quick, cost-effective manner without the need for sodium bisulfite treatment. qAMP was initially developed to confirm and further investigate results obtained by restriction landmark genomic scanning, and its efficacy in determining levels of DNA methylation in a range of sequences has been further evaluated *(8–12)*. qAMP involves three basic steps (**Fig. 20.1**): (1) the design of primers to a specific genomic region; (2) the creation of a set of PCR templates by independently digesting the DNA sample with several methylation-sensitive restrictions enzymes (MSREs) and a methylation-dependent restriction endonuclease (MDRE); (3) real-time PCR amplification. Briefly, both MSRE and MDRE restriction sites are identified in the DNA sequence of interest and primers are designed to flank these restriction sites. The set of PCR templates created for each DNA

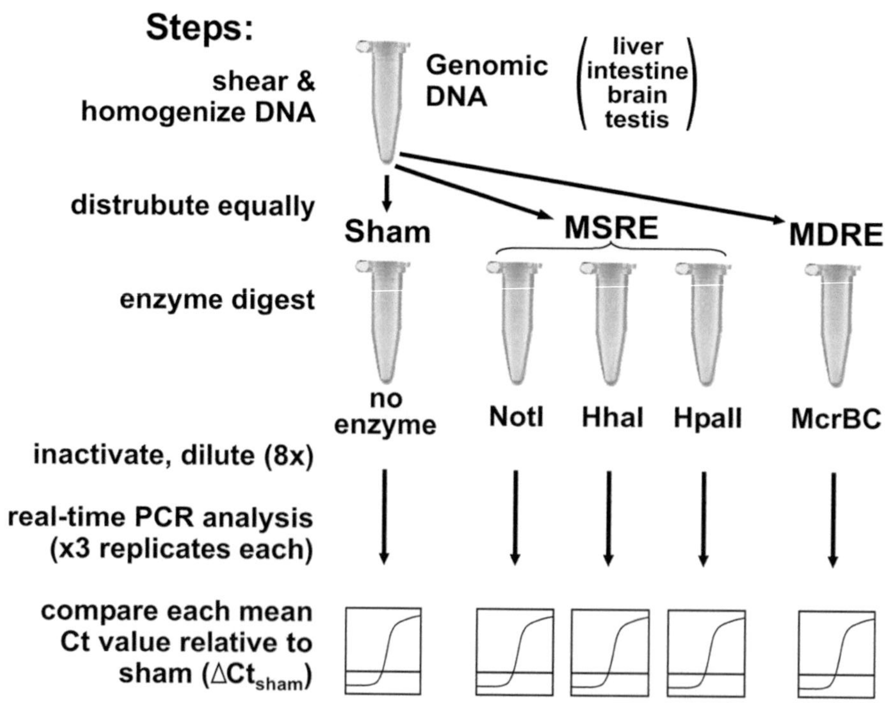

Fig. 20.1. Schematic diagram outlining the qAMP procedure (reproduced from *(8)* with permission).

sample is composed of separate tubes of DNA of equal concentration that have been digested with a single restriction enzyme. Restriction enzyme digests are divided into three categories: (1) sham group – a mock digestion with no enzyme; (2) the MSRE digests group – the DNA is cleaved if the restriction site(s) are unmethylated; and (3) the MDRE digest group – the DNA is cleaved only if methylated. Real-time PCR amplification of each template is performed, yielding cycle threshold (Ct) values that are used to calculate the percentage of DNA methylation at a given restriction site.

By generating data in a single day, qAMP is a substantially faster and more economical technique than bisulfite sequencing, though fewer CpG dinucleotides are evaluated. The simplified primer design strategy and the increased template stability are additional advantages. Overall, qAMP is a fast, cost-effective alternative to bisulfite sequencing for screening purposes of site- and region-specific levels of DNA methylation. qAMP results have been demonstrated to be comparable with those determined using bisulfite sequencing *(8, 10)*.

2. Materials

1. DNeasy Blood & Tissue kit (Qiagen, Germantown, MD), or equivalent DNA extraction kit.
2. 1 mL syringes with 27 1/2-gauge needles.
3. Restriction enzymes: *Hha*I and *Mcr*BC (New England Biolabs, Ipswich, MA); *Hpa*II and *Not*I (Invitrogen, Carlsbad, CA).
4. QuantiTect™ SYBR® Green PCR kit (Qiagen), or equivalent SYBR® Green PCR kit.
5. Mx3000P Real-time PCR machine (Stratagene, La Jolla, CA), or equivalent quantitative PCR thermal cycler.

3. Methods

Implementation of the qAMP procedure primarily depends upon the design and placement of PCR primers to flank informative restriction enzyme sites. If possible, primers are chosen to flank both MSRE and MDRE restriction sites within the region of interest. There are several factors that should be taken into consideration when designing a qAMP experiment, such as primer design (*see* **Note 1**), the special properties of *Mcr*BC (*see* **Note 2**), the number of restriction sites owing to a particular enzyme (*see* **Note 3**), and percent range-specific differences in accuracy between MSRE and MDRE digests (*see* **Note 4**). Additional primers can also be used in parallel to control for equal DNA distribution and stability within a set of templates (*see* **Note 5**). Each of these factors contributes to the successful generation and interpretation of qAMP results.

The protocol depicted below specifically describes the procedure using a five-template set: three MSREs (*Hha*I, *Hpa*II, *Not*I), one MDRE (*Mcr*BC), and sham templates. The qAMP technique can be used with any combination and number of MSREs that would cleave within a region of interest; however, some properties of various restriction enzymes make them better suited for use with the qAMP technique (*see* **Note 6**).

3.1. DNA Isolation and PCR Template Preparation

1. Isolate genomic DNA from tissues or cells using a DNA isolation kit according to the manufacturer's instructions (we use the DNeasy Blood & Tissue kit from Qiagen). Alternatively, the standard phenol–chloroform DNA isolation procedure may be used (*see* **Note 7**). RNA removal is not necessary. The amount of starting material required depends on the number of digested templates desired. Best results are obtained when 0.5–1.0 μg of DNA is used per digestion reaction. Therefore, for a five-template set, 2.5–5.0 μg of DNA are required. Highly reproducible results can also be obtained with substantially less starting material (*see* **Note 8**).
2. Increase the volume of the DNA sample to 250 μL by adding water. This is the final volume required for a five-template set; 40 μL of this DNA sample will be used per digestion reaction.
3. Homogenize the DNA by repeatedly passing the sample through a 27 1/2-gauge needle attached to a 1-mL syringe. This reduces the average DNA fragment size to ~4–5 kb.
4. Distribute 40 μL of DNA sample to five individual 1.5-mL tubes. Digest the individual templates with the appropriate enzyme for ~4–5 h in a total reaction volume of 50 μL. Use 25 units each of *Not*I, *Hpa*II, *Hha*I, and *Mcr*BC to digest the different templates according to the manufacturers' suggested conditions. For *Mcr*BC, use twice the recommended concentration of GTP. The sham template is incubated in the presence of 1 × reaction buffer (NEB buffer 2). All reactions are performed in the presence of 1 × BSA.
5. (Optional) Heat inactivate the restriction enzymes by incubating the reactions at 65°C for 20 min (*see* **Note 9**).
6. Dilute each template eightfold with 350 μL of water. The templates are ready to use for analysis using quantitative PCR (*see* **Note 10** *for storage conditions*).

3.2. Quantitative PCR and Data Analysis

1. Prepare a master mix for the desired number of reactions (the procedure described here has been optimized for the use of the QuantiTect™ SYBR® Green PCR kit; reaction conditions may vary if other kits are used). The total volume for

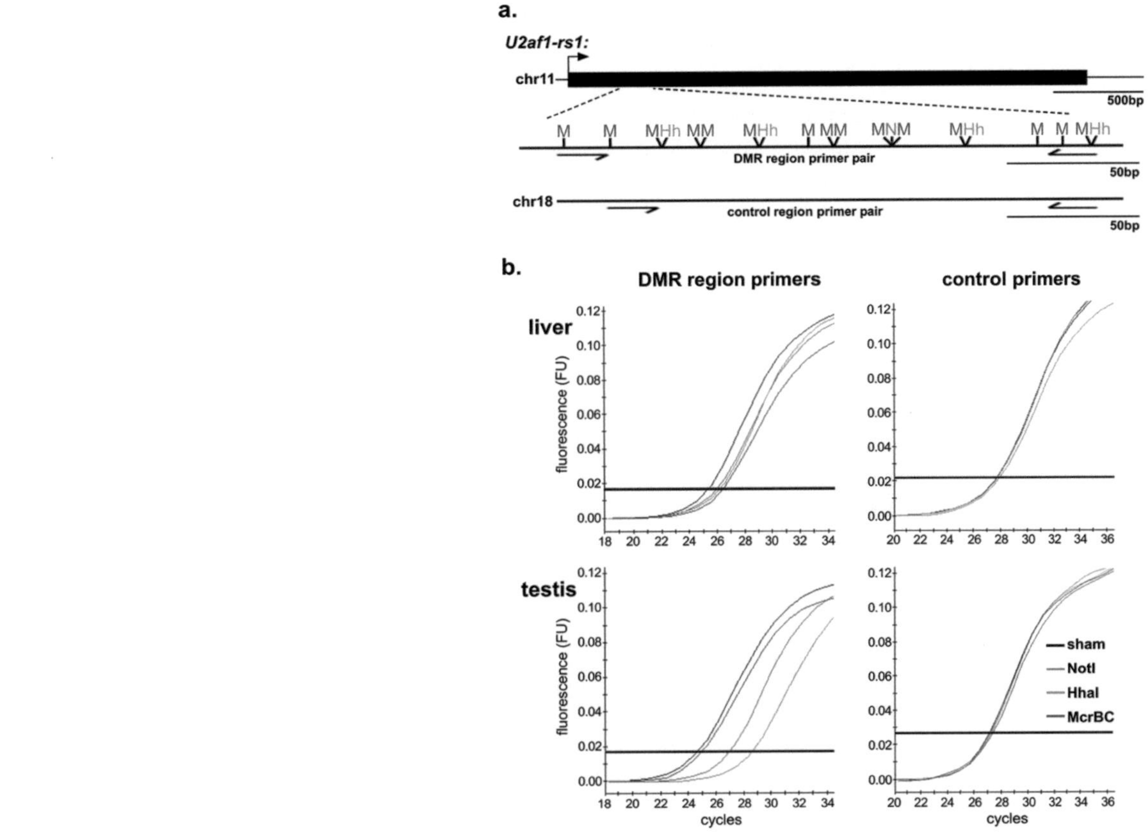

c.

liver

	Ct		ΔCt (sham)	
Template Digest	DMR region	Control	DMR region	Control
sham	25.3	27.8	\	\
NotI	26.1	27.8	0.8	0.0
HhaI	26.0	28.0	0.7	0.2
McrBC	26.3	27.8	1.0	0.0

testis

	Ct		ΔCt (sham)	
Template Digest	DMR region	Control	DMR region	Control
sham	24.8	27.1	\	\
NotI	27.1	27.3	2.3	0.2
HhaI	28.8	27.1	4.0	0.0
McrBC	25.1	27.4	0.3	0.3

Fig. 20.2. The evaluation of the differentially methylated region (DMR) of the imprinted gene, *U2af1-rs1*, in testis and liver of mouse. (**a**) Primers are designed to flank *NotI* (N), *HhaI* (Hh), and *McrBC* (M) restriction sites in the DMR region. (**b**) Liver and testis genomic DNA templates are PCR amplified following digestion with no enzyme (sham), *NotI*, *HhaI*, or *McrBC*. Amplification using a second primer pair that has been designed to a sequence devoid of restriction sites demonstrates that the templates are of equal concentration. (**c**) The difference in the Ct value of an enzyme-digested template relative to the sham-digested template determines the levels of DNA methylation in each tissue. For clarity, in (**b**) only single values are shown, whereas values in (**c**) are means of three replicates (reproduced from *(8)*; with permission). (*see* Color Plate 10)

each PCR reaction is 20 μL; each reaction contains 10 μL of 2 × PCR master mix, 1.0 μL each of the forward and reverse primers (7.5 pmol/μL), 5.5 μL water, and 2.5 μL of DNA (added separately).

2. Using the template set created in **Section** 3.1, setup the PCR reactions as follows: transfer 17.5 μL of the master mix into each well and add 2.5 μL of the appropriate template for a final reaction volume of 20 μL. Pipetting less than 2.0 μL of digested template per well can result in variability. Templates should be amplified in triplicate; however, data can be analyzed from reactions done in duplicate if the Ct values of the paired samples vary by less than 0.3 cycles.
3. Perform the following amplification reaction on the Mx3000P Real-time PCR (Stratagene) instrument: (a) initial Taq activation: 95°C for 15 min. (1 cycle); (b) denaturation: 94°C for 15 s; and annealing: 55–65°C (as determined for each primer pair) for 30 s; extension: 72°C for 30 s (40 cycles). SYBR Green is used to monitor the formation of the PCR amplicon while ROX serves as a passive reference dye to account for pipetting differences. Nonspecific amplification is monitored by a melting curve analysis performed after the amplification phase.
4. Data analysis: Ct values are obtained by real-time PCR amplification of the various templates and are calculated using the MxPro v3.00 software (Stratagene). Individually subtract the mean Ct values of the MSRE- and MDRE-digested templates from the mean Ct value of the sham-digested template to produce a ΔCt value for each digest. Calculate the percentage of methylation of a given site by using the following relationships: for MSREs, percentage of methylation = $100 \times (2^{-\Delta Ct})$; for MDRE (*Mcr*BC), percentage of methylation = $100 \times (1{-}2^{-\Delta Ct})$ (*see* **Note 11**). An example of amplification curves, the resulting ΔCt values, and associated percentage of methylation are shown in **Fig. 20.2**.

4. Notes

1. Primers are designed using the Primer3 software (http://www.frodo.wi.mit.edu/cgi-bin/primer3/primer3_www.cgi) according to the standard principles for successful quantitative PCR outlined in the QuantiTect™ SYBR® Green PCR Handbook from Qiagen. There are some key principles to bear in mind in designing primers. PCR primers should be 18–30 nucleotides in length and have a GC content of 40–60%. Product sizes range from 100 to 200 bp. To reduce primer–dimer formation, it is important to avoid at the 3′ ends of primer pairs (1) complementarity

of two or more bases, (2) runs of three or more G's or C's, (3) mismatches between the 3′ end of the primer and the target-template sequence, and (4) having a T because primers with a T at the 3′ end have a greater tolerance of mismatch. Because the relationship between ΔCt values and percent methylation assumes a reaction efficiency of 100%, efforts should be made to ensure that the PCR reaction is performing optimally (>70% efficiency). Slight reductions in efficiency do not overtly affect results. All primer pairs should be pretested to identify the optimal annealing temperature and tested for the formation of nonspecific products.

2. *Mcr*BC recognizes two half-sites of 5′-G/A^{m}C-3′ *(13)*. Because in mammals, methylated cytosines are primarily restricted to CpG dinucleotides, this sequence must be followed by a guanine residue to produce a *Mcr*BC recognition site. The recognition sequence is a nonpalindrome, thus, along with 5′-G^{m}CG-3′ and 5′-A^{m}CG-3′,*Mcr*BC will also recognize the sequences 5′-^{m}CGC-3′ and 5′-^{m}CGT-3′ (complementary to the recognition sequence). Optimal separation of the two half-sites is 55–103 bp and the enzyme cleaves the DNA in between the two sites approximately 30 bp from either site *(13)*. In regions of moderate to high CpG density there will be many *Mcr*BC sites and primer placement is not a concern. However, in regions of low CpG density, where only a few sites are present, best results are obtained when *Mcr*BC sites are separated by the optimal distance and primers are positioned such that there is at least 20 bp between the primer and the *Mcr*BC site. This ensures that cleavage will occur within the flanked region. MSRE sites can be anywhere between the primer sites. Additional note: non-CpG cytosine methylation can be present in some biological contexts *(14, 15)*, which could cause cleavage at additional G/A^{m}C sites.
3. An important aspect in the interpretation of qAMP results is the consideration of the number of restriction sites for a particular enzyme within the amplified region. Percent methylation results for primers positioned to flank a single MSRE site will reveal a true methylation percentage; however, if more than one site for the same enzyme is present, the percentage of methylation is a reflection of all sites. This is interpreted differently for the two classes of enzymes. For MSREs, if there are multiple sites for the same enzyme, the %methylation value is representative of the percent of DNA that is methylated at all of the sites. The *Mcr*BC enzyme only requires two half-sites that are methylated within the amplified region to cleave DNA despite the methylation status of other sites. As a result, the methylation value in percentage is truly representative of the percent of DNA that

is methylated at two or more sites, or, alternatively, the inverse percentage (100% methylated) represents the percentage of DNA that is totally unmethylated. Incongruent results from the two classes of enzymes reflect heterogeneity in the methylation state between neighboring restriction sites.

4. The overall sensitivity of the assay is dependent on the range of DNA methylation in the region analyzed. The relationship between the percentage of methylation and changes in the Ct value is curved such that MSREs are more sensitive to changes in methylation in the lower percent range and MDREs are more sensitive in the higher percent range *(8)*. Although all PCR reactions are run in triplicate, small variations in mean Ct values of ±0.3 cycles can occur, which can result in a significant change in the calculated level of methylation. For example, if a DNA region is 90% methylated, a 0.3 Ct variation will result in an error of 17% for MSREs; however, the same variation will result in an error of only 2% for MDREs. To evaluate a full range of methylation, a combination of data from both classes of enzymes must be considered.
5. Control primers can be designed to a region that is devoid of any of the restriction sites of the enzymes used in the experimental design. This region can be located in unique (single-copy) sequence anywhere in the genome. PCR amplification of each digested template of a given sample with the control primers ensures that the templates are of equal concentration and that nonspecific cleavage of the DNA sample has not occurred. Acceptable variability in mean template Ct values range from ±0.3 cycles. Additional note: *Mcr*BC templates will sometimes display variability despite the lack of identified recognition sites. This may be due to unavoidable background levels of non-CpG cytosine methylation.
6. Enzymes are ideally chosen to provide uniformity among templates by promoting maximal DNA stability. To limit DNA degradation during both digestion and template storage, ensure that the digestion conditions provided by the recommended reaction buffer provide a pH greater than 7.4. Lower pH reactions are conducive to DNA degradation and will produce erroneous results. It is also not ideal to use enzymes that digest at different temperatures than the sham digestion as the pH of buffers generally decreases with increasing temperature. Also, the enzyme must be thermolabile, as enzymatic activity must be extinguished during the initial step (15 min at 95°C) of the PCR reaction.
7. The use of a DNA isolation procedure that includes phenol–chloroform or any other organic solvents must include a

dialysis step prior to precipitation (minimum 3 × 4 h in 4 L of 10 mM Tris–HCl pH 8.0).

8. Consistent results can be obtained using as little as 20 ng of starting material per digestion reaction. However, there appears to be some loss of sensitivity when 2 ng or less of DNA is used *(8)*. When small amounts of DNA are used, the reaction volume, amount of enzyme, digestion time, and dilution volume should be decreased. For example, in step 4 of **Section** 3.1, DNA amounts of 200 ng, 20 ng, and 2 ng (per digestion reaction) are digested in a volume of 10 μl with 5 units of enzyme for 2 h, 40 min, and 20 min, respectively. In step 5, the templates are diluted to 150 μl if ~200 ng has been used, or to 50 μl for <20 ng. These modifications will help in maintaining the accuracy of the assay at lower DNA starting amounts. As a guideline, enough DNA should be used in each subsequent PCR reaction such that the Ct value of the sham reaction remains below 30 cycles.
9. If using small amounts of DNA (<100 ng/digestion), heat inactivation may cause some DNA degradation and is not recommended. This effect has not been observed when using higher (>100 ng/digestion) concentrations.
10. Due to the absence of a buffering agent and the dilute nature of the DNA sample, templates are subject to degradation. Degradation will produce erroneous results. Templates are best used soon after digestion and can be kept at 4°C for short-term usage. For longer-term storage, templates can be kept at either −20°C or −80°C. Consistent results have been obtained from templates that have been frozen for several months and that have undergone several freeze/thaw cycles.
11. The ΔCt versus percent methylation relationship is derived from the basic principle that each successive round of PCR amplification results in approximately a twofold increase in the amount of product. Thus, a ΔCt of 1.0 indicates that 50% of the template has been cleaved, 2.0 equals 75% cleavage, etc. The relationship for *Mcr*BC follows an inverse function.

References

1. Goll, M. G., Bestor, T. H. (2005) Eukaryotic cytosine methyltransferases. *Annu Rev Biochem* 74, 481–514.
2. Liu, L., Wylie, R. C., Andrews, L. G., et al. (2003) Aging, cancer and nutrition: the DNA methylation connection. *Mech Ageing Dev* 124, 989–898.
3. Ting, A. H., McGarvey, K. M., Baylin, S. B. (2006) The cancer epigenome – components and functional correlates. *Genes Dev* 20, 3215–3231.
4. Clark, S. J., Harrison, J., Paul, C. L., et al. (1984) High sensitivity mapping of methylated cytosines. *Nucleic Acids Res* 22, 2990–2997.
5. Frommer, M., McDonald, L .E., Millar, D. S., et al. (1992) A genomic sequencing protocol that yields a positive display of

5-methylcytosine residues in individual DNA strands. *Proc Natl Acad Sci USA* 89, 1827–1831.
6. Grunau, C., Clark, S. J., Rosenthal, A. (2001) Bisulfite genomic sequencing: systematic investigation of critical experimental parameters. *Nucleic Acids Res* 29, e65.
7. Munson, K., Clark, J., Lamparska-Kupsik, K., et al. (2007) Recovery of bisulfite-converted genomic sequences in the methylation-sensitive QPCR. *Nucleic Acids Res* 35, 2893–2903.
8. Oakes, C. C., La Salle, S., Robaire, B., et al. (2006) Assessment of a quantitative, non-bisulfite-based method for the analysis of DNA methylation using real-time PCR (qMAP). *Epigenetics* 1, 146–152.
9. Oakes, C. C., Kelly, T. L., Robaire, B., et al. (2007) Adverse effects of 5-aza-2′-deoxycytidine on spermatogenesis include reduced sperm function and selective inhibition of de novo DNA methylation. *J Pharmacol Exp Ther* 322, 1171–1180.
10. Oakes, C. C., La Salle, S., Smiraglia, D. J., et al. (2007) Developmental acquisition of genome-wide DNA methylation occurs prior to meiosis in male germ cells. *Dev Biol* 307, 368–379.
11. Oakes, C. C., La Salle, S., Smiraglia, D. J., et al. (2007) A unique configuration of genome-wide DNA methylation patterns in the testis. *Proc Natl Acad Sci USA* 104, 228–233.
12. La Salle, S., Oakes, C. C., Neaga, O. R., et al. (2007) Loss of spermatogonia and widespread DNA methylation defects in newborn male mice deficient in DNMT3L. *BMC Dev Biol* 7, 104
13. Stewart, F. J., Panne, D., Bickle, T. A., et al. (2000) Methyl-specific DNA binding by McrBC, a modification-dependent restriction enzyme. *J Mol Biol* 298, 611–622.
14. Imamura, T., Kerjean, A., Heams, T. et al. (2005) Dynamic CpG and non-CpG methylation of the *Peg1/Mest* gene in the mouse oocyte and preimplantation embryo. *J Biol Chem* 280, 20171–20175.
15. Ramsahoye, B. H., Biniszkiewicz, D., Lyko, F., et al. (2000) Non-CpG methylation is prevalent in embryonic stem cells and may be mediated by DNA methyltransferase 3a. *Proc Natl Acad Sci USA* 97, 5237–5242.

Chapter 21

MethylQuant: A Real-Time PCR-Based Method to Quantify DNA Methylation at Single Specific Cytosines

Claire Dugast-Darzacq and Thierry Grange

Abstract

MethylQuant is a cost-effective and relatively simple technique which enables quantitative analysis of the methylation status of a single cytosine at specific positions in DNA that can be assimilated to the quantitative detection of a single nucleotide polymorphism (SNP). After bisulfite conversion of DNA and PCR amplification of the region of interest, the methylation status is quantified by methylation-specific real-time PCR with one of the primers harboring the methylation status-specific nucleotide at the most 3′ end. In parallel, the amount of amplifiable DNA is quantified by a methylation-independent real-time PCR. In this protocol, we describe in detail the different stages of the MethylQuant procedure and discuss the parameters of DNA bisulfite conversion and quantitative PCR analysis with SYBR green that are crucial to achieve an accurate quantification of the methylation status of a particular cytosine. The practical aspects of DNA bisulfite conversion, primer design, and quantitative PCR analysis, discussed hereafter, should be of general interest even outside the context of the MethylQuant technique.

Key words: Epigenetics, bisulfite, DNA methylation, quantitative PCR, SYBR green, locked nucleic acid.

1. Introduction

Cytosine methylation at CpG dinucleotides contributes to the epigenetic maintenance of gene silencing. It can affect gene expression in various ways, both by favoring the establishment and maintenance of repressive chromatin structure and by interfering with the recruitment of sequence-specific activating transcription factors. The formation of repressive chromatin is favored by proteins interacting with methylated DNA and belonging in particular to the methyl-binding domain (MBD) and Kaiso family

Jörg Tost (ed.), *DNA Methylation: Methods and Protocols, Second Edition, vol. 507*

DOI 10.1007/978-1-59745-522-0_21 Springerprotocols.com

(1). In general, the efficiency of repression depends on the density of methylcytosines (^{Me}C) *(2)*. Methylation of several neighboring CpGs can be analyzed by numerous methods, ranging from the immunoprecipitation of methylated DNA with ^{Me}C-specific antibodies followed by analyses using either quantitative PCR or DNA microarrays *(3)*, to various PCR-based strategies relying on the methylation-induced change in reactivity of DNA toward enzymes or, most often, chemicals *(4, 5)*. These latter strategies can also be used or adapted to monitor the methylation status of single cytosines. Such specific methylation events can play key regulatory roles, ranging from the control of genomic imprinting to the fine-tuning of the memory of hormonal induction during development, by affecting the recognition of the surrounding DNA by sequence-specific DNA-binding proteins *(6)*.

Thus, techniques that allow the monitoring and quantification of the methylation status of individual cytosines can be extremely useful to analyze these phenomena. Most of the available techniques rely on combinations of methylation-sensitive chemical modifications and PCR analysis. C5-methylation of cytosine modifies the reactivity of the nucleotide toward hydrazine, potassium permanganate ($KMnO_4$) and sodium bisulfite, either by increasing its reactivity ($KMnO_4$) or by decreasing it (hydrazine, bisulfite) *(5)*. $KMnO_4$ and hydrazine modify cytosine so that piperidine-induced cleavage of the corresponding deoxyribose in DNA is favored, thus allowing monitoring of the original methylation status by relative quantification of the DNA backbone breaks using ligation-mediated PCR (LM-PCR) *(5, 7)*. Sodium bisulfite promotes the deamination of unmethylated cytosine in single-stranded DNA, allowing their alkali-induced conversion into uracil, which can be detected using conventional PCR-based approaches *(8)*. Unmethylated cytosines will ultimately be read as thymines, whereas methylated cytosines (and unmethylated cytosines in undenatured or readily re-annealed double-stranded DNA) will be read as cytosines. Since conventional PCR is much easier to perform than LM-PCR, bisulfite conversion became the method of choice for the analysis of DNA methylation at specific cytosines. PCR products can be cloned and sequenced, providing a detailed image of the distribution of the methylation patterns on individual molecules. Alternatively, the bulk PCR product can be directly sequenced, in particular using pyrosequencing *(9,* **Chapter 15***)*. This latter approach allows high-throughput analysis of the extent of methylation of several individual cytosines at neighboring positions, but requires specific instrumentation and is relatively costly. Less costly high-throughput real-time PCR approaches have been developed that quantify DNA methylation levels over a wider dynamic range of methylation levels, in particular, MethyLight *(10,* **Chapter 23***)*, HeavyMethyl *(11,* **Chapter 24***)*, QAMA *(12)* and MethylQuant

(13). MethyLight, HeavyMethyl, and QAMA are all similar in that they use an internal fluorescent label probe to quantify the PCR products corresponding to a specific methylation status. To achieve efficient discrimination between methylated and unmethylated DNA, and thus a good signal-to-noise ratio, the probe must cover several neighboring CpGs. These methods are thus appropriate to analyze CpG islands that can be found either in a fully methylated or fully unmethylated state. Since this kind of situation is often encountered in tumor cells, these methods are useful to study cancer epigenetics *(14)*.

MethylQuant contrasts with these methods in that it allows the quantification of the methylation levels of a single specific cytosine, and thus it parallels the quantitative detection of a single nucleotide polymorphism (SNP). MethylQuant is useful either to analyze regions with a lower density of CpGs or to study situations where the methylation of neighboring CpGs is not identical, which can be observed at regulatory sequences during gene activation *(6, 13, 15)*. MethylQuant is a real-time PCR approach that can be performed using SYBR Green I as the fluorescent probe, which renders it more economical than the aforementioned real-time approaches using fluorescent sequence-specific probes. The method is accurate and specific enough to quantify levels of methylation at a specific cytosine within a genome, even when the proportion of the analyzed species is as low as 1%. MethylQuant is based on a methylation-specific real-time PCR performed with one of the primers harboring a methylation status-specific nucleotide at the 3′ end *(13)*. In addition, the total product is quantified using a methylation-independent real-time PCR amplification, thus making it possible to calculate the proportion of a particular methylation status. MethylQuant might also be used to quantify changes in methylation levels in situations that have been characterized previously using more conventional bisulfite-PCR approaches, but we do not recommend using this technique without any preliminary characterization. Indeed, MethylQuant is not simple enough to analyze cytosines one by one. Furthermore, since quantification requires reference samples corresponding to the bisulfite-converted sequence carrying either the fully methylated or the fully unmethylated cytosine of interest, it is useful to perform first a bisulfite conversion followed by PCR, cloning, and sequencing analyses which provide the most convenient and reliable reference samples.

The principle of MethylQuant is described in detail in **Fig.** 21.1. First, genomic DNA is treated with bisulfite (*see* **Sections 3.1** and **3.2**). Next, the region containing the CpG of interest is amplified using conventional PCR with primers that cover regions that are unlikely to contain methylated cytosines (*see* **Section 3.3**). Only after this first amplification, real-time PCR reactions are performed to quantify both the total amount of

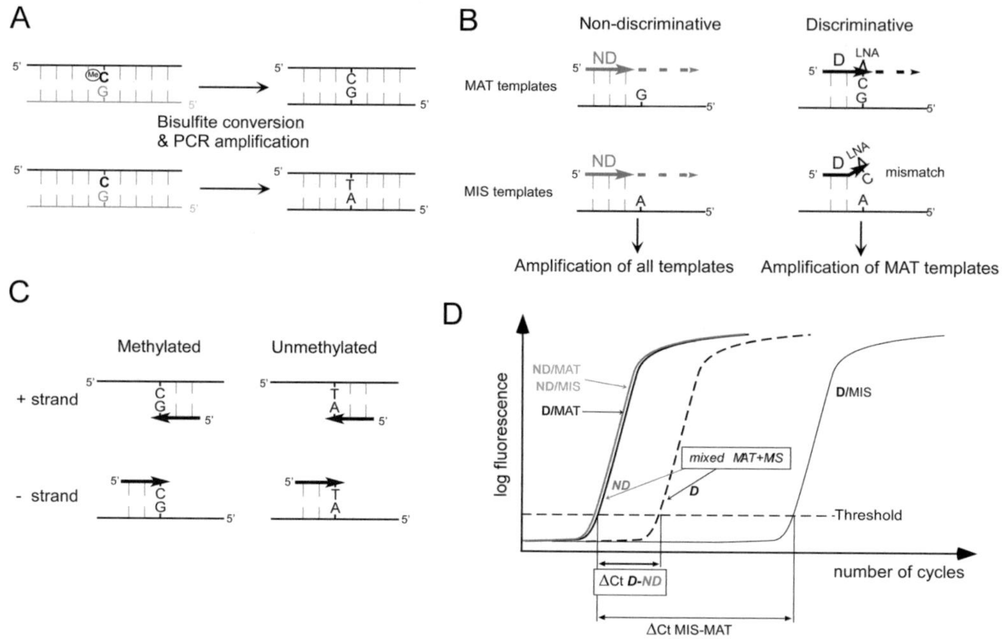

Fig. 21.1. Principle of MethylQuant (reprinted with permission from *(13)*). (**A**) Nucleotide sequence modifications resulting from bisulfite conversion and PCR amplification. Only the cytosine of interest is represented. Since, following bisulfite conversion, the two DNA strands are no longer self-complementary, different primers are required. Thus, analysis of the complementary lower strand of the target sequence (*gray* lettering) is not shown. (**B**) Design of the D primers (*black*) or ND (*gray*) between the methylation states of the cytosine of interest. The MAT templates are matched to the D primer, whereas the 3′-most base of the D primer is mismatched to the MIS templates. The LNA at the 3′-most position is represented. (**C**) The four possible designs of a D primer. Following the first round of PCR amplification of the bisulfite-converted target sequence, either of the two strands (referred to as + or −) of the PCR product can be analyzed as represented. The D primer can be matched to the sequence corresponding to either the methylated or unmethylated cytosine. (**D**) Typical real-time PCR amplification profiles obtained with the ND (*gray*) and D (*black*) primer sets and various combinations of MIS and MAT templates. The log of the amount of PCR product is plotted in function of the number of cycles. The dashed amplification curve corresponds to a sample containing a mixture of MIS and MAT templates. The threshold of PCR product synthesized allowing Ct determination is represented as a dashed line. The $\Delta Ct_{MIS-MAT}$ obtained by comparing the amplification of the MAT and MIS templates with the D primer set is represented as well as the ΔCt_{D-ND} obtained by comparing the amplification of a mixture of MIS and MAT templates with the ND and D primer sets.

amplified material and the fraction corresponding to the methylation status analyzed. The first amplification step prevents the introduction of quantification biases that can result from the use of small amounts of starting material and from the heterogeneity of the quality of the various samples. Furthermore, it makes it possible to analyze sequences that are difficult to amplify and offers a greater number of options for the design of the primers (see below). This amplification step is performed in a buffer containing tetramethylammonium chloride (TMAC), which has proved particularly useful for the amplification of the difficult sequences that are often encountered when using primers that do not cover regions containing methylated cytosine *(5)*. Indeed, in this case, the primer content is very rich in AT due to the conversion of cytosines to thymines, and TMAC improves the hybridization properties of such primers *(16)*. The PCR product is then extensively diluted and real-time PCR analyses are performed using SYBR Green I for the detection of the PCR products (*see* **Section 3.4**). For each sample, two real-time PCRs are performed using either one of the two primer sets: (1) a nondiscriminative (ND) primer set that allows quantification of the PCR product irrespective of the methylation status of the position of interest, and (2) a discriminative (D) primer set that allows quantification of the PCR product corresponding to the specific methylation status of the position of interest (**Fig. 21.1B**). The discriminative primer within the D set is designed so that the most 3′ nucleotide is complementary to the position analyzed. The presence of a locked nucleic acid (LNA) at this 3′-most end ensures efficient discrimination between the sequences corresponding to the two methylation states *(13)*. LNA is a nucleic acid analog with a 2′-O, 4′-C methylene bridge that locks the ribose moiety into a C3′-endo conformation, thus increasing the hybridization specificity of primers *(17)*. The ND primer set resembles as closely as possible the D primer set to provide similar amplification efficiencies, which facilitates the quantification analysis, and to ensure that the PCR products generated with each primer set emit fluorescence at similar intensities when bound by SYBR Green I. We generally obtain such similarity using a primer in the ND set that is almost identical to the discriminative primer but shifted by as few base pairs as possible (ideally only by one base) so that its 3′ end directly abuts the position analyzed. In addition, an identical partner primer is used in the ND and D sets.

Primers should be designed with particular care to obtain good real-time PCR data: the efficiency of PCR amplification should be maximized and almost identical for the ND and D sets and the primers should have a low tendency to produce primer dimers (*see* **Section 3.4.1**). Even though the 3′-end of the discriminative primer is constrained, there is some flexibility for primer design. In principle, four different primers can be chosen to analyze the methylation status of a specific cytosine, two for the methylated status and two for the unmethylated status on either of the two strands of the amplified product (**Fig. 21.1C**). However, not all of these four possibilities can be considered in all cases, since it is more accurate, when one of the two species is present at low levels, to measure directly the least abundant species than to deduce its levels from the measurements of the most abundant species. In situations where the cytosine of interest is flanked by another CpG that might undergo changes in methylation levels, it is possible to use primers overlapping this cytosine by introducing an inosine at the corresponding position without introducing biases in the quantification *(13)*. Indeed, inosine can base pair with any of the four conventional bases with approximately equal strength *(18)*.

Quantification is performed by comparing the PCR products originating from the bisulfite-converted DNA samples to two references, one corresponding to the template that is perfectly matched to the discriminative primer (hereafter termed the MAT template), and one corresponding to the alternative methylation status that is mismatched to the discriminative primer (referred to as the MIS template) (**Fig. 21.1B–D**). Pure MIS and MAT templates are obtained from cloned PCR products. The efficiency of PCR amplification with the ND and D primer sets is deduced from the slope of a titration curve with varying amounts of the reference MAT template. Amplification with identical amounts of the MIS template provides a measure of the background of the procedure, that is, the extent of illegitimate PCR amplification occurring with mismatched discriminative primer. This PCR amplification occurs with a delay, measured in cycle numbers, which we term $\Delta Ct_{MIS-MAT}$, where the threshold cycle (Ct), is the fractional cycle number at which one can detect a threshold level of PCR product. The value of $\Delta Ct_{MIS-MAT}$ determines the sensitivity of the approach, that is, it will prevent detection of the proportion of the MAT template that is lower than the corresponding background (Bgd) value. The background value (in %) can be calculated using the following equation:

$$\mathrm{Bgd} = \frac{100}{(1+\mathrm{E})^{\Delta Ct_{MIS-MAT}}}, \qquad (21.1)$$

where E is the efficiency of the PCR amplification varying between 0 and 1. For example, if the amplification efficiency is maximal, a $\Delta Ct_{MIS-MAT}$ of five cycles reveals that the background of the procedure is $100/2^5$ that is, about 3% of the initial mismatched population. Using discriminative primers with a 3′-LNA, we generally obtain $\Delta Ct_{MIS-MAT}$ values between 10 and 15 cycles (**Table** 21.1), whereas regular primers were, by far, less discriminative with $\Delta Ct_{MIS-MAT}$ values between 3 and 9 cycles *(13)*.

Usually, the sample PCR products contain a mixture of MAT and MIS templates because the cytosine methylation status is rarely homogeneous within the cell population. If the percentage of MAT templates is above the background value, the PCR amplification curve is found at an intermediate position (dotted lines in **Fig. 21.1D**). The delay between the amplification obtained with the ND and D primer sets, ΔCt_{D-ND}, depends on the proportion of MAT templates within the population. The percentage of MAT templates can be calculated using the following equation:

$$\mathrm{MAT} = 100\frac{(1+\mathrm{E}_{ND})^{Ct_{ND}}}{(1+\mathrm{E}_{D})^{Ct_{D}}} - \mathrm{Bgd}, \qquad (21.2)$$

where Ct_{ND} and Ct_D are determined at an identical level of detected PCR product and E_{ND} and E_D are the efficiencies of the PCR amplifications with the ND and D primer sets, respectively. In general, the background value can be neglected as long as $\Delta Ct_{D-ND} << \Delta Ct_{MIS-MAT}$. Furthermore, if the values measured are close to that of the background, one can reasonably refrain from quantifying. As seen from equation (21.2), the percentage of MAT templates is highly dependent on the efficiencies of the PCR amplification with the two primer sets, thus accurate measurement critically depends on the accuracy of these values. We recommend calculating them from the slope of at least three independently measured titration curves (*see* **Section 3.4.2**).

We describe hereafter a detailed step-by-step protocol of MethylQuant starting from a mammalian genomic DNA preparation, because the importance of a quick procedure providing clean DNA should not be underestimated. We then describe successively *(1)* bisulfite conversion using DNA embedded in agarose beads, which prevents loss of DNA during the experimental procedure and ensures optimal bisulfite reactivity by maintaining the DNA in the single-stranded form *(19)*; (2) PCR amplification in TMAC-containing buffer, which increases very significantly the success rate of PCR amplification of bisulfite-treated DNA; (3) the various stages of MethylQuant per se, with a description of key aspects of primer design and of real-time PCR quantification. We believe that many of the practical aspects we discuss here could be of general use even outside of the context of the MethylQuant technique.

2. Materials

Solutions should be stored at the temperature indicated in brackets. The * indicates the products that should be added extemporaneously. We use sterile distilled water (dH_2O) for the preparation of any solution.

2.1. Preparation of Genomic DNA

1. Wizard genomic DNA purification kit (Promega #A1120) containing Nuclei Lysis solution (RT for room temperature).
2. Trypsin solution (−20°C) (Invitrogen #15-400-054).
3. RNase solution (10 mg/mL in water) (−20°C): once the RNase is dissolved in water, boil for a few minutes in order to inactivate eventual DNase contamination.
4. PBS 10 × (RT): dissolve 80 g of NaCl, 2 g of KCl, 14.4 g of Na_2HPO_4 and 2.4 g of KH_2PO_4 in 800 mL of distilled H_2O. Adjust the pH to 7.4 with HCl. Add H_2O to 1 L.
5. PBS 1 × (4°C): 100 mL of PBS 10 × in 900 mL of H_2O.

6. 0.1 M phenylmethanesulphonylfluoride (PMSF) in isopropanol (−20°C).
7. PBS-PMSF (4°C): add PMSF* to the PBS solution at a final concentration of 1 mM (*see* **Note 1**).
8. TE 1 × (RT): 10 mM Tris–HCl, 1 mM EDTA, pH 8.0.
9. Ethanol 100% (−20°C).
10. Ethanol 70% in water (−20°C).

2.2. Bisulfite Conversion

1. 2 M NaOH solution (RT).
2. Sodium Bisulfite powder (RT) (Sigma #243973).
3. Hydroquinone powder (RT) (VWR #24704.298).
4. Low Melting Point (LMP) agarose SeaPlaque GTG (Cambrex #50111) (RT).
5. 1 M HCl (RT).
6. Heavy white mineral oil (Sigma M-5904) (RT).

2.3. PCR Amplification

1. Taq Buffer 10 × (−20°C): 670 mM Tris–HCl, pH 8.8, 67 mM $MgCl_2$, 166 mM ammonium sulphate, 5 mM tetramethylammonium chloride (Sigma # T3411), 1.7 mg/mL BSA (DNAse free).
2. Taq DNA polymerase (−20°C).
3. dNTPs: 5 mM solution (−20°C).
4. Primers 20 μM (−20°C).

2.4. Quantitative PCR Analysis

1. Primers: 3′-locked nucleic acid (LNA) discriminant primer, nondiscriminant primer, and common primer (50 μM) (−20°C).
2. LightCycler FastStart DNA Master^{PLUS} SYBR Green I mix (Roche # 03752186001) (−20°C).
3. LightCycler capillaries (Roche # 11909339001).
4. LightCycler 1.5 equipment (Roche) (*see* **Note 2**).

3. Methods

3.1. Preparation of gDNA

We detail here our own method for preparing genomic DNA, since we found it fast and simple and we obtained good results with it. Other protocols routinely performed in the laboratory might be equally suited to obtain pure DNA samples that will ensure good bisulfite-conversion rates.

1. Cultured cells were grown in 10 cm dishes and media was changed 1 day prior to collection (*see* **Note 3**).
2. Wash the cells with PBS 1 ×.
3. Add trypsin and allow it to cover all the cells.
4. Aspirate rapidly the trypsin (before the cells come off the plate).

5. Incubate the plates at RT for a few minutes until the cells come off the plate.
6. Add PBS-PMSF (*see* **Note 4**) to collect cells into a 15-mL screw cap polypropylene tube (*see* **Note 5**).
7. Centrifuge cells 2 min at 600*g* at 4°C and resuspend in 1 mL of cold PBS-PMSF.
8. Transfer cells into a 1.5-mL Eppendorf tube and centrifuge again for 2 min at 600*g* at 4°C.
9. Remove the supernatant leaving a few microliters of it in the tube and resuspend the cells in this small volume.
10. Freeze tubes in liquid nitrogen and store at −80°C.
11. Thaw cells on ice.
12. Add 600 μL of nuclei lysis solution and pipette up and down to dissociate the aggregates.
13. Add 3 μL of RNase solution, mix well by inverting the tube several times.
14. Incubate 20 min at 37°C and let the samples cool down at RT for 5 min.
15. Add 200 μL of protein precipitation solution and vortex vigorously for 20 s.
16. Leave the samples on ice for 5 min.
17. Centrifuge for 10 min at 13,000*g* at 4°C and transfer the supernatant to a new tube (*see* **Note 6**).
18. Add 560 μL of isopropanol to the supernatant and invert the tube several times until the DNA filament is clearly visible.
19. Centrifuge for 5 min at 7000*g* at RT.
20. Eliminate the supernatant and wash the pellet with 600 μL of 70% ethanol, let the pellet dry on the bench (do not overdry) and add 100 μL of TE (*see* **Note 7**).
21. Sonicate DNA in order to reduce DNA viscosity and facilitate DNA denaturation, which is crucial for efficient bisulfite conversion. In our hands, this was achieved, with fragments of an average size ranging from 1 to 5 kb (*see* **Note 8**).
22. Extract DNA with phenol/chloroform; precipitate with ethanol and wash with 70% ethanol.
23. Resuspend the digested DNA in water and measure OD_{260}(e.g., using a NanoDrop spectrophotometer). DNA is now ready to use for bisulfite conversion.

3.2. Bisulfite Conversion

3.2.1. Solution Preparation

Prepare the bisulfite solution mixed with hydroquinone on the actual day of the experiment. The sodium bisulfite and hydroquinone solutions are extremely sensitive to light and "must" be protected at all times.

1. In a 15-mL screw cap polypropylene tube, mix 3.8 g of sodium bisulfite, 4 mL of H_2O, and 1.4 mL of 2M NaOH. Protect the tube from light with aluminum foil and mix gently on a rotating wheel for 20 min in order to dissolve (*see* **Note 9**).

2. At the same time, mix 450 mg of hydroquinone with 1.28 mL H_2O in a 2-mL tube. Protect the tube from light with aluminum foil and mix gently on a rotating wheel for 20 min.
3. Add 1 mL of the saturated hydroquinone solution (*see* **Note 10**) into the bisulfite solution; shake it on a rotating wheel for 10 more minutes (be careful to always protect from light). Make sure that the pH of this solution is 5.0 and adjust the total volume to 8 mL.
4. In the dark, pass the solution through a 0.45-μm filter. Keep this solution in the dark.

3.2.2. Preparation of Agarose Beads

For bisulfite conversion, DNA is embedded in agarose beads (*see* **Note 11**). This step is delicate as agarose beads need to be firm enough to resist the subsequent treatments.

1. Weigh 40 mg of LMP agarose and mix with 2 mL of H_2O in a 2-mL Eppendorf tube. Melt the agarose by placing the tube in a boiling water bath for a few minutes, then transfer and keep the solution at 42°C to prevent the agarose from solidifying.
2. During the incubation period place several 1.5-mL Eppendorf tubes (containing 200 μL of mineral oil) at −20°C for at least 15 min (*see* **Note 12**).

3.2.3. DNA Denaturation

1. In order to ultimately obtain 20-μL beads each containing 200 ng of genomic DNA, place in low retention tubes (*see* **Note 13**) 900 ng of sonicated DNA in 27 μL H_2O. Add 3 μL of 2 M NaOH solution.
2. Incubate tubes at 50°C for 15 min and transfer the samples to the 42°C water bath containing the agarose solution prepared previously.

3.2.4. Bead Preparation

1. Add 60 μL of 2% LMP agarose to the tube containing denatured DNA. Mix well.
2. Transfer 20 μL of the DNA–agarose mix to a tube with mineral oil placed on ice (*see* **Notes 14** and **15**).

3.2.5. Bisulfite Treatment

1. Add 200 μL of bisulfite solution to each tube with beads and make sure the beads are in the aqueous phase.
2. Incubate for 4 h at 50°C in the dark in a dry bath.
3. After incubation, remove carefully all liquid with a Pasteur pipette with a very fine tip. During all the various stages of washing be careful not to touch the agarose beads with the Pasteur pipette to preserve their integrity.
4. Wash the beads 4 × 15 min with 1 mL 1 × TE (pH 8.0) (*see* **Note 16**).

3.2.6. Desulfonation

1. Incubate the beads with 500 μL of 0.2 M NaOH (2 × 15 mn).
2. Neutralize (pH 7) the NaOH by adding 50 μL of 1 M Tris–HCl (pH 7.5) and 100 μL of 1 M HCL. (The Tris–HCl buffer

can be mixed with the HCl solution prior to addition) (*see* **Note 17**).

3. Wash the beads 3 × 10 min with 1 mL TE.
4. Store the beads in a small volume of TE at 4°C (they are stable for a few days) (*see* **Note 18**).

3.3. PCR Amplification

The bisulfite treatment of DNA converts unmethylated cytosines into thymines, so that the DNA strands are no longer complementary. Either the upper or the lower strand will be PCR amplified depending on the primer used. We describe the procedure and strategy for primer design first, since it is the key to the success of MethylQuant, and to bisulfite-based DNA methylation analysis in general. Primers can be designed manually with the help of a PC primer design software like *Oligo* (http://www.oligo.net/) or automatically, with the help of an online primer design software like *primer3* (http://frodo.wi.mit.edu/primer3/input.htm, *(20)*). We often combine *primer3* design with manual refining of the various choices offered using software such as *Oligo*. We recommend searching for repeated sequences within the sequence of interest and masking these repeats with *RepeatMasker* (http://www.repeatmasker.org/cgi-bin/WEBRepeatMasker).

3.3.1. Primer Design for the First PCR

1. Convert *in silico* your sequence(s) of interest into a bisulfite-treated sequence as follows (*see* **Note 19**): convert the sequence in capital letters and replace automatically every capital C with a lower case T to obtain the conversion corresponding to the upper strand and every capital G into a lower case A to obtain the conversion corresponding to the lower strand. Perform this operation for both the repeat-masked sequence and the native sequence (*see* **Note 20**).
2. Open the *primer3* input page and paste the converted sequence of interest into the sequence window (*see* **Note 21**). Force the software to select primers that surround the CpG-containing region to be analyzed by bracketing this region in the sequence with the following symbols: [and] (e.g., ttA[CG]GGT).
3. Modify the default options of the *primer3* software as follows:
   - Product size ranges: 150–250, 100–300, 300–450 (*see* **Note 22**).
   - Number to return: 10 (*see* **Note 23**).
   - Max 3′ stability: 7.0 (*see* **Notes 24** and **25**).
   - Primer size: Min: 23, Opt: 23, Max: 23 (*see* **Note 26**).
   - Primer Tm: Min: 50, Opt:, Max: 60 (*see* **Note 26**).
   - Max 3′ self-complementarity: 2.0 (*see* **Note 27**).
   - Objective Function Penalty Weights for Primer Pairs, 3′ complementarity: 1.0 (*see* **Note 27**).

4. Select from the various primer pairs proposed by *primer3* those that satisfy your needs best (*see* **Note 23**).

3.3.2. PCR Amplification

1. Wash beads twice with 500 μL H_2O and incubate for 15 min at RT to equilibrate.
2. Prepare PCR Mix1. For each sample, mix 8 μL of 10 × Taq buffer, 2 μL of forward primer (20 μM), 2 μL of reverse primer (20 μM), and 48 μL of water.
3. Remove the maximum of supernatant from the beads and add 60 μL of PCR Mix1 to the beads.
4. Heat in a boiling water bath for a few seconds to melt the agarose beads and transfer all the liquid (about 80 μL) to a PCR tube on ice.
5. Add 20 μL of PCR Mix two per sample (containing 2 μL of 10 × Taq buffer, 10 μL dNTPs, 2 Units of Taq polymerase; adjust the volume to 20 μL with water).
6. Amplify DNA with the following program (4 min at 95°C, 32 cycles of 30 s at 95°C, 1 min at 55°C, and 1 min at 72°C, then 7 min at 72°C)
7. Analyze the products on an agarose gel (*see* **Note 28**).

Important: Due to potential carryover contamination, great care must be taken during all the steps performed after this first PCR. We recommend working in a different environment (bench, pipettes, and solutions) that is dedicated solely to post-PCR experiments. This is of particular importance when quantification of the amplified DNA is performed in subsequent experiments or when working with a small amount of DNA.

3.4. Quantitative PCR Amplification

3.4.1. Primer Design for Quantitative PCR

We use a strategy similar to the one described in **Section 3.3.1** but with different parameters reflecting the use of a different PCR buffer, different PCR conditions, and the specific constraint of performing a real-time PCR using a primer whose 3′-most nucleotide discriminates the methylation status of the cytosine of interest.

1. Manually select the discriminative primer using the virtual conversion of the sequence(s) (*see* **Section 3.3.1** and **Notes 19** and **20**) of the product(s) originating from the PCR performed in **Section 3.3.2**. **Table 21.1** presents the properties of the primers that we used for MethylQuant (*see* **Note 29**).
2. Manually select the nondiscriminating primer (ND). Usually it is sufficient to shift the discriminating primer upstream by one base, unless visual inspection reveals that it has unsatisfactory properties.
3. The amplification primer that will pair with the D and ND primers can be selected manually or with the *primer3* software as follows: Open the *primer3* input page and paste the

Table 21.1
Properties of various oligonucleotides used in the quantification step of the MethylQuant analysis (described in *(13)*). The table represents the difference between the Ct obtained with the MIS and MAT templates ($\Delta Ct_{MIS-MAT}$), the PCR amplification efficiency (E), the background percentage of misidentified methylation status (Bgd), the Tm of the primers as determined by the nearest neighbor method *(21)*, the length of the primers, the nature of the 3′-most mismatch between the D primer and the MIS template and an internal stability plot corresponding to the distribution of the ΔG (ordinate, -kcal/mole) of each pentanucleotide hybrid along the primer length (abscissa, bases)

Primer name	Δ Ct MIS-MAT	E	Bgd	Tm (°C)	Length (base)	Mismatch	3′ΔG	Internal stability
A	12	0.78	0.10	58.3	20	A:C	-7	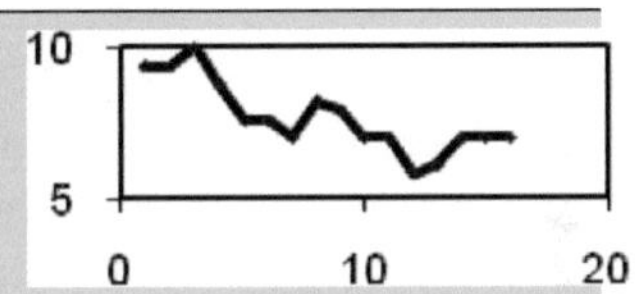
B	13.3	0.76	0.05	58.7	18	A:C	-8.2	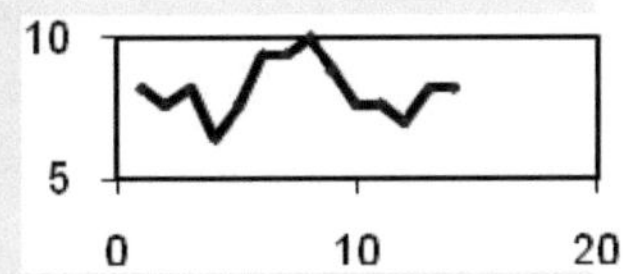
C	14.4	0.66	0.07	54.8	24	A:C	-6.9	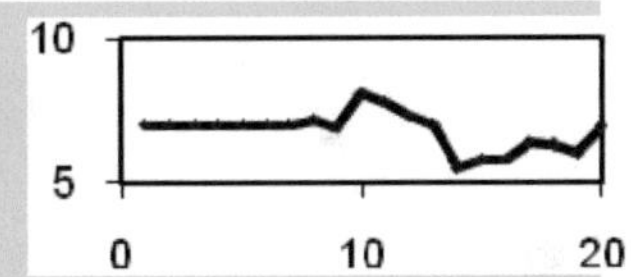
D	11.7	0.78	0.12	56.6	24	G:T	-8.6	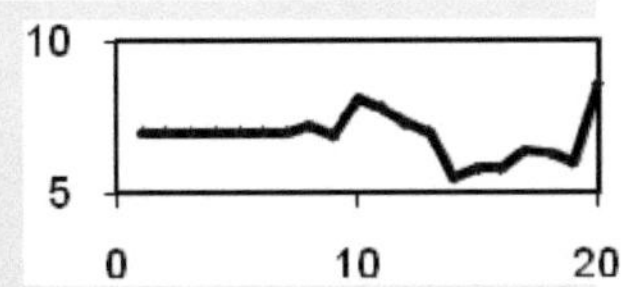
E	14	0.72	0.05	49.0	26	T:G	-7.2	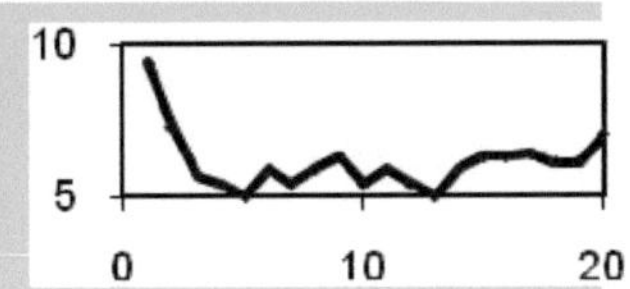
F	9.5	0.79	0.40	49.2	26	C:A	-6.9	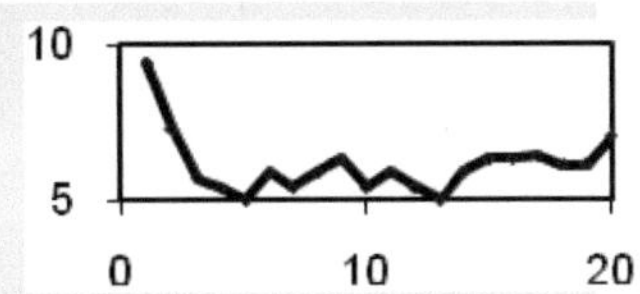

converted sequence of interest into the sequence window. Paste the sequence of the primer selected in the previous step into the appropriate "pick primer" window.

4. Modify the default options of the *primer3* software as follows:
   - Product size ranges: 80–120 60–150 50–250 (*see* **Note 30**).
   - Number to return: 10 (*see* **Note 23**).
   - Max 3′ stability: 7.0 (*see* **Note 24**).
   - Primer size: Min: 18, Opt:, Max: 26 (*see* **Note 31**).
   - Primer Tm: Min: 49, Opt: 54, Max: 60 (*see* **Note 31**).
   - Max 3′ self-complementarity: 2.0 (*see* **Note 27**).
   - Objective Function Penalty Weights for Primer Pairs, 3′ complementarity: 1.0 (*see* **Note 27**).
5. Select from the various primer pairs proposed by *primer3* those that best satisfy your needs (*see* **Note 23**).

3.4.2. Preparation of the Control Templates

In order to perform accurate quantification, the PCR analysis should be performed with control templates corresponding to either the fully methylated or the fully unmethylated cytosines of interest. These two templates are obtained after selection of sequenced-cloned, bisulfite-treated DNA (*see* **Note 32**).

1. Measure the OD_{260} of the two MET and UNMET plasmids corresponding to either the MIS or MAT templates depending on the sequence recognized by the discriminatory primer pair (DIS) (*see* **Fig. 21.1**).
2. Dilute them to obtain 10^6 molecules of plasmid DNA per μL (*see* **Note 33**).
3. Perform fivefold serial dilutions of these MIS and MAT templates (5 μL into 20 μL of water, mix, then 5 μL of the mix into 20 μL of water, mix, and repeat twice in order to obtain a serial dilution of 5 points: 10^6, $2 \cdot 10^5$, $4 \cdot 10^4$, $8 \cdot 10^3$, and $1.6 \cdot 10^3$).
4. Use the serial dilutions for the establishment of standard curves in the real-time PCR analysis.

3.4.3. Real-time PCR Quantification

1. Dilute the product from the first PCR (described in **Section 3.3.2**) 10,000-fold in water performing a serial dilution as follows: 10 μL in 90 μL of water, mix, then 10 μL of the mix into 90 μL of water, repeated three times.
2. Prepare the PCR reaction with a final reaction volume of 10 μL (if using capillaries) containing 500 nM of each primer and 2 μL of diluted DNA template. To obtain accurate measurements, it is best to minimize pipetting of reagents and samples into the capillary. Prepare a reaction mix containing primers, enzyme, and all other reagents usually provided with a Q-PCR reaction mix, so that this mix corresponds to 80% of the final volume. When using the Roche Light Cycler Master Mix Plus, combine 2 μL of 5 × Master Mix

containing the enzyme, 1 μL of forward primer (5 μM), 1 μL of reverse primer (5 μM), and 4 μL of water per sample. Prepare 10% more mix than necessary. Add to each capillary 8 μL of the reaction mix and 2 μl of the diluted template (samples, standard curves with MIS and MAT templates, or simply water to check for contamination and primer dimer formation).

3. Close the capillaries, centrifuge, and run the quantitative PCR analysis. Set up the real-time PCR cycling conditions as follow: 95°C for 8 min to activate the hot start enzyme, then 50 cycles of 10 s at 95°C, 10 s at 56°C, and 20 s at 68°C followed by fluorescence measurement. Terminate the amplification phase with a melting curve analysis at these settings: 95°C at 20°C/s, 60°C at 20°C/s, and 90°C at 0.1°C/s with continuous fluorescence measurement (*see* **Note 34**).

3.5. Data Analysis

Analyze separately the two sets of primers (D and ND), each of them should have its own standard curve, samples, and blank.

3.5.1. Analysis Using the Light Cycler Program

1. Perform a "Tm calling analysis" of the PCR and check that all the amplicons from one set of primers, except the eventual product from the blank reaction, have the same Tm (*see* **Note 35**).
2. Perform a "quantitative analysis" of the PCR and make sure the standard curves are straight (in the log scale) for each primer set. Check also that the difference in Ct for the D primer sets with the standard MIS and MAT templates is high (>10) and much greater than the difference in Ct for the analyzed samples between the D and ND primer sets (*see* **Note 36**).
3. Select the different values of Ct and concentration calculated for each sample (including standard curves) and copy them into the clipboard for further analysis using spreadsheet software.

3.5.2. Analysis and Calculation in Spreadsheet Software

1. Paste into a spreadsheet the different values of Ct and concentrations calculated for each sample, including standard curve (*see* **Note 37**).
2. Draw a graph with the different points of the standard curve representing Ct (ordinate) as a function of $\log_{10}$(concentration) (abcissa) (*see* **Note 38**).
3. Determine the slope of the standard curve.
4. Calculate the efficiency of the PCR reaction for each primer set (called E_D and E_{ND} for the D and ND sets, respectively) by using the following equation(see Note **39**).

$$1 + E = 10^{-1/\text{slope}} \tag{21.3}$$

5. If the background amplification with the D primer set and the MIS template is low (see **Section 3.5.1.2**), the background can be neglected and the percentage of MAT in

the sample can be determined using the following formula (*see* **Note 40**):

$$\mathrm{MAT} = 100\frac{(1+\mathrm{E_{ND}})^{\mathrm{Ct_{ND}}}}{(1+\mathrm{E_{D}})^{\mathrm{Ct_{D}}}} \tag{21.4}$$

4. Notes

1. The half-life of PMSF in aqueous solution is about 35 min at pH 8.0. Add to the PBS buffer just before use.
2. If not available, the LightCycler 1.5 real-time PCR system can be replaced by other equipment (i.e., for 96-wells plates). In our hands, however, LightCycler 1.5 provides the most accurate and rapid measurement.
3. We usually do the experiment with 10^7 cells. We have never felt the need to use a significantly different number of cells.
4. We use PBS containing PMSF, which acts as a serine protease inhibitor, to protect the cells from any further action of trypsin.
5. We usually add PBS-PMSF in two steps in order to collect the majority of cells. We first add 5 mL of PBS-PMSF, wash the plate, and transfer the PBS-PMSF containing the cells to the 15-mL tube. Then we add 5 mL of PBS-PMSF again to wash the plate correctly and transfer to the same tube.
6. The cellular debris is in the pellet and the genomic DNA is in the supernatant. Repeat this last step once or twice until the supernatant no longer contains any particles.
7. Leave the DNA to resuspend in TE overnight at 4°C. It is not necessary to try to resuspend the pellet manually since this would break the DNA.
8. Using the Vibro Cells VC500, genomic DNA was sonicated at 500 W (continuous output set to nine) using a cup horn, 4–6 times 30 s with intervals of 30 s between each sonication. A cup horn was used since multiple samples can be sonicated at the same time. We recommend determining, with the equipment used, the sonication settings, and times that are appropriate to obtain DNA fragments running in a gel as a smear of about 1–5 kb.
9. The sodium bisulfite solution never dissolves entirely before the addition of hydroquinone.
10. Be careful to add only the supernatant of the solution, usually the volume of liquid is just about enough.
11. Embedding in agarose beads prevents loss of DNA during the experimental procedure, facilitates subsequent manipulation and ensures optimal bisulfite reactivity by maintaining the DNA in the single-strand form and by slowing down the rapid strand re-annealing of repeated sequences.

12. It is important that the mineral oil is very cold at the time of use: take the tubes out of the freezer and put them on ice just prior to use. One tube with mineral oil is needed for each DNA sample. Extra tubes with mineral oil can be prepared. We recommend practicing the production of the beads first. Label the tubes containing mineral oil "before" placing them in the freezer, since it will not be possible to write on the frozen tubes.
13. We use Axygen low retention tubes (# 311 05 051) at this point to prevent significant material loss resulting from the use of low quantities of DNA.
14. Add the agarose to the DNA just before producing the agarose beads, since a lengthy incubation of agarose in the alkaline DNA solution causes agarose hydrolysis that adversely affects the physical properties of the beads.
15. The drops must solidify quickly in the cold oil. Leave the tubes on ice for at least 10 min to avoid disruption of the solidification process.
16. For the different wash steps, slowly invert the tubes up and down several times and then leave on the bench during the incubation time and invert again just before removing the supernatant. This must be done very gently to avoid damaging the beads.
17. It is very important to carefully determine the exact volume of HCl that is needed to neutralize the NaOH for each NaOH and HCl solution as it can be difficult to reach the proper pH.
18. Beads can be stored for a few weeks in 0.5 M EDTA or in a small volume of water at −80°C.
19. Highlight first the CpG dinucleotides to identify them after conversion.
20. The converted repeat-masked sequence should be used for the first PCR amplification because the source used for amplification will be total genomic DNA and thus primers should not hybridized to repeated DNA. The native sequence can be used for the design of the primers that are necessary for the real-time PCR step, since this latter will be performed using the PCR product of the first amplification and, thus, repeated sequences should not interfere anymore, as long as they are not repeated within the PCR product.
21. The name of the sequence can be added in the first line preceded by > with the sequence beginning on the second line (FASTA format).
22. Shorter PCR products are best, because bisulfite treatment causes some DNA degradation, but they should not be too short to ensure flexibility for the design of primers for the real-time PCR analysis. In the subsequent real-time PCR step, the size of the products should ideally range between

80 bp and 120 bp to permit optimal PCR amplification efficiency and detection with SYBR Green I. Thus, at this stage, 160–240 base long PCR fragments with the cytosine of interest in the middle will be the shortest fragments, thus leaving all options open for the real-time PCR primer design. The greater size range proposed here as a last option (450) is useful if repeated sequences surround the CpGs of interest. This length should be adapted to the size of these repeated sequences.

23. It is best to analyze the properties of the various primers manually, for example, with the help of a primer design software, to select the primer pairs that are the best placed with respect to the CpGs of interest, that have the most identical composition, the lowest tendency to form primer dimers and, if possible, the lowest 3′-end stability. The best primer pairs are not necessarily those at the top of the list and a longer list will offer more options to choose the optimal primer pair.
24. Ideally, the 5′-end of the primer should mostly contribute to the overall stability of the primer-target duplex in order to increase primer specificity and to lower the tendency to form primer dimers. This decreases the stability of the primers that are hybridized partially to nonspecific sites with a hybridized 3′-end that might be elongated by the DNA polymerase.
25. Since the PCR here is performed in a buffer that tends to render the primer's stability length dependent rather than composition dependent (*see* **Note 26**), it is not clear whether modification of this parameter is useful here, but it did prove useful in our other real-time PCR quantification experiments.
26. The 10 × Taq buffer used in this study contains tetramethylammonium chloride (TMAC). TMA^+ is a sterically bulky cation with the ability to render hybridization temperature length dependent rather than sequence dependent. It improves the efficiency and specificity of PCR from bisulfite-treated DNA, particularly since the primers have a high AT content due to the conversion of cytosines into uracils. Accordingly, we choose primers that have a uniform length rather than a uniform Tm. In the PCR conditions we describe here, primers 23-base long worked well. In contrast, the primer Tm parameters were relaxed, between 50°C and 60°C with no optimal Tm setting, since the calculations were not adapted to the buffer used.
27. To minimize the tendency to form primer dimers, we increased the stringency of the default parameters. We also forced the program to consider potential primer dimers between the reverse and forward primers by setting the value

of the penalty weight to 1 for 3′ complementarity within a primer pair.

28. Before running the samples, warm the tubes at 50°C to decrease the viscosity of the solution caused by the presence of the agarose.
29. When designing a primer, the methylation status of interest has to be chosen: methylated or unmethylated (*see* **Fig. 21.1**). This choice can be dictated by the biological questions that are being addressed. Furthermore, if there is a marked difference in the quantities of the two states, it is more accurate to measure the least abundant species and to deduce from this measurement the proportion of the most abundant one. It may thus be necessary to edit manually the converted sequence in order to replace the lower case T or A by a capital C or G at the position of interest. The discriminative primer can be either the reverse or the forward primer. This choice can be dictated by the previous primer choice. Indeed, since it is best that the length of the PCR product be comprised between 80 and 120, the position of the ends of the PCR product relative to the discriminative primer should be compatible with such a length. Furthermore, since primer dimers may be generated during the first PCR, it is important that both primers used for quantification are nested within the first PCR product so that they cannot amplify these potential dimers. PCR products that are shorter than 80 bp can be used, but the shorter the product, the most likely its Tm will be similar to that of the primer dimers. This is not necessarily a problem since good primer design should prevent the occurrence of primer dimers at the number of cycles necessary to quantify the methylation status, but when the Tm of product and dimers cannot be distinguished, the quality of the assay cannot be controlled. Once the 3′-most nucleotide and the strand analyzed with the discriminative primer have been chosen, the only remaining choice is the position of its 5′-end. This position will determine the Tm of the primer-template hybrid, and we usually select it so that the primer length ranges from 18 to 26 bases and the Tm has a value between 50 and 60°C as estimated using the nearest-neighbor method *(21)* (*see* **Table 21.1** for examples).
30. The optimal size range is between 80 bp and 120 bp, not only to allow for optimal PCR amplification efficiency and detection with SYBR Green I, but also to increase the likelihood that the Tm of the primer dimers and the PCR products will be different so that these products can be discriminated during the final melting curve analysis.

31. The primer will be selected mostly based on its Tm and the size is relaxed (no optimal size specified).
32. Prior to MethylQuant analysis, bisulfite-treated DNA of the region of interest was PCR amplified and cloned into pGEM-T (Promega # A1360) according to standard procedures (*see also* **Chapter 14**). The different clones obtained were then sequenced with one of the primers used for the PCR amplification. For the simplicity of subsequent analyses, it is best to identify plasmids corresponding to sequences where all CpGs were either methylated or unmethylated.
33. The quantity of DNA corresponding to 10^6 molecules depends on the length of the plasmid and can be calculated using the following formula: "quantity in ng for 10^6 molecules" $= 1.1 \times 10^{-6} \times$ length (bp).
34. Since PCR products from bisulfite-treated DNA are very AT rich, they often partially melt at the standard temperature used for the elongation step, that is, 72°C and thus we decreased the elongation temperature to 68°C, which in our hands was the highest temperature possible that did not cause our product to melt. Partial melting can cause erratic measurements with SYBR Green I that distort the quantification. However, it can be easily detected as the plateau of the PCR will not be flat but wavy, which is indicative of stochastic variations in the fluorescence measurements due to variations in the exact measurement temperature (**Fig.** 21.2). It is rec-

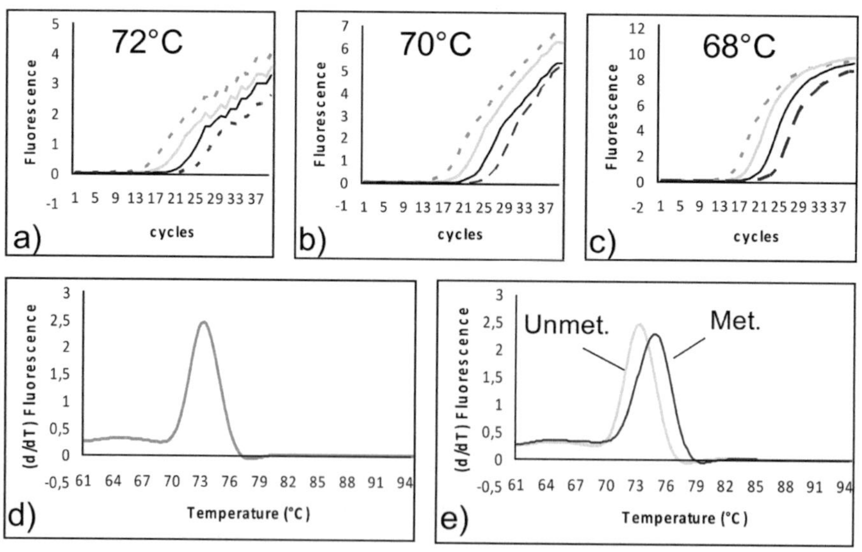

Fig. 21.2. Effect of the temperature for fluorescent measurement on real-time PCR amplification curves: 72°C (**a**), 70°C (**b**), and 68°C (**c**). (**d**) Melting curve analysis of the corresponding product. (**e**) Comparison of the melting curve analyses from two PCR products differing in their methylation status at three cytosines including the position complementary to the LNA base.

ommended to carefully check the melting curve profile of the PCR product to ensure that the product has not started to melt at the temperature of fluorescent measurement. Therefore, we recommend starting the melting curve analysis at a lower temperature than the one used for measurement, that is, 60°C. Note that almost identical products, which differ only in the methylation status of the CpGs included in the PCR product, can have distinct melting curve profiles (**Fig. 21.2e**), so that only the measurement of the unmethylated templates can be erratic at a suboptimal temperature for measurement.

35. The blank sample should yield neither product nor primer dimers. If dimers are formed, they should appear at a number of cycles significantly higher than the least abundant sample, otherwise the resulting data might be incorrect. The Tm of the dimers should be different from that of the product, but it can happen that this difference is not very significant. Check on a gel that the blank amplification product and the true product are of different sizes. Identical Tm and size is indicative of carryover contamination. If the Tm of the true product and that of the primer dimers are very similar, it will not be possible to check if the PCR quantification has not been distorted by the formation of dimers or by carryover contamination visible in the blank. Thus, if possible, it is recommended to work with primer pairs that give rise to products and dimers with distinct Tms.
36. Using the Roche software, either an automatic (second derivative) or a manual (Fit point) analysis can be performed. The automated method identifies the inflexion point of the fluorescence curve, which is determined by the second derivative maximum. The "Fit points" method uses a plot of the logarithm of the fluorescence as a function of the Ct. A fluorescence value corresponding to the beginning of the exponential amplification phase is manually set and the cycle number required to reach this threshold fluorescence level (Ct) is determined for both the samples and the standards. If there is no measurement error, then it is possible to find a fluorescence threshold that provides for all samples a very similar Ct for both the manual and the automatic procedures. Samples in which these values differ will be indicative of potential problems. Errors in the fluorescence measurements, due, for example, to pipetting errors changing the reaction volume, will affect the value determined by the "Fit points" method more than that determined by the "second derivative" method, whereas errors modifying the PCR conditions, for example, variations in primer concentration, will have a greater effect on the values determined by the "second derivative" method. We believe that errors in

fluorescence measurement are more common, but, because it is generally impossible to identify the source of error reliably, there is no robust reason to prefer data originating from one method of calculation when they differ. Depending on the accuracy of the measurements at which we are aiming, and the amplitude of the difference between the two calculation methods, these discrepancies can be treated in various ways. For very accurate measurements, or if the differences between the two calculation methods are very significant for some samples, the corresponding measurements should be ignored and the experiments redone. If differences are small or within an acceptable range given the question asked, the value obtained with both calculation methods might be averaged. If samples are quickly scanned and no precise measurements are required, the automatic "second derivative" analysis is preferred as it is less sensitive to the most common measurement errors and requires less expert and time-consuming treatment of data.

37. Two tables can be created to simplify calculations, one for the D primers set and one for the ND set.
38. For precise quantification, it is best to replicate the standard curves independently at least three times and to use the mean standard curve from this triplicate measurement for calculation.
39. If the PCR is optimal, the efficiency is 1 (100%), but this is rarely the case with bisulfite-treated DNA, and efficiency as low as 0.7 (70%) can give satisfactory results.
40. For accurate determination, perform the measurements with the ND and D primer sets in triplicate and calculate the MAT values using the nine possible pairs of Ct_{ND} and Ct_D. Then calculate the average and standard deviation of these nine MAT values.

Acknowledgements

This work was supported in part by the CNRS and by grants from the Association de Recherche sur le Cancer and the Ligue Nationale contre le Cancer. C. D. D. was supported by a fellowship from the Association de Recherche sur le Cancer. We thank E.M. Geigl, C. Brossas and A. Kropfinger for critical reading of the manuscript.

References

1. Bird, A. P. (2002) DNA methylation patterns and epigenetic memory. *Genes Dev* 16, 6–21.
2. Hsieh, C. L. (1994) Dependence of transcriptional repression on CpG methylation density. *Mol Cell Biol* 14, 5487–5494.
3. Weber, M., Hellmann, I., Stadler, M. B., et al. (2007) Distribution, silencing potential and evolutionary impact of promoter DNA methylation in the human genome. *Nat Genet* 39, 457–466.
4. Fraga, M. F., Esteller, M. (2002) DNA methylation: a profile of methods and applications. *Biotechniques* 33, 632–649.
5. Thomassin, H., Oakeley, E. J., Grange, T. (1999) Identification of 5-methylcytosine in complex genomes. *Methods* 19, 465–475.
6. Thomassin, H., Flavin, M., Espinas, M. L., et al. (2001) Glucocorticoid-induced DNA demethylation and gene memory during development. *Embo J* 20, 1974–1983.
7. Grange, T., Bertrand, E., Rigaud, G., et al. (1997) *In vivo* footprinting of the interaction of proteins with DNA and RNA. *Methods* 11, 151–163.
8. Clark, S. J., Harrison, J., Paul, C. L., et al. (1994) High sensitivity mapping of methylated cytosines. *Nucleic Acids Res* 22, 2990–2997.
9. Tost, J., Dunker, J., Gut, I. G. (2003) Analysis and quantification of multiple methylation variable positions in CpG islands by pyrosequencing. *Biotechniques* 35, 152–156.
10. Eads C. A., Danenberg, K. D., Kawakami, K., et al. (2000) MethyLight: a high-throughput assay to measure DNA methylation. *Nucleic Acids Res* 28, e32.
11. Cottrell, S. E., Distler, J., Goodman, N. S., et al. (2004) A real-time PCR assay for DNA-methylation using methylation-specific blockers. *Nucleic Acids Res* 32, e10.
12. Zeschnigk, M., Bohringer, S., Price, E. A., et al. (2004) A novel real-time PCR assay for quantitative analysis of methylated alleles (QAMA): analysis of the retinoblastoma locus. *Nucleic Acids Res* 32, e125.
13. Thomassin, H., Kress, C., Grange, T. (2004) MethylQuant: a sensitive method for quantifying methylation of specific cytosines within the genome. *Nucleic Acids Res* 32, e168.
14. Ting, A. H., McGarvey, K. M., Baylin, S. B. (2006) The cancer epigenome: components and functional correlates. *Genes Dev* 20, 3215–3231.
15. Kress, C., Thomassin, H., Grange, T. (2006) Active cytosine demethylation triggered by a nuclear receptor involves DNA strand breaks. *Proc Natl Acad Sci USA* 103, 11112–11117.
16. Melchior, W. B., Jr., Von Hippel, P. H. (1973) Alteration of the relative stability of dA-dT and dG-dC base pairs in DNA. *Proc Natl Acad Sci USA* 70, 298–302.
17. Latorra, D., Campbell, K., Wolter, A., et al. (2003) Enhanced allele-specific PCR discrimination in SNP genotyping using 3′ locked nucleic acid (LNA) primers. *Hum Mutat* 22, 79–85.
18. Martin, F. H., Castro, M. M., Aboul-ela, F., et al. (1985) Base pairing involving deoxyinosine: implications for probe design. *Nucleic Acids Res* 13, 8927–8938.
19. Olek, A., Oswald, J., Walter, J. (1996) A modified and improved method for bisulphite based cytosine methylation analysis. *Nucleic Acids Res* 24, 5064–5066.
20. Rozen, S., Skaletsky, H. (2000) Primer3 on the WWW for general users and for biologist programmers. *Methods Mol Biol* 132, 365–386.
21. Breslauer, K. J., Frank, R., Blocker, H., et al. (1986) Predicting DNA duplex stability from the base sequence. *Proc Natl Acad Sci USA* 83, 3746–3750.

Chapter 22

Methylation-Specific PCR

Julien D. F. Licchesi and James G. Herman

Abstract

Methylation-specific polymerase chain reaction (MSP) is a technique that has facilitated the detection of promoter hypermethylation at CpG islands in cell lines and clinical samples, including fresh/frozen tissues. The ability of MSP to differentiate methylated from unmethylated cytosine is dependent upon sodium bisulfite treatment of DNA which retains the methylation marks of cytosines together with the specific amplification of this modified DNA using primer sets complimentary only to the formerly methylated or unmethylated alleles. Nested-MSP (MN-MSP) is an alternative method that overcomes the limitations of MSP, especially when it comes to analyzing samples with low quality/quantity of starting DNA (e.g., paraffin-embedded specimens). MN-MSP includes a first round of amplification using primers unbiased toward the methylation status of a single (MN-MSP) or multiple (multiplex MN-MSP) genes followed by conventional MSP. Although MSP and NM-MSP are simple techniques that can easily be incorporated in most molecular biology laboratories, the ability to accurately determine the promoter methylation status of genes largely depends upon the careful design of MSP primers as well as other steps outlined in this chapter.

Key words: Methylation-specific PCR, multiplex nested-MSP, promoter hypermethylation, bisulfite modification, paraffin-embedded samples.

1. Introduction

Promoter hypermethylation in mammals occurs at cytosine residues in the context of CpG islands. These CpG islands are found at over 40% of gene promoters. In normal cells, promoters usually have unmethylated CpG islands, therefore allowing gene transcription to occur *(1)*. Methylation-specific PCR (MSP) has been essential in establishing that CpG islands at gene promoters become hypermethylated early during neoplasia *(2)*. Importantly such mechanism of gene repression affects many tumor suppressor genes whose loss of function contributes to the acquired hallmarks of cancers *(3)*, including apoptosis (*DAPK*),

Jörg Tost (ed.), *DNA Methylation: Methods and Protocols, Second Edition, vol. 507*

DOI 10.1007/978-1-59745-522-0_22 Springerprotocols.com

tumor invasiveness (*TIMP3*), cell cycle (*CDKN2A*), DNA repair (*MGMT*), and angiogenesis (*ID2*) *(4)*. MSP was first developed for assessing the methylation status of gene promoters in cell lines and fresh/frozen tissues *(5)*. Chemical modification of cytosines to uracils by bisulfite treatment is the basis of the MSP method as it is for bisulfite sequencing *(6)*. Under appropriate conditions, cytosines in DNA samples are converted to uracils, while cytosines that are methylated (5-methylcytosines) are resistant to this modification and remain cytosines *(7)*. Bisulfite-modified DNA is then amplified using specific primers that can distinguish methylated from unmethylated DNA. Because the distinction is part of the PCR amplification, extraordinary sensitivity can be achieved (typically at a detection level of 0.1% of the alleles) while maintaining specificity *(5)*. Results are obtained immediately following PCR amplification and gel electrophoresis without the need for further restriction or sequencing analysis. The detection of bands

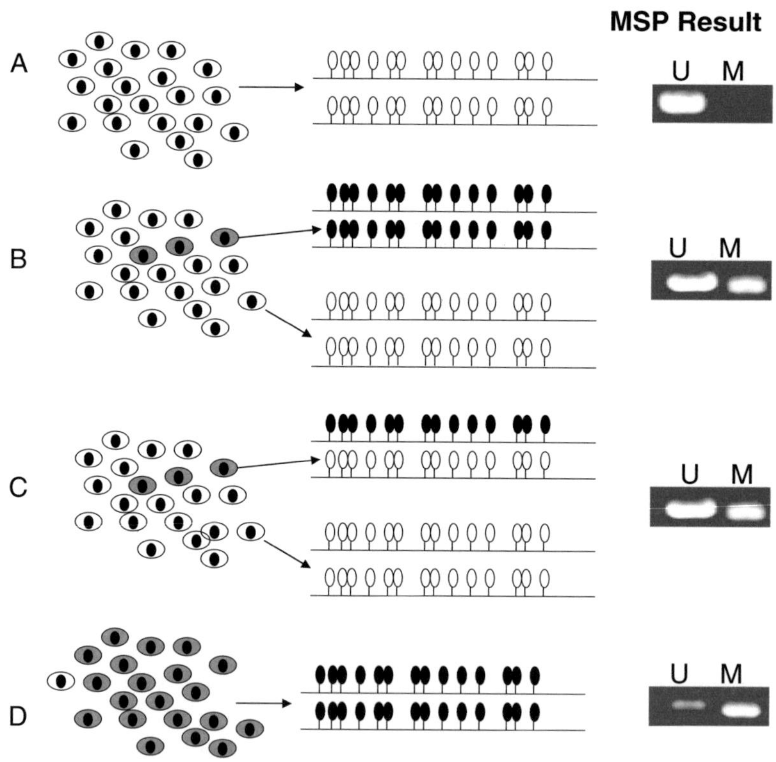

Fig. 22.1. Interpretation of MSP-based results. **(A)** This sample is composed of normal cells only (*White*). These normal cells usually have unmethylated (U) promoters for tumor suppressor gene. **(B-C)** This sample contains predominantly normal cells and a small number of tumor cells (*grey*) and will show an unmethylated as well as a methylated (M) band. **(B)** Both alleles of the normal cells are unmethylated while both alleles of the tumor cells are methylated. **(C)** Only one allele of the tumor cells is methylated. **(D)** This sample contains mostly tumor cells and will show a strong methylation signal.

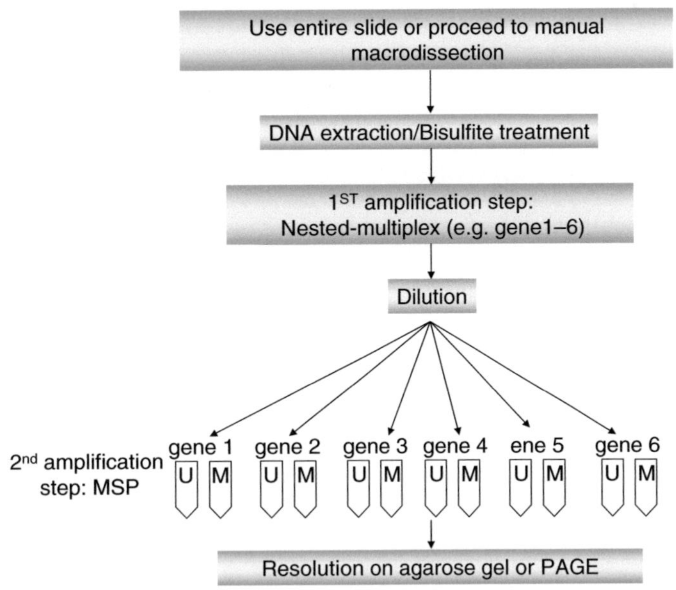

Fig. 22.2. Outlined strategy for multiplex nested-methylation-specific PCR.

with both primers sets indicates either that only one allele of the gene examined is methylated or that both alleles of the genes are methylated. The latter result is evidence that the sample analyzed is contaminated with normal tissue as indicated in **Fig. 22.1**. MSP has recently been modified in order to allow the analysis of individual (nested-MSP) or multiple (multiplex nested-MSP) promoters (**Fig. 22.2**) in samples with low quantity (e.g., macrodissected specimens) and/or quality (e.g., paraffin-embedded samples) *(8, 9)*. This strategy is particularly attractive for assessing the methylation status of gene promoters in archival specimens for which clinical outcome (e.g., response to treatment, survival) is known *(10)* as well as for sensitive molecular detection *(11)*. The use of real-time detection in MSP-based strategies should in the future help in the validation and standardization of gene promoter hypermethylation as biomarkers.

2. Materials

2.1. DNA Samples

1. Fresh, frozen, and paraffin-embedded samples as well as cell pellets can be analyzed using the instructions presented in this chapter (**Fig. 22.3**).
2. Normal lymphocytes are generally used as an unmethylated control while in vitro methylated DNA (IVD) serves as a methylated control. Cell lines for which the methylation status

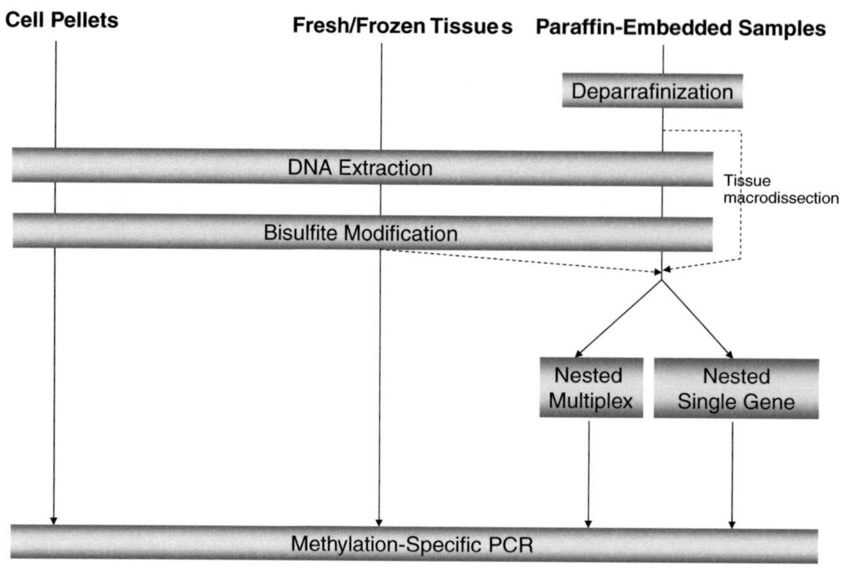

Fig. 22.3. Choosing an MSP-based strategy.

of particular genes is known can also be used as controls and will need to be determined for each gene analyzed by the user.

3. Any source of human DNA can be used for preparing the in vitro DNA methylation (IVD) control using IVD kit (available from New England Biolabs, Cat No M0226) which includes NEB buffer 2, *S*-adenosylmethionine (SAM) and the *Sss*I methylase.

2.2. Deparaffinization of Slides

This step is required in order to extract DNA from macrodissected samples obtained from paraffin-embedded slides.

1. Ethanol: 100, 95, 70 and 50% EtOH (95% grade) all at room temperature (RT).
2. Autoclave ddH_2O at RT.
3. Xylene (Sigma–Aldrich, St. Louis, MO).
4. Coplin jars.
5. Oven set at 60°C.

2.3. Common Reagents for DNA Extraction

1. 10 mM Tris–HCl, pH 8.0.
2. Phenol–Chloroform isoamyl alcohol (25:24:1) (Sigma–Aldrich).
3. 10 M Ammonium acetate.
4. Glycoblue™ 15 mg/mL (Ambion, Austin, TX).
5. 100% EtOH (ice cold).
6. 75% EtOH (RT).
7. Sterile 1.5 mL DNase/RNase free Eppendorf tubes.
8. Nanodrop-1000 (Nanodrop Technologies, Wilmington, DE). Alternatively any basic spectrophotometer can be used for DNA quantification.

2.4. DNA Extraction from Cell Lines

1. Phase lock gel^{TM} (2 mL) tubes (Qiagen Inc, Valencia, CA).
2. Digestion buffer 1 (stored at RT): 50 mM Tris–HCl, pH 8.0 (50 mL of 1 M Tris–HCl/1000 mL), 50 mM EDTA, pH 8.0 (100 mL of 500 mM EDTA/1000 mL), 2% SDS (200 mL of 10% SDS/1000 mL). Adjust to 1000 mL with 650 mL of ddH_2O.

2.5. DNA Extraction from Lymphocytes

1. Ficoll-paque plus (GE Healthcare Bio-Sciences Corp. Piscataway, NJ).

2.6. DNA Extraction from Paraffin-embedded Samples

1. Parrafin-embedded slides.
2. Scalpels/razor blades.
3. Xylene.
4. Digestion buffer 1 (stored at RT): 50 mM Tris–HCl, pH 8.0 (50 mL of 1 M Tris–HCl/1000 mL), 50 mM EDTA, pH 8.0 (100 mL of 500 mM EDTA/1000 mL), 2% SDS (200 mL of 10% SDS/1000 mL). Adjust to 1000 mL with 650 mL of ddH_2O.

2.7. DNA Extraction from Macrodissected Paraffin-Embedded Samples

1. H&E slide (reference slide).
2. Paraffin-embedded slide (used for DNA extraction).
3. Scalpels/razor blades.
4. Phase lock gel^{TM} (0.5 mL) tubes (Qiagen Inc).
5. 10 × MSP amplification buffer (100 mL): 16.6 mL of 1 M $(NH_4)_2SO_4$, 33.5 mL of 2 M Tris–HCl (pH 8.8), 6.7 mL of 1 M $MgCl_2$, 700 μL of 14.4 M of β-mercaptoethanol, and 42.5 mL of H_2O. Final concentrations are: $(NH_4)_2SO_4$ (166 mM), Tris–HCl (670 mM), $MgCl_2$ (67 mM), and β-mercaptoethanol (100 nM).
6. Digestion buffer 2 (Aliquoted and stored at −20°C): for 1 mL, 850 μL of 0.5% Tween 20, 50 μl of proteinase K (10 mg/mL) (Sigma–Aldrich), and 100 μL of 10 × MSP amplification buffer.

2.8. Bisulfite Modification

The following reagents are required if using an in-house bisulfite-treatment protocol (*see* **Note 1**). Alternatively, bisulfite modification of DNA can be carried out using commercially available kits according to the manufacturers' protocol (e.g., EZ-DNA Methylation Kit^{TM} from Zymo Research). In any case, 1 μg of starting DNA is required.

1. 2 M and 3 M NaOH.
2. 10 mM Hydroquinone (Prepare fresh before use). You will need 30 μL per sample treated.
   Dissolve 55 mg of hydroquinone (Sigma–Aldrich) in 50 mL ddH_2O.
3. 3 M Sodium bisulfite (Prepare fresh before use). You will need 520 μL per sample treated.

Dissolve 1.88 g of sodium bisulfite (Sigma–Aldrich) in 5 mL ddH_2O. Adjust pH to 5.0 with 5 M NaOH. Make sure you have enough sodium bisulfite for all your samples.

4. DNA cleanup kit (e.g., Promega DNA wizard cleanup kit, Promega, Madison, WI).
5. 10 M Ammonium acetate.
6. Glycoblue™ 15 mg/mL (Ambion).
7. 100% EtOH (ice cold).
8. 75% EtOH (RT).
9. Vacuum manifold (e.g., QIAvac 24 plus from Qiagen).
10. 80% isopropanol (RT).
11. Warm autoclaved ddH_2O.

2.9. Primer Design

Primers can either be designed manually or using the novel Internet-based computer algorithm MSPPrimer (freely accessible at http://www.mspprimer.org/cgi-mspprimer/design.cgi). Nested and MSP primers can be purchased from companies such as Sigma–Aldrich or IDTDNA (25 nM Oligonucleotide, standard desalting).

2.10. Amplification

1. Two PCR hoods: one for MSP, the other for handling and diluting nested PCR products (*see* **Note 2**).
2. DNAZAP™ (Ambion).
3. Autoclaved ddH_2O.
4. Thermocycler with gradient capabilities (e.g., MBS Satellite 0.2G from Thermo Scientific, Waltham, MA) (*see* **Note 3**).
5. 10 × MSP amplification buffer (100 mL): 16.6 mL of 1 M $(NH_4)_2SO_4$, 33.5 mL of 2 M Tris–HCl (pH 8.8), 6.7 mL of 1 M $MgCl_2$, 700 μL of 14.4 M of β-mercaptoethanol, and 42.5 mL of H_2O. Final concentrations are $(NH_4)_2SO_4$ (166 mM), Tris–HCl (670 mM), $MgCl_2$ (67 mM), and β-mercaptoethanol (100 nM).
6. 25 mM dNTP mix.
7. *Taq* DNA polymerase (e.g., JumpStart™ RED*Taq*® DNA polymerase).
8. 30 pmol/μL MSP or nested primers.
9. PCR stripes.
10. Mineral oil.

2.11. Detection

2.11.1. Sodium Borate

1. Add 61.83 g of boric acid for every 1 L of H_2O in 4 L glass beaker to make a 1 M solution. Make 4 L total (247.32 g of boric acid in a volume of 4 L H_2O).
2. Put in magnetic stir rod, and place container on a magnetic heat plate.
3. Turn on the stirrer and set heat to medium.
4. Watch carefully to make sure it does not boil (20–30 min) until the boric acid has dissolved.

5. From the 4 L solution, place 3.3 L of the 1 M boric acid solution in another glass container.
6. Add 20 M NaOH to obtain a pH of 8.15–8.2 (approximately 40 mL).
7. Add enough H_2O to equal 4 L. This will yield the desired pH of 8.
8. This is a 20 × solution sodium borate stock solution which may be diluted for making gels (10 ×) or running them (1 ×).

2.11.2. Agarose Gel

1. 2% (w/v) Agarose gel (*see* **Note 4**).
2. 10 × and 1 × sodium borate buffer (*see* **2.11.1**).
3. Gel star® (Cambrex BioSciences, Rockland, ME).
4. Multichannel pipettes (P20).
5. Gel imaging system.

2.11.3. Non-denaturing Polyacrylamide Gel Electrophoresis (PAGE)

1. 40% Bis-acrylamide (store at +4°C) (*see* **Note 4**).
2. 10 × and 1 × sodium borate buffer (*see* **2.11.1**).
3. TEMED (store at RT).
4. 10% (w/v) Ammonium persulfate (Make fresh or store aliquots at −20°C).
5. ddH_2O.
6. DNA ladder (e.g., MSP1 from New England Biolabs).
7. Ethidium bromide (Caution: carcinogenic reagent) (*see* **Note 5**).
8. Gel imaging system

3. Methods

3.1. DNA Extraction from Cell Lines

1. Pellet cells by centrifugation 5 min at ~2,500*g*.
2. Wash pellets in PBS and centrifuge. Remove supernatant.
3. Add 270 μL of digestion buffer 1 to pellet and homogenize.
4. Add 30 μL of proteinase K (10 mg/mL) to each tube and vortex.
5. Incubate samples overnight on a dry block or in a water bath set at 60°C.
6. The next day inactivate the enzyme by incubating for 10 min at 100°C.
7. Transfer the content of each tube to a 2-mL phase lock gel™ tube and gently homogenize according to the manufacturer's protocol.
8. Add an equal volume of phenol:chloroform:isoamyl alcohol (25:24:1) (in this case 300 μL) to each sample.
9. Centrifuge tubes at ~18,000*g* for 5 min.
10. The upper aqueous phase is separated from the bottom organic phase by the phase lock gel™.

11. Transfer the upper aqueous phase containing DNA into a new tube.
12. Precipitate DNA by adding 10 M ammonium acetate (1/10 of the total volume; in this case 30 μL), 2 μL of carrier such as Glycoblue™ (optional for cell lines), and two volumes of ice cold 100% EtOH.
13. Incubate at −20°C for at least 2 h.
14. Pellet DNA by centrifugation for 20 min at ~19,500*g*.
15. Wash the pellet with 150 μL of 75% EtOH, centrifuge for 5 min at ~19,500*g* and air dry.
16. Rehydrate the pellet in 50–100 μL of elution buffer (i.e., 1 M Tris–HCl pH 8.0) or nuclease-free water. To help dissolve the precipitated DNA, samples can be incubated at 50°C for 10–20 min. DNA is finally quantified using a Nanodrop or other spectrophotometer.

3.2. DNA Extraction from Normal Lymphocytes

1. Extraction of lymphocytes from blood is carried out using Ficoll as described by the manufacturer (GE Healthcare).
2. Incubate mononuclear cells overnight at 60°C in digestion buffer 1. Mix lysates with an equal volume of phenol:chloroform:isoamyl alcohol (25:24:1). Transfer into a phase lock gel™ tube and continue with step 9 **Section 3.1**.

3.3. DNA Extraction Using a Whole Paraffin-Embedded Slide

1. Cut 3–5 sections of 5–10 μ thickness and place in a single Eppendorf tube. Use razor blades to remove tissue from slide.
2. Add 1.5 mL of xylene, inverting several times (but not vortexing). Incubate at RT for a minimum of 30 min.
3. Spin down at maximum speed in a microcentrifuge for 5 min.
4. Discard xylene and add another 1.5 mL of xylene, repeat steps 2–4 two times (this is very important to ensure complete removal of paraffin).
5. Discard final volume of xylene and add 1.5 mL of 100% EtOH, mix (do not vortex) and incubate at RT for 30 min.
6. Spin, discard, add fresh EtOH, and incubate for additional 30 min.
7. Discard the EtOH wash and dry tissue in a heat block for 15 min at 90°C.
8. Add 50–200 μL of digestion buffer 1 to each specimen and 50 μL of proteinase K. Adjust the volume of digestion buffer 1and proteinase K to the richness of each sample.
9. Incubate at 55°C overnight.
10. Clear liquid indicates complete digestion. If tissue is still visible, add more proteinase K and repeat overnight digestion at 55°C.
11. Heat inactivate proteinase K for 15 min at 95°C. Punch a hole in the tube top with fine needle to avoid the lid popping.
12. Spin down quickly.

13. Samples can be stored at this stage at –80°C, or proceed to next step.
14. Transfer lysate into a phase lock gel™ tube. Continue with step 8 **Section 3.1**.
15. Store DNA at 4°C (if used on a regular basis) or −20°C (longer storage).

3.4. Manual Macrodissection of Paraffin-Embedded Slides

1. Put the paraffin-embedded slides vertically on a kimwipe tissue and incubate in an oven set at 60°C for 3–4 h or until the paraffin has melted.
2. Carefully wipe the melted paraffin off the slide using a kimwipe.
3. Wash slides three times for 30 min each with fresh xylene.
4. Rehydrate slides by incubating for 2 min in each of the following solutions: 100%, 95%, 70%, and 50% EtOH (use Coplins jars).
5. Finally, rinse slide in ddH_2O twice for 2 min and allow slides to dry.
6. Label areas of interest on the back of the de-paraffinized slide (as previously determined by the pathologist on the corresponding H&E slide).
7. Put a drop of digestion buffer 2 to each spot (around 3 μL). Scrape the area of interest using a scalpel, aspirate using a pipette and transfer into a clean tube containing 50 μL of digestion buffer 2. Incubate tubes at 60°C in dry block or water bath (optional: add a layer of mineral oil).
8. The next day incubate tubes for 10 min at 100°C to inactivate proteinase K.
9. Use the entire 50 μL of digested sample for bisulfite treatment. Alternatively samples can be further purified and DNA quantified. In that case, follow with step 7 **Section 3.1** using 0.5 mL phase lock gel™ tube.
10. Quantify DNA using a Nanodrop.
11. Store DNA as described in step 15 of **Section 3.3**.

3.5. Bisulfite Modification

Day 1 (16 h before you are ready to start cleanup)

1. Dilute 1 μg of genomic DNA in ddH_2O.
2. Add 5.5 μl of 2 M NaOH and incubate for 10 min at 37°C. This step creates single-stranded DNA, which is sensitive to reaction with sodium bisulfite.
3. Incubate at RT for 10 min.
4. Add 30 μL of freshly prepared 10 mM hydroquinone to each tube.
5. Add 520 μL of freshly prepared 3 M sodium bisulfite solution per tube.
6. Vortex.
7. Layer with mineral oil. Use enough so that you are able to see an interface between the solution and the oil.

8. Incubate at 50°C for 12–16 h and cover with foil as the solution is light sensitive (*see* **Note 1**).

   Day 2

9. Transfer the aqueous solution of each sample without the mineral oil to a clean 1.5-mL Eppendorf tube.
10. Connect columns to a vacuum manifold (e.g., QIAvac 24 Plus from Qiagen).
11. Add 1 mL of Promega DNA wizard cleanup to each barrel in addition to DNA. This step allows for the separation of the bisulfite-treated DNA from the sodium bisulfite solution.
12. Apply vacuum.
13. Fill barrel with 80% isopropanol.
14. Place empty column into labeled Eppendorf tube. Turn off vacuum.
15. Add 50 μL of heated ddH_2O (60–70°C). The warm water helps eluting DNA from the resin.
16. Spin tube/column for 1 min at ~19,500*g* and discard columns.
17. Add 5.5 μL of 3 M NaOH to each tube and incubate at RT for 5 min.

    This completes the chemical modification by desulfonating the bisulfite-treated DNA.

18. For each tube add 1 μL of Glycoblue^{TM} and 17 μL of 10 M ammonium acetate.
19. Add 200 μL of ice cold 100% EtOH and incubate tubes at −20°C for a minimum of 3 h.
20. Spin tubes for 20 min at ~19,500*g* at 4°C, discard supernatant, and add 150 μL of RT 75% EtOH.
21. Spin for 10 min at ~19,500*g* at 4°C.
22. Dry pellet for 5–10 min and resuspend in 20 μL of ddH_2O.
23. Store bisulfite-treated DNA at −80°C and avoid freeze–thaw cycles.

    Because bisulfite-treated DNA is single stranded, it is very sensitive to freeze–thaw cycles and should therefore be handled like RNA.

3.6. Manual Design of MSP Primers

1. Retrieve the promoter sequence to be analyzed (i.e., usually 1,000–2,000 bp of genomic sequence upstream of the transcription start site) using the UCSC Genome browser website (http://www.genome.ucsc.edu). Make sure to include approximately 200 bp of sequence following the transcription start site as CpG islands often continue past the transcription start site.
2. Confirm the presence of a CpG island by pasting the sequence into CpG island searcher website (http://www.uscnorris.com/cpgislands2/cpg.aspx) using default settings.
3. Paste sequence into a word document. Find and replace in the whole document CG by X, then C by T and finally X by

CG. This is the methylated-modified sequence from which the methylated set of MSP primers will be designed.

4. Paste the same original sequence in a different page of the word document. Find C and replace by T in the whole document. This is the unmethylated-modified sequence from which one can design unmethylated MSP primers sets. Coloring CGs in a distinct color allows for the fast visualization of CpG islands and helps with primer design.
5. MSP primers are designed using the same strand (either sense or antisense strand).
6. Methylated MSP primers should have the following characteristics (*see* **Note 3**):
   a. Use the methylated-modified sequence.
   b. Methylated products should be less than 200 bp.
   c. Methylated primers should amplify a region well within the CpG island rather than on the edge of the island.
   d. Methylated primers should have a minimum of 2–3 CG residues with at least 1–2 CG located at the 3′ of the primer sequence.
   e. Terminate with a C for the forward methylated primer and with a G for the reverse methylated primer.
   f. Unconverted cytosine (C), which will appear as T in the converted sequence, should be located 3′ in order to specifically amplify bisulfite-converted DNA.
   g. Tm should be approximately 60–65°C (e.g., http://www.idtdna.com/analyzer/Applications/OligoAnalyzer/).
7. Unmethylated MSP primers should have the following characteristics.
   a. Use the unmethylated-modified sequence.
   b. Use a similar region as for the methylated set of primers.
   c. Terminate with a T for the forward unmethylated primer and with an A for the reverse unmethylated primer.
   d. Tm should be approximately 60–65°C.
   e. If necessary extend the unmethylated primer sequences 5′ in order to reach an appropriate Tm.
   f. Tm for unmethylated and methylated MSP primers should be within 5°C.
   g. Unmethylated products should be less than 200 bp.

3.7. Manual Design of Nested Primers

Nested primers should amplify the region where the methylated and unmethylated MSP primers have been designed.

Nested and MSP primers may overlap by few bases pair (approximately 10 bp).

Nested primers should be void of CG residues so that the amplified products generated using these primers are neither specific for the methylated nor the unmethylated sequence. If it cannot be avoided to have a CG within the nested primer sequence minimize the number of CG and replace in the primer sequence the C of the CG by a Y (an equimolar mixture of pyrimidine C and T) for the forward nested primer, and the G of the CG by an R (an equimolar mixture of purine A and G) in the reverse nested primer sequence (*see* **Note 3**).

3.8. Primer Design with MSPPrimer

The MSPPrimer was developed by Brandes et al. *(12)* and can be used to design MSP, nested, and bisulfite-sequencing primers. Primer design is particularly critical for MSP and nested-MSP in order to accurately determine the methylation status of a gene promoter (*see* **Note 3**). Primer prediction used in this program is based on the thermodynamic characteristics of the 3′end of the proposed primer, and the number and position of individual CpG and non-CpG cytosines. Therefore, such algorithm is more accurate than the existing MSP primer design algorithms as these are primarily PCR based, have specificity issues, and do not take into account the fact that bisulfite conversion is incomplete. MSP primer includes the following considerations when predicted MSP and nested primers:

1. Primers must overlap non-CpG residues in order to specifically amplify bisulfite-modified DNA and not the residual unconverted DNA.
2. Primers must contain several CpG (usually between 2 and 4 CpG) in order to differentiate between methylated and unmethylated CpG.
3. Primer must not be prone to form secondary structure or primer–dimer pairs and must have compatible annealing temperatures.
4. Register and login at http://www.mspprimer.org.
5. Select the type of primer to design (Nested or MSP primers) and paste the unmodified 1,000 bp promoter sequence into the MSPPrimer window, use the default settings as a starting point.

3.9. Methylation-Specific PCR

3.9.1. MSP Set Up

1. Thaw 10 × PCR buffer, dNTP mix, and primers.
2. Determine the number of samples to be analyzed taking the unmethylated and the methylated amplifications as well as a water control into account. Prepare separate master mixes for both the methylated and unmethylated PCR reactions. The following amounts are for one sample and should be multiplied by the number of samples and controls analyzed.

10 × MSP amplification buffer	2.5 μL
25 mM dNTPs mix	0.5 μL
30 pmol/μL forward MSP primer	0.5 μL
30 pmol/μL reverse MSP primer	0.5 μL
Autoclaved ddH_20	18.5 μL

3. Add 0.5 units of JumpStart™ RED*Taq*® DNA polymerase (Sigma–Aldrich) or any other hot start *Taq* polymerase to each sample.

4. Pipette 23-μL aliquots of each PCR master mix into separate PCR stripes. Use one strip for the unmethylated reaction and another strip for the methylated reactions so that they are easily loaded using a multichannel pipette. Make sure the stripes are correctly labeled.
5. Add 2 μL of bisulfite-modified DNA template to each tube including the water control.
6. Add 1 or 2 drops of mineral oil to each tube, and place in a thermal cycler.

Be sure the mineral oil completely covers the surface of the reaction mixture to prevent evaporation. If the thermal cycler has a heated lid to prevent condensation, mineral oil may not be necessary, but longer run times are to be expected.

3.9.2. Amplification Conditions

1. Initiate the PCR with a 5-min denaturation at 95°C.
2. Continue PCR amplification with the following parameters (*see* **Note 3**).

35 cycles:	30 s	95°C	(denaturation)
	30 s	to be determined	(annealing: primer specific)
	30 s	72°C	(elongation)
Final step:	4 min	72°C	

Store at 4°C until analysis

3.10. Nested-MSP and Multiplex Nested-MSP

3.10.1. Nested-MSP Stage 1 Amplification Set Up

1. Thaw 10 × PCR buffer, dNTPs mix, and primers.
2. Determine the number of samples to be analyzed, including the controls (unmethylated, methylated, and water controls). The following amounts are for one sample and should be multiplied by the number of samples and controls analyzed.

10 × MSP amplification buffer	2.5 μL
25 mM dNTPs mix	0.5 μL
30 pmol/μL forward nested primer	0.5 μL
30 pmol/μL reverse nested primer	0.5 μL
Autoclaved ddH_2O	18.5 μL

3. Add 0.5 units of JumpStart™ RED*Taq*® DNA polymerase (Sigma–Aldrich) to each sample.
4. Place 23-μL aliquots of each PCR master mix into separate PCR stripes or single PCR tubes. Make sure the stripes or tubes are correctly labeled.
5. Add 2 μL of bisulfite-modified DNA template to each tube including the controls.
6. Add 1 or 2 drops of mineral oil to each tube and place in a thermal cycler. Make sure the mineral oil completely covers the surface of the reaction mixture to prevent evaporation as well as limiting contamination when opening tubes or stripes. Continue with **Section 3.10.3**.

3.10.2. Multiplex Nested-MSP Stage 1 Amplification Set Up

It is convenient to prepare a multiplex-nested primer mix. For example, if five promoter regions are analyzed, add 50 μL of each of the forward- and reverse-nested primers (100 pmol/μL) for a total volume of 500 μL. This master nested primers mix will be enough for 50 multiplex-nested reactions.

1. Prepare the following mix (for one sample, numbers should be multiplied by the number of samples and controls analyzed):

10 × MSP amplification buffer	2.5 μL
25 mM dNTPs mix	0.5 μL
Nested primer mastermix	10 μL
Autoclaved ddH_2O	9.5 μL

2. For each sample add 0.5 units of JumpStart™ RED*Taq*® DNA polymerase (Sigma–Aldrich).
3. Continue as described in step 4 **Section 3.10.1**.

3.10.3. Amplification Conditions for Stage 1 Amplification

1. Initiate the PCR with a 5-min denaturation step at 95°C
2. Continue PCR amplification with the following parameters (*see* **Note 3**).

35 cycles: 30 s	95°C	(denaturation)
30 s	to be determined	(annealing)
30 s	72°C	(elongation)
Final step: 4 min	72°C	

3. Following amplification, dilute nested amplifications 1/500 with autoclaved ddH_2O and use 2 μL as template in conventional MSP (*see* **3.9**). Each of the promoters amplified in the multiplex-nested reaction will be assayed individually in a MSP reaction (**Fig. 22.2**).

3.11. Detection

3.11.1. Agarose Gel

1. Prepare a 2% agarose gel containing GELSTAR™ using sodium borate buffer in an A3-1 Gator™ Wide Gel Electrophoresis System (ThermoFisher Scientific, Portsmouth, NH) using the 50-wells combs (*see* **Note 4**).
2. Load the DNA ladder (e.g., MSP1) in the first lane.
3. Using the multichannel pipette load 10 μL of products so that every other lane contains unmethylated samples.
4. Load 10 μL of the methylated product in the remaining lanes, also using the multichannel pipette. Make sure that MSP reactions have been done in series (e.g., one stripe is for the unmethylated samples and the next strip for the methylated samples).
5. Apply constant voltage (200 V) until desired resolution is achieved.
6. Visualize amplification products by UV using a gel imaging system. The running buffer can be reused several times.

3.11.2. Polyacrylamide Gel Electrophoresis (PAGE)

Vertical nondenaturing polyacrylamide gel electrophoresis provides better resolution that agarose gels although their preparation and loading is not amenable for running many samples (i.e., less than 50 samples maximum per gel, depending on the comb used) (*see* **Note 4**).

1. Mount PAGE plates, spacers, and cast together. We recommend placing the plates and spacers in a plastic bag and then mounting in order to prevent any leak.
2. In a 50-mL conical tube, prepare a 6% PAGE by adding the following reagents:

10 × Sodium borate buffer	4 mL
40% Bis-acrylamide	6 mL
10% (w/v) ammonium persulfate	300 μL
ddH_2O	30 mL
TEMED	30 μL

3. Pour mixture into cast, add appropriate comb, and leave for 10–15 min to allow for polymerization of the gel.
4. Upon polymerization take the cast gel out off the plastic bag, remove the comb, and transfer the cast gel into PAGE electrophoresis chamber. Rinse well with running buffer (sodium borate) and load samples individually taking care not to contaminate samples between lanes.
5. Run gel for 20–40 min at a constant voltage of 200 V. The use of sodium borate allows for a fast running time while maintaining the temperature of the gel relatively low.
6. Dismantle cast and transfer the gel into a tank containing ethidium bromide for 5 min (*see* **Note 5**). Transfer in a tank with ddH_2O. Visualize amplification products by UV using a gel imaging system.

3.12. Data Analysis

3.12.1. What to Expect?

The presence of a PCR product of the correct molecular weight (visible after gel electrophoresis) indicates the presence of either unmethylated or methylated alleles. When assessing the methylation status of tumor suppressor gene in clinical samples, expect that normal tissue and some primary tumor cells may be completely unmethylated. If the sample analyzed contains both cancerous and noncancerous cells, an unmethylated and a methylated signal will be detected. This might suggest that the cancer cells contain both unmethylated and methylated alleles. Alternatively, such result could also indicate that the promoter is methylated at both alleles in the cancer cells and that the unmethylated signal results from the normal cells (**Fig. 22.1**). This can be overcome to some extent by dissecting the sample from paraffin-embedded slides using a laser-capture microdissection.

3.12.2. No Amplification Product

If no products are obtained with U (unmethylated) or M (methylated) primers refer to **Fig. 22.3** to make sure you are using the right technique depending on the sample you are analyzing. Analyzing DNA from paraffin-embedded specimens by MSP will often fail. To exclude that the bisulfite-modified template is inadequate, use established MSP primers under conditions previously shown to produce adequate results. If standard primers and conditions fail to produce a PCR product of the appropriate size, then the DNA template is the likely source of the problem. If high-quality genomic DNA was used initially, then the problem developed during bisulfite treatment or storage after treatment. Begin again with fresh DNA and proceed according to the protocol, using the DNA soon after initial bisulfite preparation is complete, in order to avoid degradation of the modified DNA. If standard PCR conditions and primers produce satisfactory results, then a DNA template problem can be excluded, and a problem in the PCR reaction itself is the most likely possibility. Primer design and choice of annealing temperature is one of the most critical aspects of the MSP-based procedure. A possible solution is to decrease the annealing temperature by few degrees. It that fails, ensure that the primers are specific for bisulfite-modified DNA, that the size of the amplification product is below 200 bp, that face the appropriate 5′ to 3′ direction to allow amplification, that they do recognize the same DNA strand of DNA (sense or antisense) after bisulfite modification, and that their annealing temperature is compatible. If all these optimization procedures fail in yielding a specific PCR product, design a new set of PCR primers.

3.12.3. Methylated Band in All Samples

The likely source for the presence ofM products in all samples is contamination with previously amplified PCR product. Clean bench, pipettes, and racks with DNAZAP and use fresh solutions (*see* **Note 2**). Contamination can be minimized by dedicating equipment to PCR only. If methylated products are present in all samples but not in the water control, this suggests that either M is present in all samples or that the primers cannot differentiate between unmethylated and methylated alleles. In the latter case, optimize the amplification conditions (mainly the annealing temperature). Alternatively, design new MSP primers (include one or two extra CpG sites in the newly designed primer in order to increase discrimination).

3.13. Time Consideration

1. **DNA extraction:** Depending on the amount of tissue to be digested, it can take anywhere from overnight to a couple of days.
2. **Bisulfite modification day 1:** Bisulfite modification takes around 30 min to set up prior to the overnight incubation at 50°C.
3. **Bisulfite modification day 2:** DNA cleanup takes between 30 min and 2 h depending on the number of samples. Ethanol precipitation should be allowed to continue for at least 3 h before centrifugation. Similar time is required when using the EZ-DNA Methylation Kit^{TM}.

4. **MSP setup:** Assuming that the PCR primers have been synthesized and aliquoted, setting up amplification reaction takes 30–60 min. This can be speed up by aliquoting bisulfite-treated DNA in PCR stripes and adding it to the reaction with a multichannel pipette. The time for amplification depends on the thermocycler used and the number of cycles, but it is typically 2–3 h. Gel loading and running takes 1–2 h. Thus the entire diagnostic procedure can easily be completed in 1 day, provided the modified DNA samples are ready. If necessary, PCR products can be stored several weeks at 4°C or indefinitely at −20°C prior to gel analysis.
5. **Nested approaches:** Add another 1 h to set up the preliminary amplification using the nested primers, an additional 2–3 h for the amplification to occur as well as another 30 min to dilute the first round of amplification.

4. Notes

1. Bisulfite treatment parameters have been well characterized and the in-house protocol included in this chapter gives conditions which convert most of the unmethylated cytosines to uracils, while maintaining fairly good integrity of the DNA backbone. Longer times, temperature >50°C and a pH<5.0 will produce more complete conversion, but at the cost of greater degradation of template. Even under the conditions described above, bisulfite treatment fragments DNA to some extent. This makes it difficult to amplify products larger than 1 kb in length and reduces the chances of amplifying products in the 500–1,000 bp range. Also this is a concern when designing bisulfite sequencing primers. This does not interfere with MSP as the regional nature of abnormal methylation as detected by MSP is of much shorter region, normally 50–200 bp in length.
2. Take great care to ensure that the template DNA and reagents do not become contaminated with exogenous DNA or PCR products. For MSP, exogenous DNA does not pose a significant problem because modification of cytosines to uracil has not occurred, making native DNA a poor template for MSP primers. However, PCR product carryover generated in previous experiments is of great concern. At a minimum, dedicated pipettes should be maintained for sample and reagent preparation and reaction setup. Ideally, these procedures should be separated from reaction analyses. A separate PCR hood, which can be wiped down and/or UV irradiated is ideal. Filtered pipette tips should be used for all pre-PCR manipulations. Deionized water and opened boxes of plasticware can be sterilized by autoclaving. Include a negative-control

reaction with no template DNA in every round of amplification, for both the unmethylated and methylated reactions. Analyze PCR products in a separate area of the laboratory, and load gels with a dedicated set of pipettes. DNAZAP^{TM} (Ambion) can be used to clean pipettes, tube racks, bench, and other equipments in order to minimize contaminations.

3. Because MSP-based techniques utilize specific primer recognition to discriminate between methylated and unmethylated alleles, stringent amplification conditions must be maintained. The unmethylated and methylated primer sets should have a similar annealing temperature. Given the high A/T content in the unmethylated primer sets, such primers are usually longer than the methylated primer sets (extend sequence 5′). The optimum annealing temperature should be determined for each gene using a temperature gradient. Increasing the annealing temperature usually helps to reduce unspecific products. Additives such as DMSO are not necessary for these reactions, as bisulfite modification effectively reduces the GC content of even the most CG-rich CpG islands to levels which are easy to amplify. The combination of primers to be used in multiplex-nested reactions will have to be determined so that primers with similar annealing temperatures are amplified in the same reaction. The number of amplification cycles in the MSP stage that follows the nested step will also need optimization. (We routinely use between 20 and 30 cycles.) This is determined by lowering the cycle number so that both the unmethylated and the methylated controls give only an unmethylated or a methylated signal, respectively.
4. The use of agarose gels for the resolution of MSP products is only recommended once the MSP has been fully optimized and run using 6% nondenaturing polyacrylamide vertical gels which allow for the clear resolution of amplification products. Horizontal agarose gels, however, have the advantage of more efficient loading.
5. Great care should be employed when handling ethidium bromide gel and buffer as this reagent is carcinogenic. Ethidum bromide running buffer will have to be treated before being discarded using active coal column designed for this purpose (e.g., product number: 588-1140 VWR International West Chester, PA).

Acknowledgement

The authors would like to acknowledge members of the Herman laboratory, especially Dr Johan C. Brandes, Dr Joshi J. Alumkal, and Dr Mingzhou Guo for contributions to some parts of these protocols.

References

1. Bird, A.P. (1986) CpG-rich islands and the function of DNA methylation. *Nature* 321, 209–213.
2. Belinsky, S.A., Nikula, K. J., Palmisano, W.A., et al. (1998) Aberrant methylation of p16(INK4a) is an early event in lung cancer and a potential biomarker for early diagnosis. *Proc Natl Acad Sci USA* 95, 11891–11896.
3. Hanahan, D., Weinberg, R. A. (2000) The hallmarks of cancer. *Cell* 100, 57–70.
4. Herman, J. G., Baylin, S. B. (2003) Gene silencing in cancer in association with promoter hypermethylation. *N Engl J Med* 349, 2042–2054.
5. Herman, J. G., Graff, J. R., Myohanen, S., et al. (1996) Methylation-specific PCR: a novel PCR assay for methylation status of CpG islands. *Proc Natl Acad Sci USA* 93, 9821–9826.
6. Clark, S. J., Harrison, J., Paul, C. L. et al. (1994) High sensitivity mapping of methylated cytosines. *Nucleic Acids Res* 22, 2990–2997.
7. Frommer, M., McDonald, L. E., Millar, D. S., et al. (1992) A genomic sequencing protocol that yields a positive display of 5-methylcytosine residues in individual DNA strands. *Proc Natl Acad Sci USA* 89, 1827–1831.
8. House, M. G., Guo, M., Iacobuzio-Donahue, C., et al. (2003) Molecular progression of promoter methylation in intraductal papillary mucinous neoplasms (IPMN) of the pancreas. *Carcinogenesis* 24, 193–198.
9. van Engeland, M., Weijenberg, M. P., Roemen, G. M., et al. (2003) Effects of dietary folate and alcohol intake on promoter methylation in sporadic colorectal cancer: the Netherlands cohort study on diet and cancer. *Cancer Res* 63, 3133–3137.
10. Hegi, M. E., Diserens, A. C., Gorlia, T., et al. (2005) MGMT gene silencing and benefit from temozolomide in glioblastoma. *N Engl J Med* 352, 997–1003.
11. Belinsky, S. A., Liechty, K. C., Gentry, F. D., et al. (2006) Promoter hypermethylation of multiple genes in sputum precedes lung cancer incidence in a high-risk cohort. *Cancer Res* 66, 3338–3344.
12. Brandes, J. C., Carraway, H., Herman, J. G. (2007) Optimal primer design using the novel primer design program: MSPprimer provides accurate methylation analysis of the ATM promoter. *Oncogene* 26, 6229–6237.

Chapter 23

MethyLight

Mihaela Campan, Daniel J. Weisenberger, Binh Trinh, and Peter W. Laird

Abstract

MethyLight is a sodium-bisulfite-dependent, quantitative, fluorescence-based, real-time PCR method to sensitively detect and quantify DNA methylation in genomic DNA. MethyLight relies on methylation-specific priming combined with methylation-specific fluorescent probing. This combination of methylation-specific detection principles results in a highly methylation-specific detection technology, with an accompanying ability to sensitively detect very low frequencies of hypermethylated alleles. The high sensitivity and specificity of MethyLight make it uniquely well suited for detection of low-frequency DNA methylation biomarkers as evidence of disease. At the same time, the quantitative accuracy of real-time PCR and the flexibility to design bisulfite-dependent, methylation-independent control reactions allows for a quantitative assessment of these low-frequency methylation events. We describe the experimental steps of MethyLight analysis in detail. Furthermore, we present here principles and design examples for three types of quality-control reactions. QC-1 reactions are methylation-independent reactions to monitor sample quantity and integrity. QC-2 reactions are bisulfite-independent reactions to monitor recovery efficiencies of the bisulfite-conversion methodology used. QC-3 reactions are bisulfite-independent primed reactions with variable bisulfite-dependent probing to monitor completeness of the sodium bisulfite treatment. We show that these control reactions perform as expected in a time-course experiment interrupting sodium bisulfite conversion at various timepoints.

Key words: DNA Methylation, real-time PCR, TaqMan, bisulfite, epigenetics, cancer, quantitative, methylation-specific PCR.

1. Introduction

MethyLight is a sodium-bisulfite-dependent, quantitative, fluorescence-based, real-time PCR method to sensitively detect and quantify DNA methylation in genomic DNA *(1–5)*. MethyLight relies on methylation-specific priming *(6)*, combined with methylation-specific fluorescent probing *(1–5)*. This combination

Jörg Tost (ed.), *DNA Methylation: Methods and Protocols, Second Edition, vol. 507*

DOI 10.1007/978-1-59745-522-0_23 Springerprotocols.com

of methylation-specific detection principles results in a highly methylation-specific detection technology, with an accompanying ability to sensitively detect very low frequencies of hypermethylated alleles. The high sensitivity and specificity of MethyLight make it uniquely well suited for detection of low-frequency DNA methylation biomarkers as evidence of disease *(7)*. At the same time, the quantitative accuracy of real-time PCR and the flexibility to design bisulfite-dependent, methylation-independent control reactions *(5)* allows for a quantitative assessment of these low-frequency methylation events.

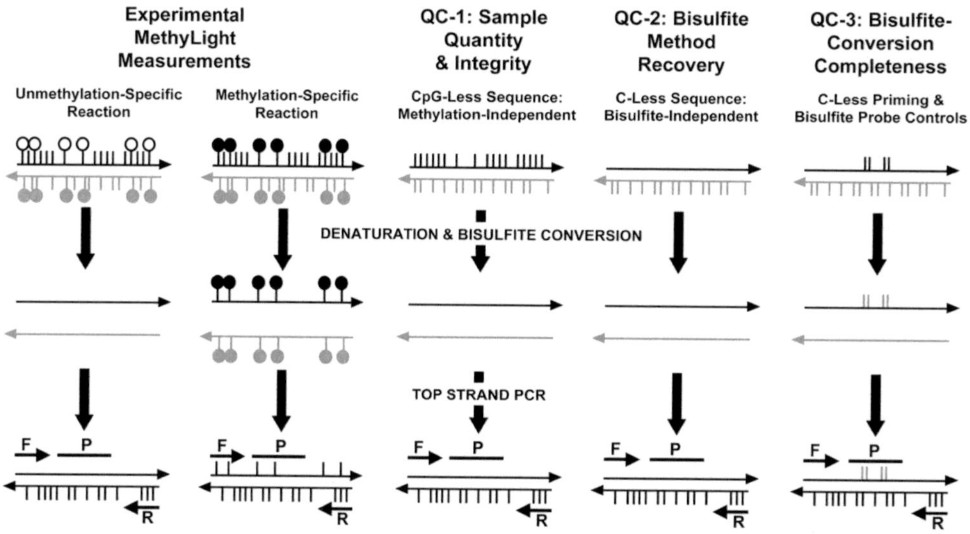

Fig. 23.1. TaqMan-based MethyLight experimental and quality-control PCR reactions used for quantitative methylation analysis of bisulfite-converted DNA. Several types of TaqMan-based PCR reactions are used for the quantitative DNA methylation analysis of bisulfite-converted DNA. Following bisulfite conversion, methylated cytosines remain unchanged, while unmethylated cytosines are deaminated to uracils. During subsequent PCR amplification of the bisulfite-converted DNA, thymine is incorporated in the place of uracil. The two strands of the bisulfite-converted DNA are no longer complementary, such that separate PCR reactions can be designed to amplify either the top or the bottom DNA strands. Each horizontal line represents one DNA strand. The tick marks represent cytosines not in the context of CpG dinucleotides while the lollipops represent cytosines in the context of CpG dinucleotides. The methylated cytosines are depicted as solid black while the unmethylated ones are open white. The bottom strand does not participate in the further analysis after bisulfite conversion, and is depicted in gray. The experimental MethyLight reactions are specific for unmethylated or methylated DNA sequences. These reactions (the first two panels) are designed to include cytosines within CpG sequences (methylation specific) as well as cytosines located outside the CpG context (bisulfite conversion specific). Reactions toward unmethylated DNA sequences are designed to amplify TG-containing sequences, while the reactions toward methylated DNA sequences are designed to amplify CG-containing sequences. Three methylation-independent reactions are used as quality controls (QC) to monitor the sample quantity and integrity (QC-1), as well as bisulfite conversion recovery (QC-2) and bisulfite conversion completeness (QC-3). The QC-1 reaction is designed toward a CpG-less sequence that still contains cytosines outside the CpG context (bisulfite conversion specific). QC-2 is a bisulfite-independent reaction in which both primers and the probes are designed toward a DNA region that does not contain any cytosines on one of the DNA strands (C-LESS). QC-3 reactions comprise a panel of 16 different reactions designed toward a single DNA sequence that have the same primer sequences but distinct probes. The DNA sequence covered by the primers lacks cytosine residues on one of the strands, while the DNA sequence covered by the probes contains four cytosines outside the CpG context (**Fig. 23.2**).

In addition to discussing in detail how to perform the experimental steps of MethyLight analysis, we present here how template, primer, and probe design flexibility can be used to develop quality-control reactions. The major principles are presented in **Fig. 23.1**. Methylation-independent, bisulfite-dependent reactions can be used as quality controls of sample quantity and integrity, as illustrated in **Fig. 23.1**: QC-1 *(5)*. We have recently developed a series of reactions to monitor recovery and completeness of the sodium bisulfite conversion step. One of the challenges in monitoring recovery during the conversion step is that the sequence changes as a result of the conversion. We therefore selected a region of the genome that did not contain any cytosine residues on one DNA strand over a short stretch (**Fig. 23.1**: QC-2). This sequence thus remains unaffected by treatment with sodium bisulfite. Therefore, we can use this C-less reaction to monitor DNA recovery at any step during the sodium bisulfite conversion. The right panel in **Figure 23.2** shows a time course of a sodium bisulfite conversion reaction. It is evident that the C-less reaction is relatively impervious to the effects of sodium bisulfite. We also designed reactions to monitor the efficacy and completeness of sodium bisulfite conversion for a

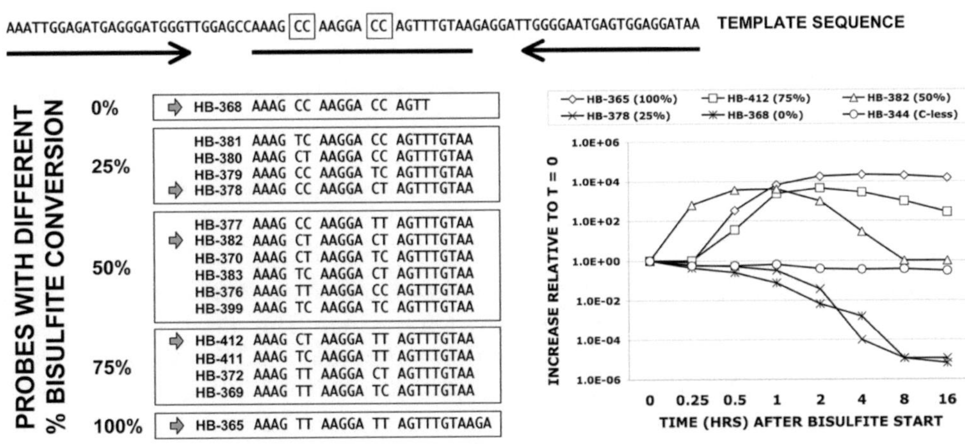

Fig. 23.2. Description of the QC-3 reactions and their performance on bisulfite-converted DNA. We designed 16 distinct bisulfite conversion control reactions that have common forward and reverse primers complementary to a DNA strand lacking cytosine residues at the positions of the primers, but have cytosine-containing unique probes that differ in their abilities to recognize various percentages of bisulfite converted DNA (0, 25, 50, 75, and 100% conversion). The genomic DNA sequence covered by these probes contains four cytosines that are normally modified to uracils after bisulfite conversion and then to thymines after a subsequent PCR amplification. The degree of conversion of these residues can be monitored by these probes, since they contain 16 different permutations of these residues to thymine reflecting possible changes that could occur in case of complete or incomplete bisulfite conversion. We tested the performance of these reactions in a time-course experiment where DNA was either left unconverted or was bisulfite-converted for different periods of time (0.25, 0.5, 1, 2, 4, 8 and 16 h). The ability of the 0% and 100% conversion reaction as well as the best 25%, 50%, and 75% conversion reactions to detect various degrees of bisulfite-converted DNA is presented in the right panel of this figure along with the performance of the C-LESS reaction that is not affected by the bisulfite-conversion process since it is designed toward a DNA sequence that contains no cytosines on one strand.

given sample (**Fig. 23.1**: QC-3). For this purpose, we selected a locus in the human genome, for which the primer locations did not cover any cytosines in one strand of the DNA. Thus, the amplification of this strand would be independent of bisulfite conversion. The region covered by the probe contains four cytosine residues. We designed and tested all 16 different permutations of the probe, assuming either conversion to uracil, or lack of conversion at each cytosine (**Fig. 23.2**). Experimental results with the probes indicated by an arrow on the left are shown for a bisulfite conversion time course on the right. We recommend using these probes to monitor completeness of the reaction. A threshold for bisulfite-conversion quality control can be implemented simply as a Δ–C(t) for each of these reactions, compared to the methylation-independent *ALU* QC-1 quantity control.

2. Materials

2.1. M.SssI Modification

1. *M.Sss*I enzyme supplied with 10 × buffer and 32 mM *S*-adenosyl methionine (SAM) (New England Biolabs, Ipswish, MA).
2. Peripheral blood leukocyte (PBL) DNA (Promega, Madison, WI).

2.2. Bisulfite Conversion and Recovery

1. Zymo EZ DNA Methylation Kit (Zymo Research, Orange, CA).

2.3. MethyLight PCR

1. TaqMan 1000 Reaction Gold With Buffer A Pack (Applied Biosystems, Foster City, CA). The kit contains the Taq enzyme, 10 × reaction buffer and the 25 mM $MgCl_2$ stock.
2. Deoxynucleotide triphosphates (dNTPs) are combined and diluted to a stock concentration of 10 mM for each nucleotide.
3. Primers and Black-Hole Quencher containing probes are obtained from Biosearch Technologies Inc. (Novato, CA). The primers and probes are prepared as 300 and 100 μM solutions, respectively, in H_2O. The probes containing the Minor Groove Binder Non Fluorescent Quencer (MGBNFQ) are obtained from Applied Biosystems, and are prepared as 100 μM solution in H_2O.
4. TaqMan stabilizer (10 ×): 0.1% Tween-20, 0.5% gelatin. Prepare a 20% working stock of Tween-20 in nuclease-free H_2O. Heat 0.2 g of gelatin in 20 mL of H_2O and after it is all melted, add 0.2 mL of 20% Tween-20 and bring the final volume to 40 mL with nuclease-free water. Store at +20°C.

3. Methods

3.1. DNA Isolation

Highly purified DNA is not a requirement for MethyLight analysis. Crude DNA extraction protocols involving lysis of the cells or tissues followed by DNA precipitation, or just crude lysates can be used in conjunction with MethyLight analysis. These approaches are desirable when limited quantities of DNA are available, such as in tissues embedded in paraffin slides, tissues from biopsies, or in bodily fluids such as blood (plasma/serum) containing small amounts of free circulating DNA. After biopsy tissues or microdissected cells from paraffin slides are lysed, an aliquot of this lysis solution can be directly used in bisulfite conversion. For the sensitive detection of DNA methylation in plasma or serum, the plasma/serum DNA needs to be concentrated from a larger initial volume. This can be achieved by using commercially available kits for DNA extraction from blood or various biological fluids or by precipitation of the DNA.

3.2. M.SssI Modification

M.*Sss*I is a CpG methylase and therefore each CpG dinucleotide is a target of the enzyme, which uses *S*-adenosyl methionine (SAM) as a methyl donor. M.*SssI*-treated DNA is used as a universally methylated reference sample in most MethyLight implementations. PBL DNA is used as a substrate in this protocol. A dilution of bisulfite-converted M.*SssI*-treated DNA will be used for normalization and is the basis for the ALU-C4 standard curves.

1. The M.*Sss*I treatment is carried out overnight at 37°C in a solution containing H_2O, 0.05 μg/μL of PBL DNA, 0.16 mM of SAM, 1 × reaction buffer, and 0.05 units/μL of M.*Sss*I enzyme.
2. Next day, add an extra boost of M.*Sss*I enzyme and SAM (1/3 of the original volume) for both components together with H_2O in a total volume representing 1/50 of the initial treatment volume.
3. In order to achieve complete methylation at all the genomic CpG sites, multiple rounds of M.*SssI* treatment can be performed.
4. M.*SssI*-treated DNA can be stored at +4°C, and 20 μL (~1 μg) is used for each bisulfite conversion.

3.3. Bisulfite Conversion and Recovery

1. First, prepare the CT Conversion Reagent and M-Wash Buffer included in the Zymo EZ DNA Methylation kit. Add 750 μL of water and 210 μL of M-Dilution Buffer to one tube of CT Conversion Reagent and mix by vortexing every 1–2 min for a total of 10 min. Each tube of CT Conversion Reagent is designed to treat ten DNA samples. For best results, the prepared CT Conversion Reagent should be used

immediately. Add 24 mL of 100% ethanol to the M-Wash Buffer Concentrate.

2. Start the bisulfite conversion protocol by adding 5 μL of the M-Dilution Buffer to the DNA sample and adjust the total volume to 50 μL with sterile H_2O. Mix the sample by flicking or pipetting up and down.
3. Incubate the sample at 37°C for 15 min.
4. Add 100 μL of the prepared CT Conversion Reagent to each sample and vortex gently.
5. Incubate the samples in a thermocycler using the following conditions: 95 c for 30 s, 50 c for 60 min repeated for 16 cycles.
6. Incubate the sample on ice for 10 min.
7. Add 400 μL of M-Binding buffer to the sample and mix by pipetting up and down.
8. Load the sample onto a Zymo-Spin I column and place the column into a 2-mL collection tube.
9. Centrifuge at full speed (>10,000*g*) for 30 s. Discard the flowthrough.
10. Add 200 μL of M-Wash Buffer to the column. Spin at full speed for 30 s.
11. Add 200 μL of M-Desulphonation Buffer to the column and let column stand at room temperature for 15 min. After incubation, spin at full speed for 30 s.
12. Add 200 μL of M-Wash Buffer to the column. Spin at full speed for 30 s.
13. Add another 200 μL of M-Wash Buffer and spin at top speed for 2 min.
14. Add 10 μL of M-Elution Buffer directly to the column matrix. Place into a 1.5 mL tube. Spin at top speed for 1 min to elute the DNA.
15. Bisulfite-converted DNA is stored at −20°C.

3.4. TaqMan PCR Reaction Setup for MethyLight Analysis

The MethyLight assay makes use of the TaqMan PCR principle, which requires forward and reverse primers as well as an oligomeric probe which emits fluorescence only after it is degraded by the 5′→3′ exonuclease activity of the *Taq* polymerase.

1. Each PCR reaction uses the same basic reaction setup. The choice of primer/probe sets is the only variable in these reactions.
2. Each individual PCR reaction contains 10 μL DNA, 15.4 μL PreMix Solution, 4.5 μL OligoMix Solution (1.5 μL of each primer and probe), and 0.1 μL *Taq* polymerase in a total of 30 μL PCR volume.
3. The combined PreMix Solution, OligoMix Solution, and *Taq* Gold polymerase for each reaction is referred to as the

MasterMix Solution (*see* **Note 1**). Load 10 μL of bisulfite-converted DNA and 20 μL of the MasterMix Solution in each PCR well.

4. The PreMix Solution contains all the TaqMan components, except Taq polymerase. These components and their final concentration in a PCR reaction are: $MgCl_2$ (3.5 mM), 1 × TaqMan Buffer, 1 × TaqMan stabilizer, and 0.2 mM of each dNTP. Each TaqMan reaction Kit is sufficient for 2000 MethyLight reactions. Therefore, to prepare a PreMix Solution for 2000 reactions, mix 8.4 mL of the $MgCl_2$ stock with 6 mL of 10 × TaqMan Buffer, 6 mL of 10 × TaqMan Stabilizer, 1.2 mL of 10 mM combined dNTPs, and 9.2 mL of H_2O. Small aliquots are stored at +4°C. For each PCR reaction, use 15.4 μL of the PreMix Solution in a total PCR volume of 30 μL.
5. The OligoMix Solution is specific for each MethyLight and quality control reactions, and represents a mixture of both primers and the probe. From the working stock of primers (300 μM) and probe (100 μM) prepare an OligoMix Solution by combining both primers and the probe in one tube. The concentrations of each the forward and reverse primers in the OligoMix Solution are 2 μM and the probe concentration is 0.67 μM. For each PCR reaction, use 4.5 μL of the OligoMix Solution in a total PCR volume of 30 μL.
6. For example, to determine the DNA methylation status of a specific gene of interest such as MLH1, first prepare an MLH1 OligoMix Solution by combining 2 μL of the MLH1 forward primer (300 μM), 2 μL of the MLH1 reverse primer (300 μM), and 2 μL of the MLH1 probe (100 μM) with 294 μL water. The MethyLight primers and probe sequences for MLH1 have previously been published *(8)*. In each individual MLH1 PCR reaction, combine 4.5 μL of this MLH1 OligoMix Solution with 15.4 μl of PreMix Solution, 0.1 μL *Taq* polymerase, and 10 μL of the bisulfite-converted DNA sample to be analyzed.
7. Individual OligoMix Solutions are prepared for any other gene investigated by MethyLight or any other quality control reactions used in the analysis, and 4.5-μL aliquots are then combined with the PreMix Solution, *Taq* polymerase, and bisulfite-converted DNA as described above.
8. Each MethyLight-based data point is the result of the combined analysis of a methylation-dependent PCR reaction (Experimental MethyLight reaction, *see* **Fig. 23.1**) and methylation-independent PCR reaction (CpG-less sequence, *see* **Fig. 23.1**) on reference (M.*Sss*I-treated DNA) and experimental DNA samples. The MethyLight assay setup is described in **Section 3.7**.

9. All PCR reactions are carried out as follows: one cycle at 95°C for 10 min followed by 50 cycles at 95°C for 15 s, and 60°C for 1 min.

3.5. Initial Quality Control

1. QC-1: Sample quantity and integrity. Samples vary in the initial template quantity and integrity. The most reliable measure of amplifiable DNA quantities after bisulfite conversion is a bisulfite-dependent, methylation-independent reaction for a multicopy number sequence well distributed throughout the genome. For this purpose, use the ALU-C4 bisulfite control reaction *(5)* (*see* **Table 23.1** for primer and probe sequences) to perform a preliminary TaqMan PCR test to check the Ct of each sample. Following the Zymo kit purification, the DNA is contained in 10 μL. Dilute the sample 1:10 (final volume 100 μL) and test 2 μL by PCR using the ALU-C4 bisulfite control reaction and the PCR conditions described in **Section 3.4**. The The Ct value generated from this 1:5 dilution will give an indication of the amount of bisulfite-converted DNA available for further analysis (*see* **Note 2**).
2. QC-2: Sample recovery. If problems are regularly encountered with recovery of samples after bisulfite conversion, then C-less bisulfite-independent reactions can be used to monitor recovery of each step. Primer and probe sequences of the C-less reaction are given in **Table 23.1**, and the cycling conditions are described in **Section 3.4.**
3. QC-3: Bisulfite conversion completeness. The efficacy and completeness of the bisulfite conversion of the sample can be assessed using a panel of bisulfite conversion reactions (**Fig. 23.2**) (*see* **Table 23.1** for the primer and probe sequences of these reactions and **Section 3.4** for cycling PCR conditions). These reactions are specific for unconverted DNA (0% conversion), fully converted DNA (100% conversion), or DNA with various degrees of conversion (25%, 50%, and 75% conversion) (*see* **Note 3**).

3.6. MethyLight Reactions

Two types of reactions are used in the MethyLight protocol that both use the bisulfite-converted DNA as a substrate: methylation-dependent reactions (CpG-based) specific for methylated DNA and methylation-independent control reactions (CpG-less), as described for the QC-1 quality control reaction above.

1. The methylation-dependent reactions are both bisulfite- and methylation-specific reactions, that is, they cover CpGs as well as C's not in a CpG context in their sequence (methylated CpGs will remain CpGs, other C's and unmethylated CpGs will become T's or TpGs, respectively, after bisulfite conversion and PCR).

Table 23.1
Primer and probes sequences for the quality-control PCR reactions used for the analysis of bisulfite-treated DNA. All probes contain a 6FAM fluorophore at the 5′ end and a minor groove binder nonfluorescent quencher at the 3′ terminus. The genomic coordinates for the QC-2 reaction are; chromosome 20, 19199387–19199455, and for the QC-3 reactions; chromosome 11, 17649485–17649620 obtained from the NCBI Build 36.2. The QC-1 reaction is based on an ALU consensus sequence and therefore does not have precise genomic coordinates

HB-number	Reaction ID	Description	Forward primer sequence (5′ to 3′)	Reverse primer sequence (5′ to 3′)	Probe sequence (5′ to 3′)
HB-313	ALU-C4	QC-1	GGT TAG GTA TAG TGG TTT ATA TTT GTA ATT TTA GTA	ATT AAC TAA ACT AAT CTT AAA CTC CTA ACC TCA	CCT ACC TTA ACC TCC C
HB-344	C-LESS-C1	QC-2	TTG TAT GTA TGT GAG TGT GGG AGA GA	TTT CTT CCA CCC CTT CTC TTC C	CTC CCC CTC TAA CTC TAT
HB-365	CONV-C1	QC-3: 100%	AAA TTG GAG ATG AGG GAT GGG T	TTA TCC TCC ACT CAT TCC CCA A	TCT TAC AAA CTA ATC CTT AAC TTT
HB-368	CONV-C2	QC-3: 0%	AAA TTG GAG ATG AGG GAT GGG T	TTA TCC TCC ACT CAT TCC CCA A	AAC TGG TCC TTG GCT TT
HB-369	CONV-C3	QC-3: 75%	AAA TTG GAG ATG AGG GAT GGG T	TTA TCC TCC ACT CAT TCC CCA A	TTA CAA ACT GAT CCT TAA CTT T
HB-370	CONV-C4	QC-3: 50%	AAA TTG GAG ATG AGG GAT GGG T	TTA TCC TCC ACT CAT TCC CCA A	TTA CAA ACT GAT CCT TAG CTT T
HB-372	CONV-C5	QC-3: 75%	AAA TTG GAG ATG AGG GAT GGG T	TTA TCC TCC ACT CAT TCC CCA A	TTA CAA ACT AGT CCT TAA CTT T
HB-376	CONV-C9	QC-3: 50%	AAA TTG GAG ATG AGG GAT GGG T	TTA TCC TCC ACT CAT TCC CCA A	TTA CAA ACT GGT CCT TAA CTT T
HB-377	CONV-C10	QC-3: 50%	AAA TTG GAG ATG AGG GAT GGG T	TTA TCC TCC ACT CAT TCC CCA A	TTA CAA ACT AAT CCT TGG CTT T
HB-378	CONV-C11	QC-3: 25%	AAA TTG GAG ATG AGG GAT GGG T	TTA TCC TCC ACT CAT TCC CCA A	TTA CAA ACT AGT CCT TGG CTT T

(continued)

Table 23.1 (continued)

HB-number	Reaction ID	Description	Forward primer sequence (5′ to 3′)	Reverse primer sequence (5′ to 3′)	Probe sequence (5′ to 3′)
HB-379	CONV-C12	QC-3: 25%	AAA TTG GAG ATG AGG GAT GGG T	TTA TCC TCC ACT CAT TCC CCA A	TTA CAA ACT GAT CCT TGG CTT T
HB-380	CONV-C13	QC-3: 25%	AAA TTG GAG ATG AGG GAT GGG T	TTA TCC TCC ACT CAT TCC CCA A	TTA CAA ACT GGT CCT TAG CTT T
HB-381	CONV-C14	QC-3: 25%	AAA TTG GAG ATG AGG GAT GGG T	TTA TCC TCC ACT CAT TCC CCA A	TTA CAA ACT GGT CCT TGA CTT T
HB-382	CONV-C16	QC-3: 50%	AAA TTG GAG ATG AGG GAT GGG T	TTA TCC TCC ACT CAT TCC CCA A	TTA CAA ACT AGT CCT TAG CTT T
HB-383	CONV-C17	QC-3: 50%	AAA TTG GAG ATG AGG GAT GGG T	TTA TCC TCC ACT CAT TCC CCA A	TTA CAA ACT AGT CCT TGA CTT T
HB-399	CONV-C15	QC-3: 50%	AAA TTG GAG ATG AGG GAT GGG T	TTA TCC TCC ACT CAT TCC CCA A	TTA CAA ACT GAT CCT TGA CTT T
HB-411	CONV-C20	QC-3: 75%	AAA TTG GAG ATG AGG GAT GGG T	TTA TCC TCC ACT CAT TCC CCA A	TTA CAA ACT AAT CCT TGA CTT T
HB-412	CONV-C21	QC-3: 75%	AAA TTG GAG ATG AGG GAT GGG T	TTA TCC TCC ACT CAT TCC CCA A	TTA CAA ACT AAT CCT TAG CTT T

2. The methylation independent control reaction (CpG-less) (ALU-C4) is used to normalize for differing quantities and quality of DNA samples (*see* **Note 4**). This reaction is not methylation specific since there are no CpGs in the primers/probe sequences, but is specific for bisulfite-converted DNA since it covers C's not in a CpG context.

3.7. MethyLight Assay Setup

In order to determine the methylation status of a specific gene using the MethyLight assay, four PCR reactions are required. Two types of samples are needed: the bisulfite-converted DNA of the sample of interest and the M.*Sss*I-converted DNA as a reference sample. For each of these DNAs we perform a PCR reaction for the gene of interest (Experimental MethyLight measurement, *see* **Fig. 23.1**) and one control PCR reaction to measure the amount of amplifiable DNA sample (QC-1, ALU-C4 reaction, *see* **Fig. 23.1**). The use of M.*Sss*I-converted DNA as a reference helps to control for variations in reagent batches, including primers and probes, reaction efficiency, machine performance, and various other sources of batch effects (*see* **Note 5**).

1. Dilute the bisulfite-converted M.*Sss*I-DNA (1:100) and use 10 μL of this sample for each PCR reaction. Use 10 μL of the bisulfite-converted experimental sample DNA (diluted based on the ALU-C4 Ct value from the 1:5 dilution test, see **Section 3.5.1.**). Perform each MethyLight reaction as well as each control reaction in duplicate.
2. Two independent four-point standard curves using the ALU control reaction and bisulfite–converted, M.*Sss*I-modified DNA (diluted 1:100) are required for quantification. From this initial stock of bisulfite–converted, M.*Sss*I-modified DNA, perform 1:25 serial dilutions. A volume of 10 μL of each dilution should be used for each amplification.

3.8. MethyLight Percentage of Methylated Reference (PMR) Calculations

1. The formula to calculate PMR values represents the quotient of two ratios (× 100). Thus, the formula is: $100 \times [(GENE\text{-}X \text{ mean value})_{sample}/(ALU \text{ mean value})_{sample}] \;/\; [(GENE\text{-}X \text{ mean value})_{M.SssI} \;/\; (ALU \text{ mean value})_{M.SssI}]$.
2. Once the real-time PCR program is finished, the Ct values are converted to mean values/copy numbers using the standard curve for each plate (*see* **Note 6**)
3. One PMR value per sample will be calculated based on the mean values derived from each of the two standard curves. The two PMRs obtained will be averaged at the end of the procedure.
4. Using the data generated with the first standard curve, divide the mean/copy value for the methylation reaction of the sample of interest by the mean/copy value of the ALU reaction

for the very same sample. Average this Quotient for duplicate reactions.

5. Divide the mean/copy value for M.*SssI* sample for the same methylation reaction as in step 4 by the mean/copy value for the ALU reaction of the M.*Sss*I sample. Average this quotient for duplicate reactions.
6. Divide the value from step 4 by the value from step 5 and multiply that value by 100. This is the first PMR value.
7. Calculate the second PMR value by the same procedure using the data generated based on the second standard curve. This can be achieved by simply reassigning the second ALU-C4 standard curve wells as standards. Then redetermine the values from steps 4 and 5. The PMR values from each quantitation can then be averaged to generate the final PMR value for each sample.

4. Notes

1. Uracil DNA glycosylase (AMPerase) should not be included in MethyLight reactions. This is a component used in some TaqMan reaction kits, and this poses a complication in MethyLight reactions as uracil is a product following bisulfite conversion.
2. It should be noted that low Ct values are always preferred to achieve the best possible data. An ALU-C4 Ct $\leq$ 17 is usually desirable, but data can also be generated from samples with Ct $\leq$ 22 when the samples are precious.
3. The lack of complete decline of the mean values obtained for the 75% and 50% reactions is likely due to some degree of cross-hybridization with fully converted sequence. Nevertheless, these two reactions are likely to be the most sensitive probes for detecting incomplete bisulfite conversion of a sample. It should be noted that these reactions assess completeness of bisulfite conversion only at this one locus. To the extent that sequences differ in their resistance to denaturing or bisulfite conversion, this locus may not be representative for other parts of the genome.
4. Cancer DNA can contain copy number alterations, which can affect the quantitation of both the locus of interest as well as the methylation-independent reaction. Ideally, one would design a CpG-less methylation-independent reaction as close to the MethyLight reaction as possible to adjust for such events. This significantly increases the cost of reaction design and experimental implementation. As a next-best solution, we avoid the influence of copy-number alterations of single loci for the methylation-independent QC-1 control reaction by using a repetitive element. Even in samples with aneuploidy,

this control reaction will yield a reasonable approximation of total DNA quantities due to the distributed nature of the target sequence.

5. We describe here the standard procedure in our laboratory to calculate the PMR value as a universal measure for the percentage of fully methylated alleles of a DNA sample, regardless of origin or locus being assessed. We find this a useful measure for most instances. Although our implementation is based on a comparison to a reference sample, it utilizes the absolute method of quantitation for real-time PCR, which is based on mean values derived from a standard curve of defined initial template quantities. By comparing to a control reaction and to a reference sample, we turn this absolute method into a relative method. We also sometimes implement relative methods of analysis, including the calculation of Δ–Ct values. We use this most frequently, when we are interested in a separate within-sample comparison, such as the comparison of different bisulfite=conversion control reactions. We prefer the PMR method as a general measure of DNA methylation, since it controls for many other sample-independent sources of experimental variation and error.
6. Under usual real-time PCR conditions, the standard curve is based on dilutions of known absolute quantities of template. Although this could be implemented for each reaction, using synthetic or cloned template, we prefer to avoid this, in part to limit sources of high-concentration PCR contamination. Since the PMR calculation is a relative measure, it is sufficient to use unknown quantities of standard DNA, but with precisely defined dilutions. This will yield mean values that do not have any absolute meaning, but which can be used to derive the ratios in the PMR calculation.

References

1. Eads, C. A., Danenberg, K. D., Kawakami, K., et al. (1999) CpG island hypermethylation in human colorectal tumors is not associated with DNA methyltransferase overexpression. *Cancer Res* 59, 2302–2306.
2. Eads, C. A., Danenberg, K. D., Kawakami, K., et al. (2000) MethyLight: a high-throughput assay to measure DNA methylation. *Nucleic Acids Res* 28, e32.
3. Trinh, B. N., Long, T. I., Laird, P. W. (2001) DNA methylation analysis by MethyLight technology. *Methods* 25, 456–462.
4. Ogino, S., Kawasaki, T., Brahmandam, M., et al. (2006) Precision and performance characteristics of bisulfite conversion and real-time PCR (MethyLight) for quantitative DNA methylation analysis. *J Mol Diagn* 8, 209–217.
5. Weisenberger, D. J., Campan, M., Long, T. I., et al. (2005) Analysis of repetitive element DNA methylation by MethyLight. *Nucleic Acids Res* 33, 6823–6836.
6. Herman, J. G., Graff, J. R., Myohanen, S., et al. (1996) Methylation-specific PCR: a novel PCR assay for methylation status of CpG islands. *Proc Natl Acad Sci USA* 93, 9821–9826.
7. Laird, P. W. (2003) The power and the promise of DNA methylation markers. *Nat Rev Cancer* 3, 253–266.
8. Fiegl, H., Gattringer, C., Widschwendter, A., et al. (2004) Methylated DNA collected by tampons – a new tool to detect endometrial cancer. *Cancer Epidemiol Biomarkers Prev* 13, 882–888.

Chapter 24

Quantification of Methylated DNA by HeavyMethyl Duplex PCR

Jürgen Distler

Abstract

The HeavyMethyl (HM) assay is a real-time PCR assay suitable for the qualitative and quantitative DNA methylation analysis of fresh, frozen, or formalin-fixed paraffin-embedded tissues and remote samples, such as serum, plasma, and urine. The HM uses a methylation-specific oligonucleotide blocker and a methylation-specific probe to achieve methylation-specific amplification and detection. A protocol for a duplex real-time PCR for the analysis of the methylation status of the *GSTP1* exon1 in prostate tissue samples is presented. This chapter describes the preparation and analysis of a combined HM *GSTP1* Exon1 and *GSTP1* reference assay.

Key words: Real-time PCR, methylation-specific PCR, HeavyMethyl assay, *GSTP1*, prostate cancer.

1. Introduction

The glutathione S-transferase gene (*GSTP1*) is silenced in the majority of prostate cancers and high-grade prostatic intraepithelial neoplasias as a result of methylation of CpGs in the promoter and the 5′-region of the gene *(1)*. In relation to the *GSTP1* transcription start, CpGs sites −42 to 31, comprising the promoter, exon 1, and intron 1 region of the *GSTP1* gene, are differentially methylated in cancerous and healthy prostate tissue *(2)*. Using quantitative methylation-specific real-time PCR, elevated methylation levels in the region of *GSTP1* were found in >90% of prostate cancer tissues *(1)*. As *GSTP1* methylation analysis significantly improves the diagnostic sensitivity of histopathological analysis of prostate biopsies, a quantitative *GSTP1* methylation

Jörg Tost (ed.), *DNA Methylation: Methods and Protocols, Second Edition, vol. 507*

DOI 10.1007/978-1-59745-522-0_24 Springerprotocols.com

assay might be a highly valuable tool to improve the prostate cancer diagnosis *(3)*.

In HeavyMethyl (HM) PCR, methylation specificity is not provided by the amplification primers, but by a nonextendable blocker oligonucleotide. The blocker binds to bisulfite-treated unmethylated DNA. As the binding sites of the blocker overlap with the binding sites of the amplification primer, the primer cannot bind to the unmethylated template. Therefore, the amplicon is not generated. In contrast, the blocker does not bind to methylated sequences, and therefore the primer-binding site is accessible and an amplicon is generated. In real-time PCR the amplification is monitored by a methylation-specific detection probe (4, 5; *see also* **Chapter 23**). HM real-time PCR assays allow the quantitative detection of methylated sequences in the background of an 8000-fold excess of unmethylated sequences *(5, 6)*. Based on the HM assay format, highly sensitive real-time assays were developed for the detection of several different DNA methylation markers in body fluid samples, such as urine and plasma. This chapter provides protocols for a HM *GSTP1* Exon1, a *GSTP1* reference real-time PCR, and a duplex PCR assay simultaneously analyzing the exon1 and the reference assay.

2. Materials

2.1. DNA Extraction and Bisulfite Conversion Kit

1. QIAamp DNA Mini Kit (Qiagen).
2. EpiTect Bisulfite Kit (Qiagen).

2.2. Real-Time PCR

1. LightCycler FastStart DNA Master HybProbe Kit (Roche Diagnostics).
2. PCR grade water (Roche Diagnostics).
3. LightCycler capillaries (Roche Diagnostics).
4. LightCycler 1.0 or 2.0 instrument (Roche Diagnostics).
5. Universally methylated bisulfite-treated human male DNA (Epigenomics AG, Berlin, Germany).
6. PolyA solution dissolve 100 mg PolyA (Roche Diagnostics) in 1 mL PCR grade water.
7. LightCycler Colour Compensation Set (Roche Diagnostics, Mannheim Germany).
8. 25 mM $MgCl_2$ solution as delivered with the FastStart DNA Master HybProbe Kit.

2.3. PCR Oligonucleotides for HM GSTP1, GSTP1 Reference and GSTP1/Reference Duplex Real-Time PCR

1. Reference forward primer: GGAGTGGAGGAAATTGAGAT.
2. Reference reverse primer: CCACACAACAAATACTCAAAAC.
3. Reference donor probe: GTTTAAGGTTAAGTTTGGGTGTTTGTA-FL (where FL is fluorescein).

4. Reference acceptor probe: red705-TTTTGTTTTGTGTTAG GTTGTTTTTTAGG-PH (where PH is a 3′-phosphate modification and red705 is a LCred705 dye).
5. *GSTP1*-HM forward primer: GGGATTATTTTTATAAGGTT.
6. *GSTP1*-HM reverse primer: CCATACTAAAAACTCTAAAC CC.
7. *GSTP1*-HM blocker: CCCATCCCCAAAAACACAAACCACAC-PH (where PH is a 3′-phosphate modification).
8. *GSTP1*-HM donor probe: TTCGTCGTCGTAGTTTTCGTT-FL (where FL is fluorescein).
9. *GSTP1*-HM acceptor probe: red640-TAGTGAGTACGCGC GGTT-PH (where PH is a 3′-phosphate modification and red640 is a LCred640 dye).

3. Methods

The total amount of amplifiable bisulfite DNA is quantified using the *GSTP1* reference PCR, which amplifies a 130 bp fragment within the *GSTP1* gene (GenBank accession no. X08058; nucleotides 2273–2402). The amount of methylated DNA is determined by HM *GSTP1* exon1 PCR generating a 123 bp (GenBank accession no. X08058, nucleotides 1183–1304) amplificate *(4, 6)*.

The amounts of total and methylated DNA are calculated against an external standard curve prepared with known concentrations of bisulfite-converted, universally methylated human male DNA. The methylation status in the *GSTP1* exon1 is expressed as percent methylation reference value (PMR, *see* also **Chapter 23**) determined as ratio of the amount of methylated DNA and total DNA multiplied by 100. The PMR values can be determined either by performing two individual real-time PCRs, HM *GSTP1* Exon1 and *GSTP1* reference PCR, or by one HM GSTP Exon1/*GSTP1* reference assay duplex real-time PCR. The results of the methylation analysis of prostate cancer tissue obtained by prostatectomy and benign prostatic hyperplasia (BPH) using *GSTP1* Exon1 and *GSTP1* reference single (**Fig. 24.1**) and duplex real-time PCR are shown (**Figs. 24.2** and **24.3**).

3.1. Sample DNA Extraction and Bisulfite Treatment

1. Extract DNA from 20 mg to 25 mg of fresh frozen prostatectomy samples using the QIAamp DNA Mini Kit according the recommendation of the manufacturer.
2. Treat 1 μg of the extracted DNA with sodium bisulfite as described in detail in **Chapter 26** (*see* **Note 1**) or the EpiTect Bisulfite Kit is used according the recommendation of the manufacturer.

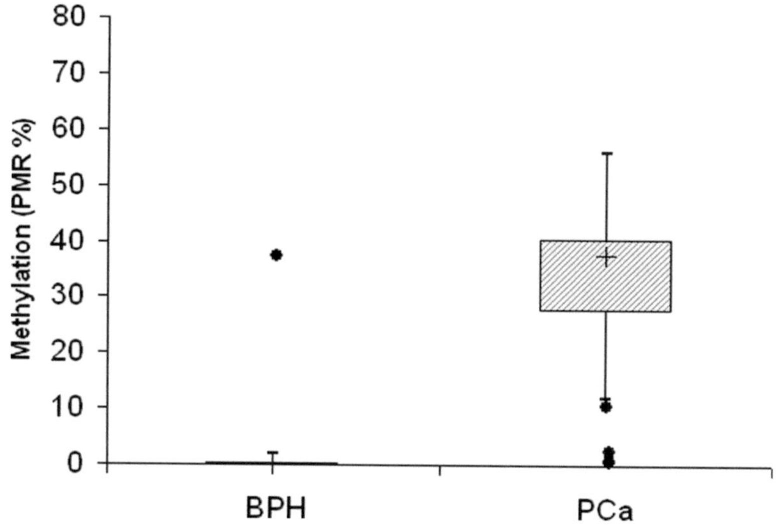

Fig. 24.1. Box-whisker plot of the *GSTP1* exon1 PMR values of prostate cancer tissue obtained by prostatectomy (PCa, n = 39) and benign prostatic hyperplasia (BPH, n = 37) tissue samples. The PMR values are determined performing HM *GSTP1* exon1 and *GSTP1* reference single-plex assays in three replicates. The box comprises 50% and the whiskers 95% of the data. The median of the PMR are indicated by a cross. Setting the PMR threshold to 2.5% prostate cancers are identified with a sensitivity of 92.5% (95% CI: 79–100) and a specificity of 97.6% (95% CI: 85–100).

3.2. Preparation of Quantification Standard DNA Solutions and Oligonucleotide Solutions

1. Prepare six quantification standard DNA solutions containing 5 ng/μL, 1 ng/μL, 0.2 ng/μL, 0.04 ng/μL, 0.008 ng/μL, and 0 ng/μL universally methylated and bisulfite-treated human male DNA dissolved in PCR grade water. In addition, each solution contains PolyA as carrier at a concentration of 5 ng/μL. Store at −20°C (*see* **Note 1**).
2. Prepare a 10 mM solution of the oligonucleotides: reference forward primer, reference reverse primer, reference donor probe, reference acceptor probe, *GSTP1*-HM forward primer, *GSTP1*-HM reverse primer, *GSTP1*-HM donor probe, and *GSTP1*-HM acceptor probe using PCR grade water. Store at −20°C (*see* **Note 2**).
3. Prepare a 100 mM solution of the *GSTP1*-HM blocker oligonucleotide using PCR grade water. Store at −20°C (*see* **Note 2**).

3.3. Setup of LightCycler 1.0/2.0 Real-Time Instrument

1. Generate a color-compensation file using the LightCycler Colour Compensation Set according to the procedure described by the manufacturer (*see* **Note 3**).
2. Define the capillaries containing the DNA quantification standards and the samples in the LightCycler Software 3.5. Enter the amount of DNA in the PCR of the individual quantification standards (50 ng, 10 ng, 2 ng, 0.4 ng, 0.08 ng, and 0 ng).

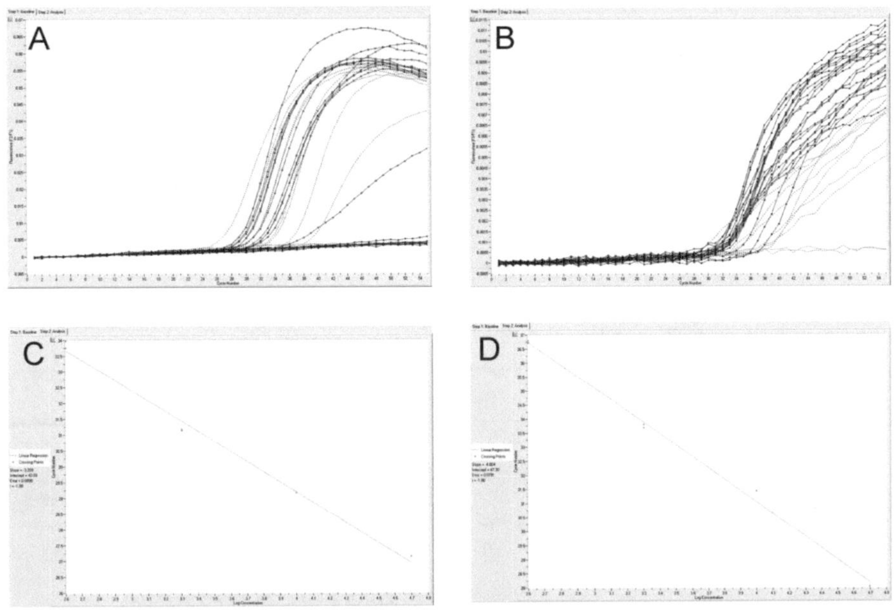

Fig. 24.2. **A-B** Screen shots of the HM *GSTP1* Exon1/*GSTP1* reference duplex real-time PCR on prostatectomy cancer (n = 10) and benign prostatic hyperplasia (n = 10) tissue samples. The detection of HM *GSTP1* Exon1 (**A**) and *GSTP1* reference (**B**) was carried out during the annealing step monitoring the ratio of fluorescence at 640 nm and 530 nm (channel F2/F1) and the ratio of fluorescence at 705 nm and 530 nm (channel F3/F1), respectively. Amplification curves of prostate cancer (open circles), BPH samples (black circles) and DNA quantification standards (broken lines) are shown. **C-D** Linear calibration standard curves of HM *GSTP1* exon1/*GSTP1* reference duplex real-time PCR are shown. The curves are based on the analysis of HM *GSTP1* Exon1 (**C**) and *GSTP1* reference (**D**) of the duplex PCR using channel F2/F1 and F3/F1, respectively. The x-axis indicates the log DNA amounts, the y-axis the threshold cycle number of the real-time PCR.

3. Enter the PCR cycling program using the following cycling conditions: activation 95°C for 10 min, 50 cycles at 95°C for 10 s (denaturation), 56°C for 30 s (annealing) and 72°C for 10 s (extension). The detection is carried out during the annealing step monitoring the ratio of fluorescence at 640 nm and 530 nm (HM *GSTP1* exon1 assay), at 705 nm and 530 nm (*GSTP1* reference assay) or both (duplex PCR).

3.4. GSTP1 Exon1 HM PCR

1. Thaw, mix, and centrifuge the $MgCl_2$, oligonucleotide, and DNA-quantification standard solutions in a bench centrifuge for 5 s.
2. Prepare the FastStart DNA Master HybProbe Kit according to the manufacturer's recommendations.
3. For the preparation of the PCR master mix, combine the following solutions in the indicated order: 122.4 μL water, 72 μL $MgCl_2$, 21.6 μL *GSTP1*-HM forward primer, 21.6 μL *GSTP1*-HM reverse primer, 28.8 μL *GSTP1*-HM blocker, 10.8 μL *GSTP1*-HM donor probe, 10.8 μL *GSTP1*-HM acceptor probe, and 72 μL FastStart DNA Master HybProbe (*see* **Note 4**).

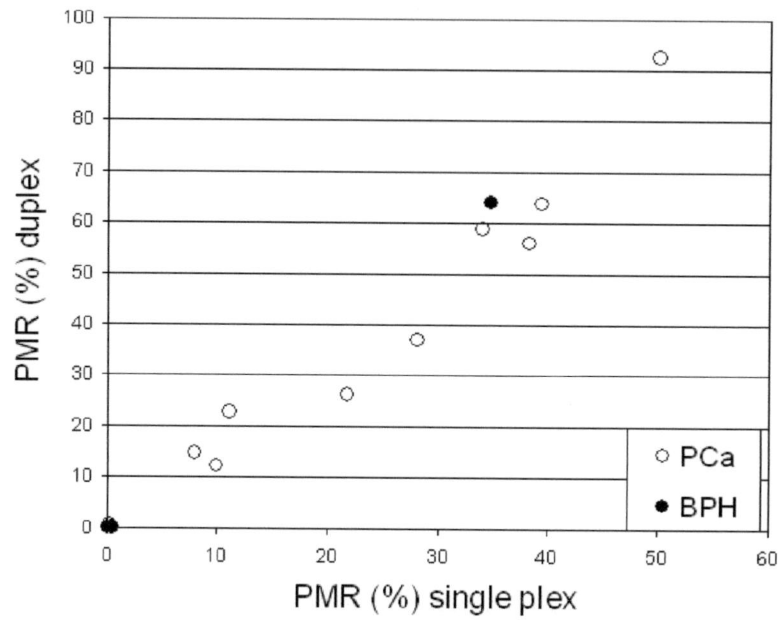

Fig. 24.3. Comparison of the performance of HM *GSTP1* exon1/*GSTP1* reference duplex and single-plex real-time PCRs. The correlation of PMR values generated with single-plex PCR and duplex PCR is shown for ten prostatectomy cancer (open circle) and ten benign prostatic hyperplasia (black circle) tissue samples.

4. Carefully mix the master mix and spin for 5 s.
5. Transfer 10 μL of the master mix to each of the 32 LightCycler capillaries.
6. Add 10 μL of template DNA to each capillary (*see* **Note 5**).
7. Spin the LightCycler capillaries according the recommendation of the manufacturer and put the capillaries into the LightCycler carousel according the predefined order (*see* **3.3**).
8. Set the carousel into the LightCycler instrument and start the real-time PCR.

3.5. GSTP1 Reference PCR

1. The first two steps are as described in 3.4
2. For the preparation of the PCR master mix combine the following solutions in the indicated order: 151.2 μL water, 72 μL $MgCl_2$, 21.6 μL reference forward primer, 21.6 μL reference reverse primer, 10.8 μL reference donor probe, 10.8 μL reference acceptor probe, and 72 μL FastStart DNA Master HybProbe (*see* **Note 4**).
3. The next steps are as described in 3.4.

3.6. HM GSTP1 Exon1/GSTP1 Reference Assay Duplex Real-Time PCR

1. The first two steps are as described in 3.4
2. For the preparation of the PCR master mix, combine the following solutions in the indicated order: 75.6 μL water, 57.6 μL $MgCl_2$, 5.4 μL reference forward primer, 5.4 μL reference reverse primer, 10.8 μL reference donor probe,

10.8 μL reference acceptor probe, 43.2 μL *GSTP1*-HM forward primer, 43.2 μL *GSTP1*-HM reverse primer, 14.4 μL *GSTP1*-HM blocker, 10.8 μL *GSTP1*-HM donor probe, 10.8 μL *GSTP1*-HM acceptor probe, and 72 μL FastStart DNA Master HybProbe (*see* **Note 4**).

3. The next steps are as described in 3.4.

3.7. Data Analysis and Determination of the Percent Methylation Reference (PMR) Values

1. The ratio of fluorescence at 640 nm and 530 nm (channel F2/F1) and the ratio of fluorescence at 705 nm and 530 nm (channel F3/F1) are used to monitor the *GSTP1* Exon1 HM and *GSTP1* reference PCR, respectively.
2. Calculate the cycle thresholds (Cts) according to the second derivative maximum method of the LightCycler software 3.5.
3. Based on the Cts of the quantification standards, the amount of DNA (total DNA, channel F3/F1 and methylated DNA, channel F2/F1) is determined using the LightCycler software 3.5.
4. The PMR value of the samples are determined according the following equation: amount of methylated DNA in PCR (HM *GSTP1* exon1 assay readout) divided by the total amount of amplifiable DNA in PCR (*GSTP1* reference assay readout) multiplied by 100.

4. Notes

1. Bisulfite-treated DNA can be stored for up to 6 months at −20°C. However, the DNA should not be frequently thawed and refrozen.
2. As the ratio of primer and blocker oligonucleotide is critical for an optimal performance of HM assays, the concentrations of the oligonucleotide solution should be carefully adjusted and tested by UV measurement. If available, the oligonucleotides should be quality checked by mass spectrometry. Oligonucleotide solutions can be stored in aliquots at −20°C for up to 3 months. Thaw and refreeze individual aliquots three times at maximum.
3. The color-compensation file is a prerequisite to perform dual-color PCR such as the HM *GSTP1* exon1 / *GSTP1* reference assay duplex PCR, in order to avoid fluorescence cross-talk between the channels. The color-compensation procedure must be performed on each instrument and should be rerun every 6 months.
4. The master mix is sufficient for 32 PCRs comprising a 10% pipetting margin.
5. In each run (32 capillaries total) 22 sample DNAs and 10 quantification standards (5, 1, 2 × 0.2, 2 × 0.04, 2 × 0.008, 2 × 0 ng/μL) can be analyzed. This setup should be run three

times in order to obtain reliable results. If the bisulfite DNA is derived from formalin-fixed and paraffin-embedded tissue, 5 μL of template DNA and 5 μL of water should be used.

Acknowledgement

The author would like to thank Reimo Tetzner for many productive scientific discussions and Matthias Schuster for critically reviewing the manuscript.

References

1. Meiers, I., Shanks, J. H., Bostwick, D. G. (2007) Glutathione S-transferase pi (*GSTP1*) hypermethylation in prostate cancer: review 2007. *Pathology* 39, 299–304.
2. Millar, D. S., Paul, C. L., Molloy, P. L., et al. (2000) A distinct sequence (ATAAA)n separates methylated and unmethylated domains at the 5′-end of the *GSTP1* CpG island. *J Biol Chem* 275, 24893–24899.
3. Harden, S. V., Sanderson, H., Goodman, S. N., et al. (2003) Quantitative *GSTP1* methylation and detection of prostate adenocarcinoma in sextant biopsies. *J Natl Cancer Inst* 21, 1634–1637.
4. Cottrell, S. E., Distler, J., Goodman, N.S., et al. (2004) A real-time PCR assay for DNA-methylation using methylation-specific blockers. *Nucleic Acids Res* 32, e10.
5. Cottrell, S. E., Laird, P. W. (2003) Sensitive detection of DNA methylation. *Ann NY Acad Sci* 983, 120–130.
6. Tetzner, R., Dietrich, D., Distler J. (2007) Control of carry-over contamination for PCR-based DNA methylation quantification using bisulfite treated DNA. *Nucleic Acids Res* 35, e4.

Part V

Special Applications

Chapter 25

Analysis of Methylated Circulating DNA in Cancer Patients' Blood

Eiji Sunami, Anh-Thu Vu, Sandy L. Nguyen, and Dave S. B. Hoon

Abstract

Circulating extracellular nucleic acids derived from body fluids such as blood are commonly analyzed to assess malignant diseases. Efficient isolation, extraction, quantification, modification, and analysis methods remain important for utilizing circulating nucleic acids as potential molecular biomarkers. Our refined techniques of DNA isolation from serum, sodium bisulfite modification of extracted DNA, and methylation analysis provide a robust approach for quantitative analysis of circulating tumor-related DNA. The approach allows direct comparison of methylated and nonmethylated genomic sequences in a specimen.

Key words: Methylation, circulating DNA, methylation-specific PCR (MSP), capillary array electrophoresis (CAE), serum.

1. Introduction

Extracellular nucleic acids from body fluids have been targeted to investigate the status of malignant diseases. These include various types of carcinomas, melanomas, and lymphomas *(1–10)*. Extracellular nucleic acids and tumor-derived DNA are now used as biomarkers and are observed in body fluids such as blood and bone marrow *(11)*.

There is no convincing evidence as to whether serum or plasma is a better analytical tool for studying circulating nucleic acids. The most significant difference between the two is the presence of clotting factors and their related proteins. The amount of circulating DNA that is available in body fluids is very limited; however, the circulating DNA from serum is higher than from

Jörg Tost (ed.), *DNA Methylation: Methods and Protocols, Second Edition, vol. 507*

DOI 10.1007/978-1-59745-522-0_25 Springerprotocols.com

plasma *(12, 13)*. We demonstrated that serum potentially has up to a sixfold higher amount of free-circulating DNA than plasma. Contaminated extraneous DNA from leukocytes can play a role in the difference in amount of circulating DNA found in serum versus plasma *(13)*. Because of this finding, we selected serum as the specimen source for our circulating nucleic acids studies. Another critical factor in choosing serum or plasma for our studies is the specificity of the biomarker.

The characteristics of free-circulating DNA in blood are summarized as follows: (1) low DNA quantity, (2) fragmented DNA of different sizes, and (3) DNA bound to lipids and proteins. Depending on the method of DNA extraction from serum, quantification methods, and patient characteristics such as cancer type and stage, the total DNA amount isolated from 1 mL of serum ranges from 10 ng to 1,000 ng *(11, 13)*. Serum from cancer patients or patients with inflammation tends to contain higher amounts of DNA than healthy blood donors *(14, 15)*. Circulating tumor DNA is released by means of apoptosis, necrosis, and active shedding. The main source of circulating DNA in healthy individuals is apoptosis. DNA released from apoptotic cells is usually truncated into roughly <200 bp fragments, whereas those released from tumor cells vary in size *(15,16)*. Biomarkers assessed in these samples may be limited to the fragment size of the circulating DNA.

Several experimental approaches have been taken to establish the utility of nucleic acids obtained from the body fluid of diseased patients for molecular biomarkers. DNA quantification/integrity analysis, detection of point mutations, analysis of microsatellite alterations, and methylation of CpG island sequences in the promoter region are the most commonly used analyses for circulating DNA. Methylation of CpG dinucleotides in a gene's promoter region can cause gene silencing *(17)*. The extent of promoter region CpG island hypermethylation observed depends on the specific gene investigated and the tumor type. When the appropriate combination of tumor-related genes are selected, circulating methylated DNA can be a major tool in monitoring cancer patients *(7, 18)*. Assessment of circulating DNA allows a single point or repetitive assessment of blood in absence of the tumor material. Methylation studies of circulating nucleic acids are usually performed on the basis of methylation-specific PCR (MSP, **Chapter 22**). Often, circulating DNA contains a small quantity of low-quality DNA and may be contaminated with DNA from normal cells, thus it is essential to introduce sensitive and quantitative methylation analysis. To accurately assess the methylation status of circulating DNA, we use capillary array electrophoresis (CAE) for semi-quantitative methylation analysis *(7)* and quantitative analysis of methylated alleles (AQAMA) as the absolute quantitative method *(19)*.

To date, there are no standard protocols for isolating DNA from serum/plasma. Optimization and standardization are necessary. This step is paramount for the accurate interpretations of the results. In this protocol, we demonstrate our optimized and standardized methods of DNA isolation from serum, sodium bisulfite modification (SBM) of extracted DNA, and DNA methylation analysis.

2. Materials

2.1. Serum Isolation

1. Serum Filter System 13 mm ×4″ (Fisher Scientific).
2. 7-mL Red-Top Vacutainer tubes (Fisher Scientific).
3. 2-mL Transfer Pipettes (Fisher Scientific).
4. 2-mL Cryovials (Corning Inc).

2.2. DNA Extraction

1. Sterile 0.9% NaCl (0.45 g in 50 mL final volume, filter with 0.45-μm syringe).
2. Proteinase K (20 mg/mL).
3. 10% Sodium Dodecyl Sulfate (SDS).
4. 2-mL Eppendorf microcentrifuge tube (Axygen Scientific Inc).

2.3. DNA Purification

1. Phenol–Chloroform–Isoamyl (PCI) 25:24:1, pH 8.

2.4. DNA Precipitation

1. Pellet paint NF (Novagen).
2. 100% Isopropanol.
3. 90% Ethanol, store at 4°C (200 proof stock).
4. Disposable fine-tip transfer pipettes (Fisher Scientific).

2.5. DNA Quantification

1. TE Buffer (1 ×): 10 mM Tris–HCl, pH 8.0, 1 mM EDTA.
2. Pico-Green Assay Kit (Molecular Probes).
3. 96 round-well bottom microplate (Coster).

2.6. Sodium Bisulfite Modification

1. EpiTect Bisulfite Kit (Qiagen).
2. Oli-Green Assay Kit (Invitrogen).

2.7. Methylation Analysis

1. CEQ8000XL (Beckman Coulter).
2. *Sss*I Methylase (New England Biolabs).
3. GenomeLab Separation Buffer (Beckman Coulter).
4. GenomeLab Separation Gel-LPAI (Beckman Coulter).
5. GenomeLab Sample Loading Solution (Beckman Coulter).
6. GenomeLab DNA Size Standard Kit (Beckman Coulter).
7. Genomi*Phi* DNA Amplification Kit (GE healthcare).

3. Methods

Methylation analysis of serum DNA consists of several procedures: serum processing (from blood draw to serum storage), DNA isolation from serum (DNA extraction, purification, and precipitation), sodium bisulfite modification of DNA, and methylation analysis. In this section, we introduce our standard operating procedures.

3.1. Serum Isolation

1. Blood specimens must be drawn using a serum separator tube and kept at room temperature before processing (*see* **Note 1**).
2. Centrifuge the blood samples in the serum separator tubes at 1,000*g* (RCF) at 22–25°C for 15 min.
3. Obtain a 7-mL red-top vacutainer tube. Remove the red-top cap while keeping the 7-mL vacutainer inside the fume hood.
4. Using a transfer pipette, transfer serum from the centrifuged serum separator tube to the 7-mL red-top vacutainer. Avoid disturbing the gel layer with the transfer pipette.
5. Discard transfer pipette and serum separator tube.
6. Obtain a serum filter. Avoid touching the bottom filter portion.
7. Insert the serum filter at an angle into the mouth of the 7-mL vacutainer and push down on it.
8. Use a new transfer pipette to transfer the filtered serum to 2 mL cryovials for storage. Use as many cryovials as necessary so that the volume for each cryovial does not exceed the 1.8-mL mark. Cap tightly.
9. Discard the serum filter, 7-mL vacutainer tube, and transfer pipette.
10. Store all 2-mL serum cryovials temporarily (<30 days) at –30°C. Transfer serum cryovials to –80°C for long-term storage.

3.2. DNA Extraction

1. Frozen serum samples need to be thawed; this can be carried out in a 37°C incubator for approximately 20 min (*see* **Note 2**).
2. Aliquot 0.5 mL serum into a 2-mL microcentrifuge tube. Dilute aliquot with 0.3 mL of 0.9% NaCl.
3. Prepare a premix consisting of Proteinase K (20 μL per sample) and 10% SDS (95 μL per sample). Aliquot 115 μL of premix into each sample. Vortex each sample for 2 s.
4. Incubate all samples at 50°C for 3 h. Briefly vortex for a few seconds every hour or place samples on a continuously rocking platform.

3.3. DNA Purification

1. After 2.5 h, remove PCI from refrigerated storage and allow reagent to equilibrate to room temperature in fume hood before proceeding to the next step (*see* **Note 3**).
2. Under a fume hood, pipette only the bottom aqueous layer of PCI reagent, and add 1 mL of PCI to each sample (*see* **Note 4**). Vortex for 10 s three times.
3. Incubate at room temperature for 10 min, allowing each sample to separate into two distinct phases.
4. Centrifuge the microcentrifuge tubes for 10 min at 16,000*g* (RCF) at 22°C (*see* **Note 5**).
5. Remove 75–85% of the upper (aqueous) phase from the microcentrifuge tube and aliquot into a new, labeled 2-mL centrifuge tube (*see* **Note 6**). Discard remaining PCI according to protocol into hazardous waste containers.

3.4. DNA Precipitation

1. Add 2 μL of pellet paint NF to the aqueous aliquot and thoroughly mix by vortexing for a few seconds.
2. Add 700 μL of 100% isopropanol. Vortex well and then quickly centrifuge the samples.
3. Precipitate the DNA by storing all samples at −30°C for at least 3 h. Serum DNA samples can be stored overnight.
4. Three hours later (or the next day), sediment the DNA by centrifuging tubes at 16,000*g* for 30 min at 4°C. A visible pellet should be seen after centrifuging.
5. Remove a majority of the isopropanol supernatant using a disposable fine-tip transfer pipette. It is not pertinent to remove all of the supernatant; just remember not to pipette and remove the DNA pellet from the centrifuge tube.
6. Add a 1-mL aliquot of 90% 4°C ethanol using a pipette to each sample in order to wash the DNA pellet. Invert tubes 2–3 times to mix.
7. Then, centrifuge tubes at 16,000*g* for 10 min at 4°C. Remove the 90% ethanol supernatant with disposable fine-tip transfer pipette. Once again, do not remove or disturb the pellet during removal of supernatant.
8. Air-dry the pellet remaining in centrifuge tube or dry in a speed vacuum, without heat.
9. After the pellet is completely dry in the centrifuge tubes, resuspend DNA in 50 μL H_2O (*see* **Note 7**). Vortex thoroughly.

3.5. DNA Quantification

1. For quantification of double-stranded DNA (dsDNA), reserve 5 μL (of the total 50 μL) and mix it with a 95 μL aliquot of TE Buffer. Use this diluted sample in the PicoGreen quantification assay, according to the manufacturer's protocol.

3.6. Sodium Bisulfite Modification (SBM)

1. The maximum DNA solution volume that can be applied for the Qiagen EpiTect kit is 40 μL per tube. Take up to

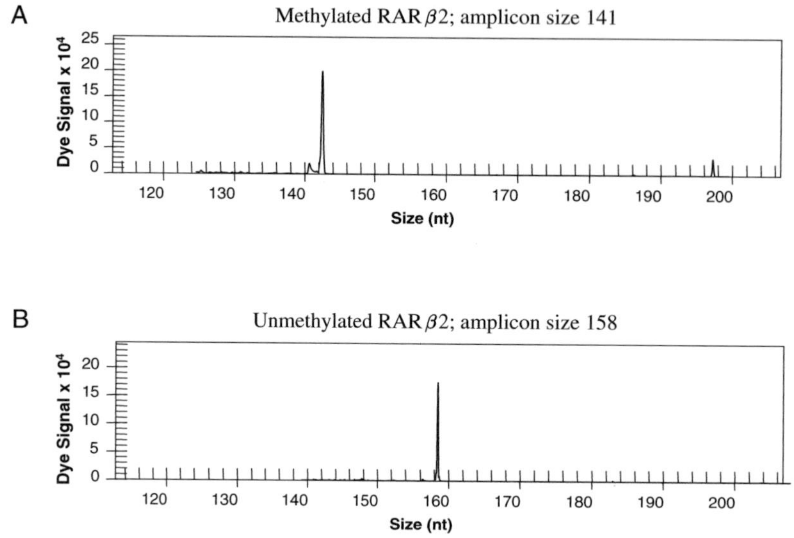

Fig. 25.1. Differentiation of methylated and unmethylated MSP product by color and size. **(A)** Result for the methylated RARβ2 allele analysis by MSP is shown (D4 dye) and amplicon size is 141 bp. **(B)** Result for the unmethylated RARβ2 allele analysis by MSP is shown (D3 dye) and amplicon size is 158 bp.

40 µL DNA solution (from step 3.4.9) for SBM. For SBM DNA modification and cleanup of bisulfite-converted DNA, see manufacturer's recommendations (*see* **Note 8**).

2. For quantification of sodium bisulfite-modified DNA, reserve 2 µL (of the total 20–40 µL) and mix it with a 98 µL aliquot of TE Buffer. Use this diluted sample in the Oli-Green quantification assay, according to the manufacturer's protocol.

3.7. Methylation Analysis

1. After SBM, the methylation status of serum DNA is assessed using methylation-specific PCR (MSP) using two sets of fluorescent-labeled primers specifically designed to amplify methylated and unmethylated DNA sequences.
2. Bisulfite-modified DNA is subjected to PCR amplification in a final reaction volume of 10 µL containing PCR buffer, 2.5–4.5 mM $MgCl_2$, dNTPs, 0.3 µM primers, 0.5 U Ampli*Taq* gold polymerase, and 50 pmol of each forward (F) and reverse (R) primer set specific for methylated (M) and unmethylated (U) primers. Forward MSP primers of methylated and unmethylated were each labeled with different color dyes (i.e., D3 and D4).
3. PCR is carried out using the optimized annealing temperatures for each primer set (*see* **Note 9**). An initial 10 min incubation step at 95°C is followed by 40 cycles of denaturation at 95°C for 30 s, annealing for 30 s, extension at 72°C for 30 s, and a final hold of 7 min at 72°C. Sodium bisulfite modified-lymphocyte DNA from healthy donors serves as unmethylated negative control DNA for each PCR reaction, whereas

*Sss*I Methylase-treated lymphocyte DNA followed by sodium bisulfite modification serves as positive-methylated control DNA. Molecular biology grade water without DNA serves as a blank control for reagent contamination (*see* **Note 10**).

4. Post-PCR products are separated and assessed using CAE run on a CEQ 8000XL system. CAE was performed in a 96-well microplate, combining 1 μL of each methylated and unmethylated PCR reaction product with 40 μL of loading buffer and 0.5 μL of dye-labeled size standard in a single well. Forward MSP primers were each labeled with different color dye to permit discrimination of the two peaks for direct comparison of post-PCR methylated and unmethylated products in a single run. CEQ 8000XL software was used to determine PCR product signal intensity (**Fig. 25.1**).

4. Notes

1. All serum specimens should be processed as soon as possible under a fume hood. The long-time storage of blood samples may cause an increase in serum DNA due to contamination from leukocyte DNA.
2. We have confirmed that these methods can also be applied for plasma.
3. PCI should be used at room temperature.
4. Do not shake bottle of PCI reagent prior to use. There should be two distinct layers: an upper aqueous buffer solution layer and a lower organic PCI reagent layer.
5. Do not disturb the separated phases when removing tubes from the centrifuge.
6. Extract the top aqueous phase only. The extracted aqueous phase is approximately 450–700 μL in total volume. Do not disturb the middle, turbid (emulsion) layer. To avoid disturbing, pipette three 150 μL aliquots of the aqueous phase into a newly labeled centrifuge tube.
7. Resuspension volume should be calculated from the volume that can be used for SBM (see manufacturer's instructions). If DNA quantification is not needed for the study, the entire volume of extracted DNA can be used. Total DNA amount from 500 μL of serum ranges between 50 ng and 500 ng, depending on patient characteristics.
8. We tested another SBM kit from Zymogen and confirmed that both kits work equally well.
9. Primer design and optimization of MSP conditions are very important. For optimization of MSP conditions, completely methylated and unmethylated controls are necessary (*see* also **Chapter 22**).

10. Synthesized universal unmethylated control by nested whole genome amplification with *phi29* DNA polymerase can be used *(20)*. Extracted DNA from cell lines previously known as methylated or unmethylated can be used as controls.

References

1. Allan, J. M., Hardie, L. J., Briggs, J. A., et al.(2001) Genetic alterations in bronchial mucosa and plasma DNA from individuals at high risk of lung cancer. *Int J Cancer* 91, 359–365.
2. Anker, P., Mulcahy, H., Stroun, M. (2003) Circulating nucleic acids in plasma and serum as a noninvasive investigation for cancer: time for large-scale clinical studies? *Int J Cancer* 103, 149–152.
3. Chiang, P. W., Beer, D. G., Wei, W. L., et al. (1999) Detection of erbB-2 amplifications in tumors and sera from esophageal carcinoma patients. *Clin Cancer Res* 5, 1381–1386.
4. Coulet, F., Blons, H., Cabelguenne, A., et al. (2000) Detection of plasma tumor DNA in head and neck squamous cell carcinoma by microsatellite typing and p53 mutation analysis. *Cancer Res* 60, 707–711.
5. Goessl, C., Muller, M., Heicappell, R., et al. (2001) DNA-based detection of prostate cancer in blood, urine, and ejaculates. *Ann N Y Acad Sci* 945, 51–58.
6. Mayall, F., Fairweather, S., Wilkins, R., et al. (1999) Microsatellite abnormalities in plasma of patients with breast carcinoma: concordance with the primary tumour. *J Clin Pathol* 52, 363–366.
7. Mori, T., Martinez, S. R., O'Day ,S. J., et al. (2006) Estrogen receptor-alpha methylation predicts melanoma progression. *Cancer Res* 66, 6692–6698.
8. Swisher, E. M., Wollan, M., Mahtani, S. M., et al. (2005) Tumor-specific p53 sequences in blood and peritoneal fluid of women with epithelial ovarian cancer. *Am J Obstet Gynecol* 193, 662–667.
9. Whitney, D., Skoletsky, J., Moore, K., et al. (2004) Enhanced retrieval of DNA from human fecal samples results in improved performance of colorectal cancer screening test. *J Mol Diagn* 6, 386–395.
10. Yamashita, K., Kida, Y., Shinoda, H., et al. (1999) K-ras point mutations in the supernatants of pancreatic juice and bile are reliable for diagnosis of pancreas and biliary tract carcinomas complementary to cytologic examination. *Jpn J Cancer Res* 90, 240–248.
11. Fleischhacker, M., Schmidt, B. (2007) Circulating nucleic acids (CNAs) and cancer – a survey. *Biochim Biophys Acta* 1775, 181–232.
12. Lee, T. H., Montalvo, L., Chrebtow, V., et al. (2001) Quantitation of genomic DNA in plasma and serum samples: higher concentrations of genomic DNA found in serum than in plasma. *Transfusion* 41, 276–282.
13. Umetani, N., Hiramatsu S., Hoon D. S. (2006) Higher amount of free circulating DNA in serum than in plasma is not mainly caused by contaminated extraneous DNA during separation. *Ann N Y Acad Sci* 1075, 299–307.
14. Umetani, N., Giuliano, A. E., Hiramatsu, S. H., et al. (2006) Prediction of breast tumor progression by integrity of free circulating DNA in serum. *J Clin Oncol* 24, 4270–4276.
15. Umetani, N., Kim, J., Hiramatsu, S., et al. (2006) Increased integrity of free circulating DNA in sera of patients with colorectal or periampullary cancer: direct quantitative PCR for ALU repeats. *Clin Chem* 52, 1062–1069.
16. Widschwendter, M., Menon, U. (2006) Circulating methylated DNA: a new generation of tumor markers. *Clin Cancer Res* 12, 7205–7208.
17. Hoon, D. S., Spugnardi, M., Kuo, C., et al. (2004) Profiling epigenetic inactivation of tumor suppressor genes in tumors and plasma from cutaneous melanoma patients. *Oncogene* 23, 4014–4022.
18. Müller, H. M., Widschwendter, A., Fiegl, H., et al. (2003) DNA methylation in serum of breast cancer patients: an independent prognostic marker. *Cancer Res* 63, 7641–7645.
19. de Maat, M. F., Umetani, N., Sunami, E., et al. (2007) Assessment of methylation events during colorectal tumor progression by absolute quantitative analysis of methylated alleles. *Mol Cancer Res* 5, 461–471.
20. Umetani, N., de Maat, M. F., Mori, T., et al. (2005) Synthesis of universal unmethylated control DNA by nested whole genome amplification with phi29 DNA polymerase. *Biochem Biophys Res Commun* 329, 219–223.

Chapter 26

Prevention of PCR Cross-Contamination by UNG Treatment of Bisulfite-Treated DNA

Reimo Tetzner

Abstract

Amplification of sodium bisulfite-treated DNA is widely used to study DNA methylation. The proportion of methylated sequences of a specific DNA region in a sample can be determined by the analysis of PCR products or directly calculated from real-time PCR amplification of bisulfite-treated DNA. At the same time, PCR based methods always involve the risk of false positive or incorrect quantitative results due to the unintended reamplification of contaminating PCR products. The incubation of PCR reactions with Uracil-DNA Glycosylase (UNG) prior to the thermal cycling in combination with the use of dUTP in the PCR amplification is a commonly used technology to prevent such cross-contamination. Since sodium bisulfite treatment converts unmethylated cytosine bases into uracil residues, not only contaminating PCR products but also the converted template DNAs would be degraded as well. This chapter describes a modified bisulfite treatment procedure to generate sulfonated DNA enabling the application of UNG-based carryover prevention to DNA methylation analysis. The high efficiency of the decontamination procedure as well as the universal applicability of this simple method is shown.

Key words: DNA methylation, polymerase chain reaction, real-time PCR, sodium bisulfite-treated DNA, contamination prevention, carryover prevention, Uracil-DNA Glycosylase (UNG)

1. Introduction

PCR products derived from prior amplifications can serve as a template for subsequent PCR experiments. This may lead to false-positive reactions, positive no template controls, and invalid results. In practice, this problem can be so persistent that scientists may need to move to a new location, because contamination of the work environment makes it impossible to obtain reliable PCR results. In a clinical laboratory, the main concern is that contaminating DNA may cause incorrect results in diagnostic assays.

Jörg Tost (ed.), *DNA Methylation: Methods and Protocols, Second Edition, vol. 507*

DOI 10.1007/978-1-59745-522-0_26 Springerprotocols.com

The use of the enzyme Uracil-DNA Glycosylase (UNG) reduces the risk of false-positive reactions due to PCR product carryover *(1)*. This contamination protection method is based on the substitution of dTTP by dUTP in all PCR reactions. The resulting PCR products contain uracil in place of thymine and are therefore distinguishable from the template DNA. Before starting the PCR cycling, the reaction mixture is incubated in the presence of UNG to degrade specifically PCR products from former reactions preventing its unintended reamplification. Unincorporated dUTP and the DNA template in the mixture remain unaffected. The initial denaturation step of the PCR leads to an inactivation of the UNG and therefore allows for the formation of new PCR product. This procedure is considered to be the standard method for carryover prevention and is widely used in PCR-based research and diagnostic laboratories *(2, 3)*. However, the method is not applicable to assays that are based on the amplification of uracil containing template DNA. Most techniques for the analysis of DNA methylation patterns depend on bisulfite treatment of the template DNA leading to a deamination of all unmethylated cytosines to uracils, leaving only methylated cytosines unaltered *(4)*. Therefore, Uracil-DNA Glycosylase cannot be applied to techniques using bisulfite conversion of genomic DNA. In the modified bisulfite treatment described in this chapter, the desulfonation step at the end of the procedure is omitted, leading to sulfonated DNA (SafeBis DNA), containing 6-sulfonyl uracil. SafeBis DNA is resistant to UNG cleavage and therefore enables the application of UNG to prevent PCR contamination *(5)*. After treatment of the reaction mixture with UNG, the PCR is carried out with an extended initial denaturation step. During this step, the bisulfite-treated DNA becomes desulfonated, the Taq DNA polymerase is activated and the UNG is inactivated simultaneously. This chapter describes how this procedure can be applied to sensitive and quantitative methylation analysis. In a study, 24 samples from colorectal cancer and normal tissue were analyzed with and without the carryover prevention procedure and the results were compared. The study comprised the use of a reference real-time PCR and a methylation-specific real-time HeavyMethyl (HM) PCR *(5,6,* **Chapter 24)**. The percentage of methylated reference (PMR) of TMEFF2 was determined by analyzing bisulfite DNA obtained by a standard procedure and compared to the results obtained by applying the carryover prevention workflow. The TMEFF2 promoter was confirmed to be differentially methylated in colorectal cancers *(7, 8)*. To demonstrate the efficiency of the decontamination, the analysis was repeated after addition of PCR products, simulating a carryover contamination. The use of UNG to prevent PCR contamination is not limited to real-time PCR methods and can easily be applied to any bisulfite-based DNA methylation analysis.

2. Material

2.1. DNA Isolation

1. QIAamp DNA Mini Kit (Qiagen).
2. Ethanol (purity > 99.8%).
3. Proteinase K.
4. Water for molecular biology (DEPC treated).
5. RNase A (Qiagen).
6. 1.5-mL and 2-mL reaction tubes, thermoblock, and centrifuge (Eppendorf).
7. NanoDrop spectrophotometer (NanoDrop Technologies).

2.2. Sodium Bisulfite Treatment

1. Sodium bisulfite ($Na_2S_2O_5$).
2. Sodium sulfite, anhydrous (Na_2SO_3).
3. Water for molecular biology (DEPC treated).
4. 1,4-dioxane, stabilized (Riedel de Haën).
5. 6-hydroxy-2,5,7,8-tetramethylchroman-2-carboxylic acid (Sigma-Aldrich).
6. Sodium hydroxide pellets.
7. Microcon Centrifugal Filter devices (Microcon YM-30, Millipore).
8. 2.0-mL safe-lock tubes and safe-lock clamps (Eppendorf).
9. 15- and 50-mL Falcon tubes.
10. Thermomixer, waterbath, centrifuge.
11. Universally methylated DNA (Chemicon).

2.3. Real-Time PCR

1. LightCycler 2.0 system (Roche Diagnostics).
2. LightCycler capillaries (20 μL, Roche Diagnostics).
3. LightCycler FastStart DNA Master HybProbe (Roche Diagnostics).
4. Water, PCR grade (purified, double-deionized, autoclaved, not DEPC treated, Roche Diagnostics).
5. Uracil-DNA Glycosylase (recombinant, *E.coli* K12, Roche Diagnostics).
6. Polyadenylic acid (Poly(A), Roche Diagnostics).

3. Methods

3.1. DNA Isolation

The stage TN1–TN4 tumor samples were obtained from five male and seven female patients. Normal adjacent tissues were also available from all 12 patients. The DNA was extracted using the QIAamp DNA mini kit following the manufacturer's recommendations:

1. Equilibrate the fresh frozen tissue samples to room temperature.

2. Dissect the samples and transfer up to 25 mg into 1.5-mL microcentrifuge tubes.
3. Add 180 mL of lysis buffer ATL and 20 μL of proteinase K.
4. Lyse the samples overnight at 56°C in a thermomixer.
5. After overnight lysis, add 4 μL of RNase A, vortex the tubes, and incubate for 2 min.
6. After adding 200 μL of AL buffer, vortex the tubes again, and incubate at 70°C for 10 min to inactivate all enzymes.
7. Add 200 μL of ethanol, vortex the tubes, and apply all of the solution to the provided spin columns.
8. After centrifugation at 6,000*g* for 1 min, discard the filtrate with the collection tube.
9. Place the spin columns into clean collection tubes and wash by adding 500 μL of AW1 buffer and centrifugation at 6,000*g* for 1 min.
10. Perform a second washing step with 500 μL of AW2 buffer.
11. Remove residues of the AW2 buffer by additional centrifugation in a clean microcentrifuge tube at 14,000*g* for 1 min.
12. For the elution of the DNA, transfer the spin columns into clean 1.5-mL microcentrifuge tubes.
13. After adding 50 μL of prewarmed sterile water (40°C), incubate the columns for 5 min and elute the DNA by centrifugation at 6,000*g* for 1 min.
14. Repeat the elution step 13 to obtain 100 μL of DNA solution.
15. Measure 2 μL of the purified DNA twice in the NanoDrop instrument to obtain the mean absorption at 260 nm.
16. Calculate the DNA concentration based on the average extinction of double-stranded DNA to be 1.0 A_{260} (50 μg/mL).

3.2. Bisulfite Conversion Without Desulfonation Generating "SafeBis DNA"

1. Prepare the bisulfite solution by dissolving 1.13 g sodium sulfite and 4.71 g sodium disulfite in 10 mL of water. Rigorous vortexing is required to completely dissolve the salt.
2. Prepare the radical scavenger solution by dissolving 197 mg of 6-hydroxy-2,5,7,8-tetramethyl-chroman-2-carboxylic acid in 5 mL of 1,4-dioxane.
3. Dilute 0.5 μg DNA of each sample into 100 μL of water in safe-lock tubes.
4. Add 354 μL of sodium bisulfite solution and 146 μL of dioxane containing the radical scavenger to each sample. Close the tubes carefully and additionally secure with safe-lock clamps.
5. Mix the tubes containing 600 μL of reaction mixture thoroughly by pulse-vortexing for 10 s.
6. Incubate the tubes initially in a 99°C (or boiling) water bath for 3 min.

7. Then, incubate reactions at 50°C at 1,000 rpm/min for 5 h in a thermomixer.
8. During the incubation, subject samples to thermospikes at 30 and 90 min incubation by transferring the tubes again to the 99°C water bath for 3 min.
9. Following the 50°C incubation, centrifuge the tubes to dry the lid, and cool the solution to room temperature.
10. Add 200 μL of water to each reaction mixture to redissolve potential precipitates.
11. Load 400 μL of the reaction mixture onto the Microcon YM30 membrane filter device.
12. Centrifuge the membrane filter devices for 15 min at 14,000*g*, discard the flowthrough and reload the device with the second 400 μL of the reaction mixture and centrifuge again for 15 min at 14,000*g*. To ensure effective desalting of the sample, all liquid must pass through the membrane. Additional centrifugation time might be required depending on individual sample characteristics. Filters should be damp, but not dry.
13. Wash the membrane four times by adding 400 μL of water and centrifuge for 12 min at 14,000*g*.
14. Then, transfer the microcon filter unit to a new collection tube.
15. Apply 75 μL of prewarmed water (50°C) to the membrane and incubate in a thermomixer for 10 min at 50°C and 1,000 rpm.
16. For elution, invert the membrane into a collection tube and retrieve the DNA solution by centrifugation at 1,000*g* for 5 min.

3.3. UNG Treatment for Carryover Prevention

1. Only SafeBis DNA is used as template in experiments where the UNG treatment is carried out (**Fig. 26.1**, *see* **Note 1**).
2. Perform the carryover prevention procedure by adding 0.2 U of Uracil-DNA Glycosylase to the PCR master mix (*see* **Notes 2** and **3**).
3. To ensure the efficient degradation of uracil containing nucleic acids, incubate the closed PCR tubes at 37°C for 10 min before the initial activation of the reaction (*see* **Note 4**).

3.4. Standard DNA and Calibration Curves

1. Treat 2 μg universal methylated DNA with sodium bisulfite as described in **Section 3.2**.
2. Measure 2 μL of the purified DNA twice in the NanoDrop instrument to obtain the mean absorption at 260 nm.
3. Calculate the DNA concentration based on the average extinction of single-stranded DNA to be 1.0 A_{260} (33 μg/mL).

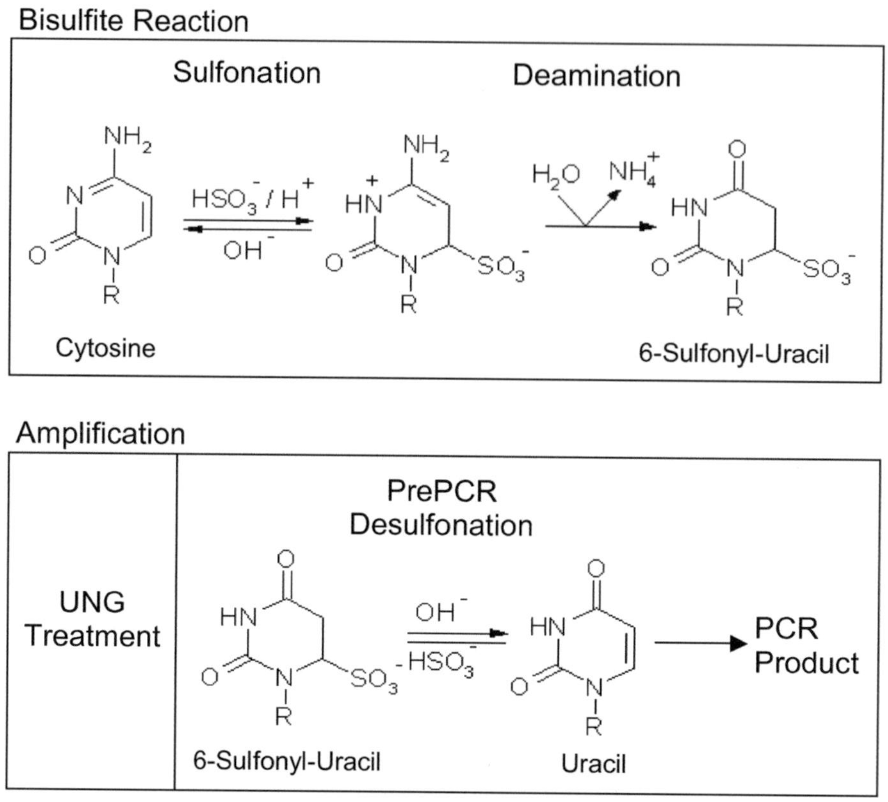

Fig. 26.1. Schematic illustration of the SafeBis bisulfite-conversion reaction enabling the use of UNG for carryover prevention. To prevent the bisulfite-converted DNA from hydrolysis by UNG, the desulfonation step of the bisulfite conversion reaction is not performed. Unmethylated cytosines are converted to 6-sulfonyl uracil by sodium bisulfite treatment (upper part). As long as the sulfon group is not released, the DNA is protected from the hydrolysis activity of the Uracil-DNA Glycosylase (UNG). The DNA can therefore be used directly in PCR reactions with UNG treatment for carryover prevention. The desulfonation of the 6-sulfonyl uracil containing DNA (SafeBis DNA) is performed prior the PCR by an extended initial activation step at 95°C in the PCR buffer. At the same time, the UNG is inactivated and will not interfere with the amplification of the bisulfite DNA.

4. Prepare the calibration standards from standard DNA with final concentrations of 20 ng, 5 ng, 2 ng, 0.8 ng, and 0.2 ng in 10 μL (*see* **Note 5**).

3.5. Quantification of Total Amount of SafeBis DNA by Reference PCR

The total amount of bisulfite-converted DNA was quantified using a real-time reference PCR, which is not specific for the methylation status. The primers amplify a 130 bp fragment within the GSTP1 gene (accession no. X08058; nucleotide 2,273 to 2,402) and do not cover any CpG. The PCR was performed in the LightCycler instrument 2.0 in a total volume of 20 μL containing 10 μL of template DNA.

1. The reaction mix contains 2 μL of the LightCycler FastStart DNA Master HybProbe per reaction and a final concentrations of 3.5 mM $MgCl_2$, 0.60 μM forward primer (**Table 26.1, no. 1**), 0.60 μM reverse primer (**Table 26.1, no. 2**) and 0.2 μM hydrolysis probe (**Table 26.1, no. 3**).

Table 26.1
Oligonucleotides used in the described experiments, provided by TibMolBiol (Berlin, Germany)

No	Name	Sequence	5′-Mod.	3′-Mod.
1	Ref-F	GGAGTGGAGGAAATTGAGAT		
2	Ref-R	CCACACAACAAATACTCAAAAC		
3	Ref-P	TGGGTGTTTGTAATTTTTGTTTTGTGTTAGGTT	FAM	BHQ1
4	TMEFF2-F	AAAAAAAAAAAACTCCTCTACATAC		
5	TMEFF2-R	GGTTATTGTTTGGGTTAATAAATG		
6	TMEFF2-B	ACATACACCACAAATAAATTACCAAAAACATCAACCAA		PH
7	TMEFF2-D	TTTTTTTTTTCGGACGTCGTT		FL
8	TMEFF2-A	TCGGTCGATGTTTTCGGTAA	Red640	PH

Notes: *Abbreviations: PH = phosphate, FL = fluorescein, red640 = LightCycler fluorescence label for 640 nm, FAM = carboxyfluorescein, and BHQ1 = black whole quencher 1

2. Mix the master mix carefully and dispense 10 μL per capillary.
3. Prepare two replicates of each calibration standard with 20 ng, 5 ng, 2 ng, and 0.8 ng by pipetting 10 μL to the capillaries (*see* **Note 5**).
4. Dilute samples 1:5 by adding 40 μL of water to 10 μL of DNA sample.
5. Use 10 μL of the diluted DNA sample per reaction.
6. Perform the PCR under the following cycling conditions: 95°C for 30 min, followed by 50 cycles at 95°C for 10 s, and 56°C for 30 s and 72°C for 10 s.
7. Carry out the detection during the annealing step at 56°C at a wavelength of 530 nm.
8. Calculate the cycle thresholds (Cts) according to the second derivative maximum method of the LightCycler software.
9. The amount of the sample DNA is automatically calculated when the calibration standards are defined in the LightCycler software.
10. The final concentration is obtained by the mean of two replicates each of which was measured in a separate PCR amplification reaction.

3.6. Quantification of Methylated TMEFF2 Promoter by Real-Time HeavyMethyl (HM) PCR

The TMEFF2 HM PCR amplifies a 113 bp fragment of methylated bisulfite-converted DNA in the promoter region of TMEFF2 (Accession no. AF242221, nucleotide 1,102 to 1,214). The selectively amplification of methylated sequences is mediated by a blocker oligonucleotide, binding to unmethylated DNA of the region *(5,6,* **Chapter 24***)*. The PCR efficiency was calculated to be 1.85 (**Fig. 26.2**).

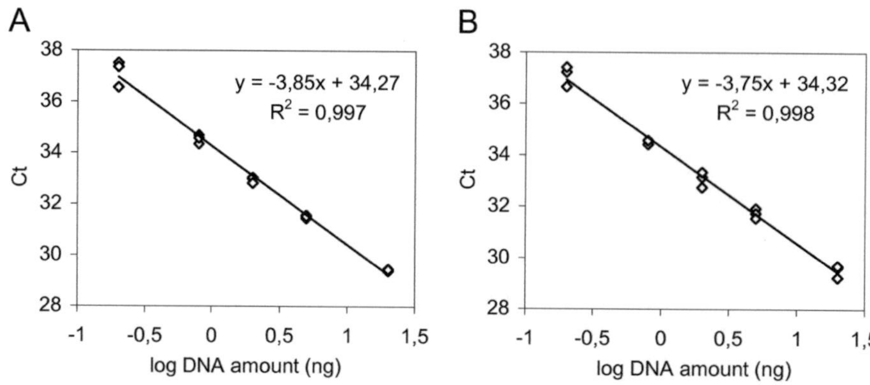

Fig. 26.2. Examples of standard curves of the TMEFF2 HM PCR. The 20 ng, 5 ng, 2 ng, 0.8 ng, and 0.2 ng methylated bisulfite-converted DNA were amplified in triplicates. The PCR efficiency $E = 10^{(-1/\text{slope})}$ was determined for desulfonated bisulfite DNA being 1.82 (**A**) and for SafeBis DNA being 1.85 (**B**).

1. The reaction mix contains 2 μL of the LightCycler FastStart DNA Master HybProbe per reaction and final concentrations of 3.5 mM $MgCl_2$, 0.30 μM forward primer (**Table 26.1, no. 4**), 0.30 μM reverse primer (**Table 26.1, no. 5**), 4.0 μM blocker (**Table 26.1, no. 6**), 0.15 μM hybridization donor probe (**Table 26.1, no. 7**), and 0.15-μM hybridization acceptor probe (**Table 26.1, no. 8**).
2. Additionally, add 0.2 units/reaction UNG to the master mix (*see* **Note 6**).
3. Mix the master mix carefully and dispense 10 μL into each capillary.
4. Prepare two replicates of each calibration standard with 20 ng, 5 ng, 2 ng, and 0.8 ng by pipetting 10 μL into the capillaries (*see* **Note 5**).
5. Use 10 μL of the same 1:5 dilution of the sample DNA (*see* **Section 3.5.4**) per reaction.
6. Perform the PCR under the following cycling conditions: 37°C for 10 min, 95°C for 30 min, followed by 50 cycles at 95°C for 10 s, 56°C for 30 s, and 72°C for 10 s (**Fig. 26.3**, *see* **Notes 1 and 7**).
7. Carry out the detection during the annealing step at 56°C at a wavelength of 640/530 nm.
8. Calculate the Cts according to the second-derivative maximum method of the LightCycler software.
9. The amount of sample DNA is automatically calculated, using the calibration standards defined in the LightCycler software.
10. The final concentration of the methylated DNA of the TMEFF2 promoter region is obtained by the mean of two replicates each of which is measured in a separate PCR amplification reaction.

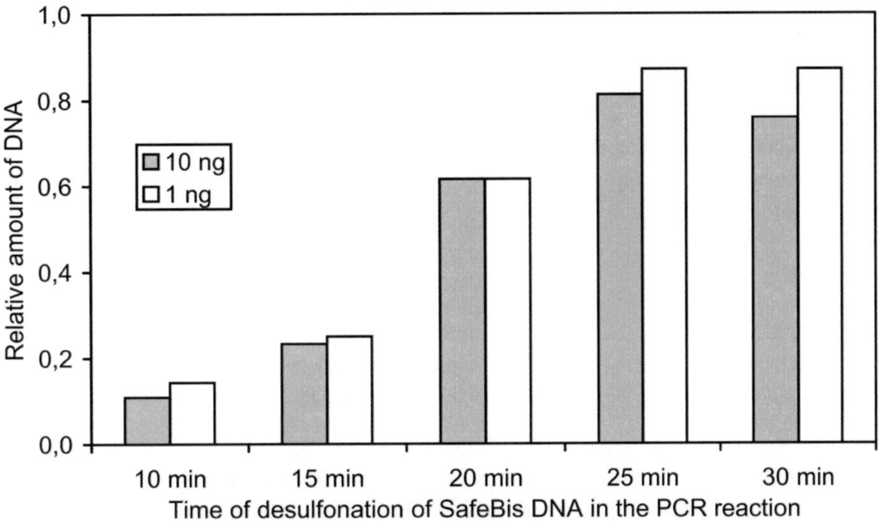

Fig. 26.3. The conditions for the desulfonation step of the PCR were optimized by a time course. The Ct of a GSTP1-HM PCR using 10 ng and 1 ng SafeBis were measured at different times of initial incubation of the real-time PCR at 95°C. The relative amount of amplified SafeBis DNA was quantified from the delta Ct compared to desulfonated DNA of the same concentration by the following equation: $R = 1/(2^{\Delta Ct})$. If SafeBis DNA was desulfonated for only 10 min at 95°C, the amplification was delayed for about three cycles compared to bisulfite DNA, which refers to relative amount of approximately only 10–15%. With increasing desulfonation time, the ΔCt of SafeBis and Bisulfite DNA was reduced to 0.2 cycles. Thirty minutes initial incubation of the real-time PCR was found as sufficient desulfonation time. Although the Ct of SafeBis DNA does not reach the level of bisulfite DNA, the desulfonation of the DNA was complete, because no further improvements were obtained for elevated incubation times. The final Ct delay can be explained by the slightly changed reaction conditions using SafeBis DNA.

3.7. Calculation of the Percentage Methylation Reference (PMR)

The relative level of methylation of the TMEFF2 promoter region was calculated according to the PMR value method *(9)*.

1. Determine the total amount of sample DNA per 10 μL sample using the reference PCR as described in **Section 3.5**.
2. The amount of methylated TMEFF2 DNA per 10 μL sample is determined using the HM PCR as described in **Section 3.6** using the same calibration standard DNA (*see* **Note 5**).
3. Calculate the percentage of methylated copies in relation to the total DNA using the following equation: PMR = 100 × (ng methylated DNA/ng total DNA).
4. For a comparison, the PMR values of the same samples were also obtained from standard bisulfite DNA, which was desulfonated (*see* **Note 8**). Standard bisulfite DNA was analyzed without UNG treatment and a different PCR cycling program (**Fig. 26.4,** *see* **Notes 8–10**).

4. Notes

1. If higher Ct values were obtained using SafeBis DNA compared to desulfonated bisulfite DNA, the following reasons

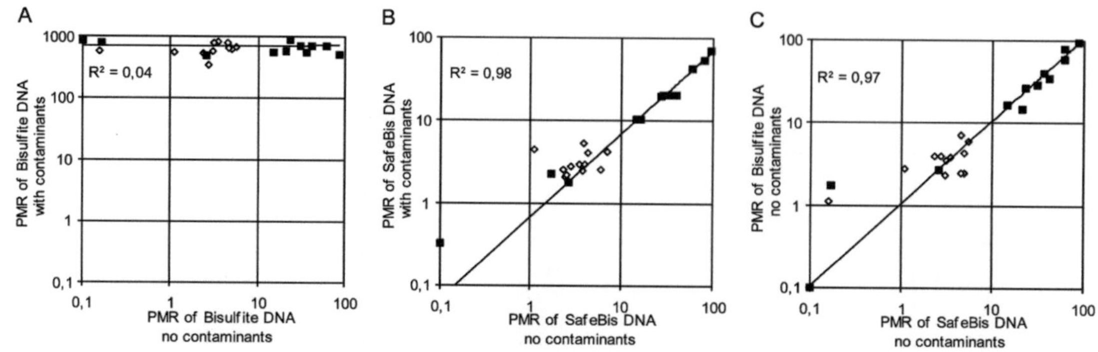

Fig. 26.4. Percent methylation reference (PMR) of TMEFF2 promoter calculated from the TMEFF2 HM PCR and the reference PCR using different work flows. In the first part of the study, the samples from colorectal cancers and normal tissue were processed with a standard work flow without carryover prevention as described in **Notes 3** and **4.** Afterwards the same samples were intentionally contaminated by the addition of 10,000 copies of PCR product of the TMEFF2 HM PCR to the final reaction. The calculated PMR values are therefore dramatically different compared to the noncontaminated samples (**A**). Next, the same samples were processed applying the described carryover prevention workflow resulting in SafeBis DNA. Again the samples were measured twice, once uncontaminated and secondly spiked with 10,000 copies of PCR product from the TMEFF2 HM PCR. No significant differences were obtained for the PMR values from contaminated and noncontaminated reactions (**B**). The results of the two workflows without contamination were plotted against each other and showed a good correlation with a regression coefficient of $R^2 = 0.97$ (**C**). These results demonstrate that the elimination of contaminating PCR product was highly efficient and that the use of SafeBis DNA gives the same results as the standard work flow without carryover prevention.

can be considered: (a) The DNA was not properly desulfonated in the PCR buffer. Check the initial activation step to be 30 min or change the PCR reaction buffer. Some PCR buffers are not suitable for the desulfonation process. The pH has to be >8. Accordingly, real-time PCR kits should be tested for their desulfonation capability prior to their usage. (b) The DNA used was partially desulfonated before the PCR preparation and therefore not fully protected from the UNG treatment. It is necessary to ensure that the bisulfite treatment results in sulfonated DNA (SafeBis DNA) as described. It is essential to skip the NaOH treatment step which is usually carried out. (c) Elevated pH of the DNA solution combined with an increased storage temperature can initiate the desulfonation process. It is absolutely essential to elute the SafeBis DNA in a pH neutral solvent and to store it at low temperatures *(5)*.

2. The UNG method can be easily applied to all PCR-based methods, including conventional thermocycler PCR with endpoint analysis. It only needs the consistent substitution of dTTP with dUTP in all PCR reactions and a decontamination step using UNG. The use of SafeBis DNA therefore can be recommended for all methylation analysis methods, where PCR products are reanalyzed and therefore involve a high risk of PCR product carryover. Clone sequencing (**Chapter 14**), direct sequencing, COBRA (**Chapter 19**),

MS-SNuPE (**Chapters 16 and 18**), and other methods can easily be adapted to the described carryover prevention procedure reducing the risk of false results caused by cross-contamination (*see also* **Note 6**).

3. In general, the described method is universally applicable, but several PCR kits might not be suitable for the amplification of SafeBis DNA. As described, the critical parameter is the pH of the buffer, which needs to be above 8 for an efficient desulfonation *(5)*. The desulfonation efficiency can easily be tested with a side-by-side comparison of SafeBis DNA with conventionally desulfonated bisulfite DNA.
4. If the contaminating PCR products are not completely inactivated, the following reasons can be considered: (a) Ensure that the master mix, containing UNG and potential contaminating PCR products, was mixed properly. Spin down the reaction tubes to ensure that all potential PCR contaminants will be in contact with the UNG. Ensure that the master mix was incubated for at least 10 min at 37°C. (b) Very high copy numbers of PCR product might not be inactivated completely. Dilute out the PCR product to check the individual performance of the UNG treatment in a reamplification experiment. (c) The UNG activity could be affected by any impurity of the sample. Ensure that the sample DNA is pure and dissolved in water without any enzyme inhibitors like EDTA. (d) The carryover prevention method requires the consistent use of dUTP in all PCRs in the laboratory amplifying the same sequence. Any persistent contamination might be caused by earlier work with the same sequence using dTTP.
5. The preparation of calibration curves is one of the most critical steps of quantitative analysis methods, especially in real-time PCR. In studies, where the described carryover prevention procedure is applied, the standard DNA should be sulfonated similarly to the analyzed samples. The standard DNA needs to be well characterized for its complete conversion. Preferably, the calibration curve should be established by direct pipetting from one stock. Serial dilutions should be avoided because of the risk for propagation of pipetting errors. DNA dilutions should always be done with poly(A) as carrier to prevent loss of DNA due to unspecific binding to the surfaces of vials and tips. Several bisulfite specific real-time PCR were tested with 5 ng/μL poly(A) carrier, resulting in 50 ng RNA per PCR. In contrast to poly(dA) the RNA carrier does not affect the real-time PCR results. For most precise PMR values, the identical calibration curve should be used in the reference assay and methylation-specific PCR.
6. The inactivation of PCR products by UNG depends on several factors. First, enzyme activity can differ in different PCR buffers. Furthermore, individual handling and preparation

Table 26.2
Cycle thresholds (Ct) of the reamplification of PCR products with and without UNG treatment for different PCR amplicons, showing different decontamination efficiencies

PCR	UNG[(1)]	Copies of PCR-products per reaction				
		100,000	30,000	10,000	3,000	1,000
TMEFF2-HM	–	*20.0*	*21.8*	*23.5*	*25.4*	*26.8*
	+	*31.1*	*38*	–	–	–
HM-Assay 2	–	*20.2*	*21.9*	*24.0*	*25.6*	*26.9*
	+	–	–	–	–	–
HM-Assay 3	–	*20.3*	*22.1*	*24*	*25.8*	*27.2*
	+	*35.3*	*36.2*	*36.6*	–	–
HM-Assay 4	–	*22.9*	n.d.	n.d.	27.3	n.d.
	+	35.7	n.d.	n.d.	–	n.d.

Notes: [(1)] + : PCR reaction contained 0,2 Unit UNG and were incubated for 10 min at 37°C before cycling; – : no UNG was added; n.d.: not determined.

times after the addition of UNG can cause differences in the decontamination efficiency. Therefore, in contrast to the manufacturer's recommendation, we applied an additional incubation step at 37°C for 10 min to ensure a reproducible procedure. Also, impurities introduced by sample preparation, could cause lower enzyme activity *(10)*. Finally, the sensitivity of the PCR products against the UNG treatment depends directly on the uracil content and thus on its length. Several HM PCR products were tested and 3,000 copies were found to be the amount of PCR products that was still completely inactivated as indicated by the absence of a signal after 50 cycles (**Table 26.2**).

7. Although the SafeBis DNA is not amplifiable, because 6-Sulfonyl-Uracil is not a substrate for polymerases, it can be directly used as template DNA in the PCR *(5)*. To ensure the efficient desulfonation of the SafeBis DNA in the PCR buffer, the cycling program must start with an extended initial activation step at 95°C for 30 min (**Fig. 26.3**). No further changes to the cycling program or analysis procedure are required. Up to 10 μL SafeBis DNA can be added to the master mix without impairing PCR performance.
8. To receive desulfonated bisulfite-converted DNA basically the same protocol was used as described under **Section 3.2** with following changes:

1–14. No changes

15. After discarding the flowthrough, perform the desulfonation by adding 400 μL of a 0.2 M NaOH solution to the membrane and centrifugation at 14,000*g* for 12 min.
16. Repeat step 15 with 400 μL of a 0.1 M NaOH solution.
17. Wash the membrane two times by adding 400 μL of water followed by centrifugation at 14,000*g* for 12 min.
18. Transfer the filter device to a new collection tube.
19. Apply 75 μL of prewarmed water (50°C) to the dry membrane and incubate in a thermomixer for 10 min at 50°C.
20. For the elution, invert the membrane in a clean collection tube and retrieve the DNA solution by centrifugation at 1,000*g* for 5 min.

9. The cycling conditions for experiments using desulfonated bisulfite DNA without UNG are as follows: 95°C for 10 min, followed by 50 cycles at 95°C for 10 s, 56°C for 30 s, and 72°C for 10 s.
10. The efficiency of the carryover prevention method was examined by treating PCR products with UNG. To obtain PCR products 10 ng methylated bisulfite-converted DNA was amplified using the TMEFF2 HM PCR assay as described in 3.6. The PCR products were purified using the QIAquick PCR Purification Kit (Qiagen) according to the manufacturer's recommendation and analyzed in a 2% agarose gel to ensure a single product. 2 μL of a 1:10^8 dilution of purified PCR product were used as template in the TMEFF2 HM real-time PCR. In the quantification, the determined Ct was equivalent to 33 ng bisulfite-converted universal methylated DNA. Assuming that 3.3 pg human DNA represents one genome equivalent, the prepared 1:10^8 dilution of PCR product can be calculated as having 10,000 copies in 2 μL. If PCR products were spiked into the reactions, the PCR master mix was recalculated to have the same final amount of all components in 8 μL instead of 10 μL to allow the addition of 2 μL contaminating PCR products (**Table 26.2**).

Acknowledgement

The author would like to thank Jürgen Distler, Theo DeVos, and Dimo Dietrich for critically reviewing the manuscript. The work was supported by a grant from the Bundesministerium für Bildung und Forschung (BioChancePlus AkZ 0313166).

References

1. Longo, M. C., Berninger, M. S., Hartley, J. L. (1990) Use of uracil DNA glycosylase to control carry-over contamination in polymerase chain reactions. *Gene* 93, 125–128.
2. Pang, J., Modlin, J., Yolken, R. (1992) Use of modified nucleotides and uracil-DNA glycosylase (UNG) for the control of contamination in the PCR-based amplification of RNA. *Mol Cell Probes* 6, 251–256.
3. Pennings, J. L., van de Locht, L. T., Jansen, J. H., et al. (2001). Degradable dU-based DNA template as a standard in real-time PCR quantitation. *Leukemia* 15, 1962–1965.
4. Frommer, M., McDonald, L. E., Millar, D. S., et al. (1992) A genomic sequencing protocol that yields a positive display of 5-methylcytosine residues in individual DNA strands. *Proc Natl Acad Sci USA* 89, 1827–1831.
5. Tetzner, R., Dietrich, D., Distler, J. (2007) Control of carry-over contamination for PCR-based DNA methylation quantification using bisulfite treated DNA. *Nucleic Acids Res* 35, e4.
6. Cottrell, S. E., Distler, J., Goodman, N. S., et al. (2004) A real-time PCR assay for DNA-methylation using methylation-specific blockers. *Nucleic Acids Res* 32, e10.
7. Young, J., Biden, K. G., Simms, L. A., et al. (2001) HPP1: a transmembrane protein-encoding gene commonly methylated in colorectal polyps and cancers. *Proc Natl Acad Sci USA* 98, 265–270.
8. Ebert, M. P., Mooney, S. H., Tonnes-Priddy, L., et al. (2005) Hypermethylation of the TPEF/HPP1 gene in primary and metastatic colorectal cancers. *Neoplasia* 7, 771–778.
9. Eads, C. A., Lord, R. V., Wickramasinghe, K., et al. (2001) Epigenetic patterns in the progression of esophageal adenocarcinoma. *Cancer Res* 61, 3410–3418.
10. Pruvost, M., Grange, T., Geigl, E. M. (2005) Minimizing DNA contamination by using UNG-coupled quantitative real-time PCR on degraded DNA samples: application to ancient DNA studies. *Biotechniques* 38, 569–575.

Chapter 27

Profiling DNA Methylation from Small Amounts of Genomic DNA Starting Material: Efficient Sodium Bisulfite Conversion and Subsequent Whole-Genome Amplification

Jonathan Mill and Arturas Petronis

Abstract

Sodium bisulfite modification-based fine mapping of methylated cytosines represents the gold standard technique for DNA methylation studies. A major problem with this approach, however, is that it results in considerable DNA degradation, and large quantities of genomic DNA material are needed if numerous genomic regions are to be profiled. This chapter describes a method for profiling DNA methylation from small amounts of genomic-DNA starting material utilizing an efficient sodium bisulfite conversion method followed by whole-genome amplification (WGA). WGA is a useful method to overcome the problem of low initial amount of DNA and/or severe DNA degradation during conventional sodium bisulfite treatment in studies investigating DNA methylation. WGA is a relatively inexpensive process that can be optimized for high-throughput application and enables the thorough investigation of methylation at numerous genomic locations on samples for which DNA availability is low. Data from our lab has demonstrated that bisulfite-treated DNA amplified using WGA can be used for a range of downstream DNA methylation mapping procedures, including bisulfite-primer optimization, the sequencing of cloned PCR products, MS-SNuPE, and Pyrosequencing.

Key words: DNA methylation, epigenetics, sodium bisulfite, whole-genome amplification, method.

1. Introduction

Epigenetics refers to the heritable, but reversible, regulation of various genomic functions mediated through partially stable modifications of DNA and chromatin *(1)*. DNA methylation – that is,. cytosine methylation at CpG and occasionally non-CpG sites – is one of the best-understood epigenetic mechanisms and has been investigated using a myriad of laboratory techniques *(2)*. Over the last 15 years, the gold standard technique for fine

Jörg Tost (ed.), *DNA Methylation: Methods and Protocols, Second Edition, vol. 507*

DOI 10.1007/978-1-59745-522-0_27 Springerprotocols.com

mapping of methylated cytosines (^{me}C) has been based on the treatment of genomic DNA with sodium bisulfite, which converts unmethylated cytosines to uracils (and subsequently, via PCR, to thymidines), while methylated cytosines are resistant to bisulfite and remain unchanged *(3)*. After sodium bisulfite treatment, DNA regions of interest are amplified and sequenced to identify C→T transitions or stable C positions, respectively, corresponding to unmethylated and methylated cytosines in the native DNA. Typically, PCR amplicons are either sequenced directly to provide a strand-specific average sequence for the population of DNA molecules or cloned and sequenced to provide methylation maps of single DNA molecules *(3,* **Chapter 14***)*. An alternative approach to quantitatively assess the methylation level at specific cytosine sites within an amplicon is methylation-sensitive single nucleotide primer extension (Ms-SNuPE) using SNaPshot or Pyrosequencing *(4,5,* **Chapters 15 and 18***)*.

Conventional sodium bisulfite treatment is beset by a number of problems that result from the fact that the full conversion of unmethylated cytosines requires harsh reaction conditions which cause large-scale degradation of genomic DNA *(6)*. Alternative bisulfite-conversion protocols have been developed in which DNA is embedded in agarose during treatment to reduce DNA loss and ensure efficient conversion *(7)*, but these procedures only partially reduce the degradation and are not suited for high-throughput analysis. In many epigenetic studies the amount of genomic DNA starting material is limited, especially in experiments utilizing valuable clinical samples such as oocytes, laser-capture microdissected cells, and microscope slides. Following bisulfite treatment, converted DNA is single stranded, and prone to further degradation unless stored at −80°C. The degradation of DNA during and after sodium bisulfite treatment is a major hurdle to successful studies of DNA methylation. In addition, given the degenerate nature of sodium bisulfite-treated DNA, downstream applications such as bisulfite-PCR often require considerable optimization, further eroding valuable DNA stocks if several loci are to be interrogated.

Whole-genome amplification (WGA) methods are routinely employed on genomic DNA for genotyping and sequence analysis when the amount of starting template is extremely low. One common application of WGA, for example, is in forensic analyses where it is used to improve both the quality and quantity of DNA, and allows accurate genetic profiling from single cells *(8)*. Two commonly used WGA strategies are (i) primer extension preamplification (PEP), a *Taq* DNA polymerase PCR-based reaction first described by Zhang et al. *(9)*, and (ii) multiple displacement amplification (MDA), an isothermal genome amplification using *Phi 29* DNA polymerase *(10)*. Both methods are widely utilized for genotyping, with several studies demonstrating the reliability of data produced from WGA templates *(11, 12)*. This chapter describes an efficient sodium bisulfite conversion method, suitable for high-throughput analyses using 96-well microtitre plates, and the subsequent application of WGA to sodium bisulfite-treated DNA.

WGA is a useful method to overcome the problem associated with a low amount of DNA-starting material and/or severe DNA degradation during conventional sodium bisulfite treatment in studies investigating DNA methylation. WGA is a relatively inexpensive process that can be optimized for high-throughput application and enables the thorough investigation of methylation at numerous genomic locations on samples for which DNA availability is low. Data from our lab has demonstrated that bisulfite-treated DNA amplified using WGA can be used for a range of downstream DNA methylation mapping procedures including bisulfite–primer optimization, the sequencing of cloned PCR products, MS-SNuPE, and pyrosequencing. While it should be acknowledged that WGA could potentially introduce biases into quantitative estimates of CpG methylation, our data suggest that such biases may not be a major problem in the profiling methods we have tested. Data from our laboratory indicates that the DNA methylation profiles obtained from WGA of sodium bisulfite-treated DNA are generally consistent with those obtained from non-WGA DNA *(13)*. Examples of direct sequencing and Ms-SNuPE analysis of WGA sodium bisulfite DNA are shown in **Fig. 27.1**. Even if not utilized for final DNA methylation data collection, given the large amount of template generated in each reaction, WGA is a useful tool for laboratories optimizing a large number of bisulfite-PCR reactions.

2. Materials

2.1. Sodium Bisulfite Treatment

1. Sterile water, preferably freshly degassed under a vacuum (*see* **Note 1**).
2. Fresh 3 M NaOH solution. Dissolve 3 g NaOH pellets in 25 mL of degassed water.
3. Fresh 0.1 M NaOH solution made from a dilution of above.
4. Fresh hydroquinone solution. Dissolve 0.22 g hydroquinone (Sigma) in 10 mL degassed water. Keep this solution shielded from light.
5. Saturated sodium bisulfite solution. Bring 10.8 g sodium bisulfite (Sigma) to 16 mL final volume in preheated degassed water (55°C). Invert to mix until solution is fully saturated.

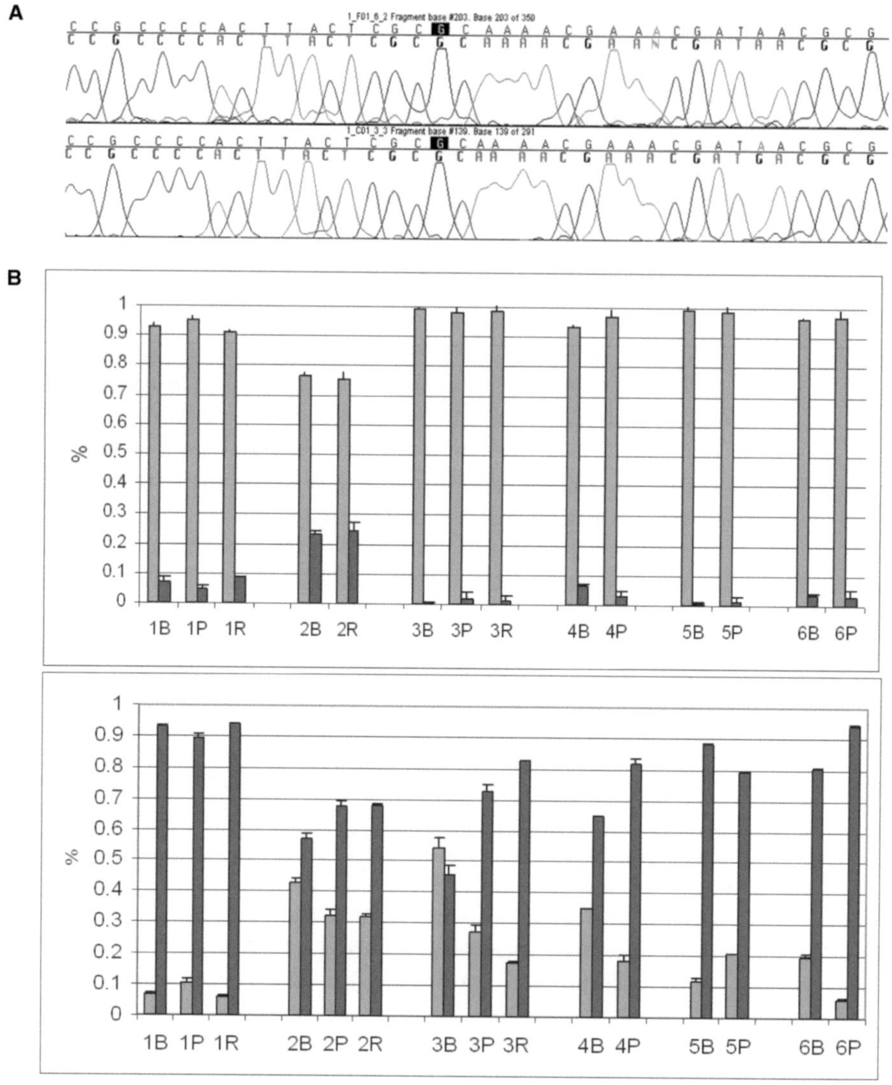

Fig. 27.1. DNA methylation profiling on WGA bisulfite DNA. (**A**) Direct sequencing chromatograms of PCR amplicons generated from normal bisulfite-treated DNA (WGA–) and PEP-amplified bisulfite-treated DNA (WGA+) from the same individual. (**B**) DNA methylation estimates obtained using Ms-SNuPE for two CpG sites on templates generated from six DNA samples (1–6). Methylation estimates derived from WGA templates are generally very similar to those produced by standard bisulfite-treated DNA templates. Light bars denote the percentage of methylated cytosines, and dark bars denote the percentage of unmethylated cytosines at each CpG site. B = non-WGA bisulfite-treated DNA; P = PEP-amplified bisulfite-treated DNA; R = bisulfite-treated DNA amplified with the Qiagen REPLI-g MDA kit.

Add 2.6 mL 3 M NaOH solution and 1.0 mL hydroquinone solution. Mix well (*see* **Note 2**).

6. Microcon YM-50 columns (Millipore) or (for high throughput) Montage PCR96 96-well filtration plate (Millipore) (*see* **Note 3**).
7. TE buffer: 10 mM Tris–HCl, 1 mM EDTA, pH 7.5.

2.2. Primer Extension Preamplification (PEP) of Sodium Bisulfite DNA

1. Sterile, autoclaved water (*see* **Note 4**).
2. 100 pmol/μL 15-mer degenerate primers (5′-NNNNNNN-NNNNNNNN-3′).
3. *Taq* polymerase [5 U/μL] (NEB) or a mix of *Taq* polymerase and proofreading *Pwo* polymerase (e.g., high fidelity PCR enzyme mix [5 U/μL] (Fermentas)) (*see* **Note 5**).
4. 10 × PCR buffer.
5. 25 mM $MgCl_2$ solution.
6. 10 mM dNTP mix.
7. MiniElute PCR-purification kit (Qiagen).
8. TE buffer: 10 mM Tris–HCl, 1 mM EDTA, pH 7.5.

2.3. Multiple Displacement Amplification on Sodium Bisulfite DNA (Alternative Procedure)

1. RELPI-g Mini kit (Qiagen).

2.4. Quality Control Analysis of WGA Products

1. Biotechnology-grade agarose.
2. 1 × TBE buffer: 90 mM Tris–boric acid, 2 mM EDTA.
3. 5 mg/mL Ethidium Bromide.
4. Agarose gel loading buffer.

3. Methods

3.1. Sodium Bisulfite Treatment

1. Take DNA sample (50 ng to 2 μg) and adjust volume to 10 μL (*see* **Note 6**).
2. Transfer DNA sample(s) to PCR tubes (or 96-well plate for high-throughput processing). Add 1.1 μL of fresh 3 M NaOH solution. Spin down and seal tubes (or plate).
3. Place in a thermocycler for 20 min at 42°C.
4. Spin down tubes/plate to catch condensation and carefully open seal. Add 120 μL of fresh sodium bisulfite solution, seal plate/tube with a fresh lid, invert a few times to mix, and then spin down.
5. Place in a thermocycler for 4–5 h at 55°C (*see* **Note 7**).
6. Remove plate/tubes from thermocycler, spin down, and carefully remove lid.
7. Add 100 μL of sterile water.
8. Transfer each sample to a Microcon YM-50 column (Millipore) or (for high-throughput) a well into a Montage PCR96 96-well filtration plate (Millipore).
9. Draw solution through filtration matrix by either centrifugation at maximum speed (Microcon YM-50 column) or vacuum (Montage PCR96 96-well filtration plate), until wells

are visibly empty (~4–5 min). The DNA remains on the matrix.

10. Desalt DNA by adding 175 μL of sterile water to each well and draw the solution through the matrix (via centrifugation or vacuum) as before. Repeat this step two more times.
11. Desulfonate by adding 175 μL fresh of 0.1 M NaOH. Draw solution through the matrix (via centrifugation or vacuum).
12. Perform a final wash step by drawing 175 μL of sterile water through the matrix.
13. Recover DNA by adding 50 μL of 1 × TE Buffer and incubate for 2 min at room temperature.
14. If using Microcon YM-50 columns, carefully vortex the Microcon unit for 30 s, separate sample reservoir from filtrate cup and place sample reservoir upside down into a new vial. Spin for 3 min at 1,780*g* in invert spin mode to elute DNA. If using a Montage PCR96 96-well filtration plate, use a plate shaker to release DNA from filtration matrix for 10 min at 500 rpm. Remove eluted DNA solution from individual wells using a pipette and filter tips.
15. Remove an aliquot for WGA if needed. Otherwise, store bisulfite-treated DNA at −20°C (or −80°C for long-term storage).

3.2. Primer Extension Preamplification of Sodium Bisulfite DNA

1. PEP reactions should be setup on ice, taking care to maintain a clean and sterile laboratory environment to ensure that no contamination occurs (*see* **Note 8**).
2. Each PEP reaction uses 2-μL sodium bisulfite template (from the 50 μL final volume eluted in the protocol described above). Assuming sodium bisulfite conversion was performed on 250 ng starting genomic DNA, this corresponds to ~10 ng original genomic DNA template (*see* **Note 9**).
3. Each PEP reaction is performed in triplicate to increase the likelihood that all genomic regions are adequately enriched. The total volume of sodium bisulfite DNA required for each PEP-based WGA reaction is 6 μL (corresponding to ~30 ng of the original genomic DNA template assuming bisulfite conversion was performed on 250 ng) (*see* **Note 9**).
4. Add bisulfite DNA to reaction well. Remember to add appropriate negative control samples. Use filter tips and pipette carefully. Be careful to avoid aerosol contamination.
5. Make up a master mix of the PEP amplification reagents, mix well, and aliquot 48 μL (to a total volume of 50 μL) to each bisulfite-treated DNA sample. Each 50 μL reaction contains 2 μL 100 pmol/μL PEP primer, 2 μL 10 mM dNTPs, 5 μL 10 × PCR buffer, 10 μL 25 mM $MgCl_2$, 0.8 μL *Taq* polymerase (NEB) or high fidelity PCR enzyme mix (Fermentas), and 28.2 μL sterile H_2O (*see* **Note 10**).

6. Carefully seal the PCR tubes and centrifuge the reaction mix to collect it at the bottom of the reaction well.
7. Transfer samples to a thermalcycler, preprogrammed to amplify using 50 cycles, with each cycle consisting of a denaturing step for 1 min at 95°C, a annealing step for 2 min at 37°C, a programmed ramping step of 10 s/°C up to 55°C, and an incubation step for 4 min at 55°C.
8. Following amplification, ensure the reaction is cooled to 4°C (to avoid condensation on the lid) and centrifuge to collect the product at the bottom of the PCR tube.
9. Carefully open the PCR tube/plate ensuring that no cross-contamination of wells occurs (*see* **Note 11**).
10. Carefully combine the three replicates for each sample into one tube (→150 μL total volume) and mix well.
11. (Optional – *see* **Note 12**). Add 1,000 μL of Buffer EB (Qiagen) and mix. Transfer 600 μL to a labeled MinElute column (Qiagen) and centrifuge at maximum speed for 1 min. Discard flowthrough and repeat step with the remaining 600 μL. Discard flowthrough. Add 750 μL of Buffer PE (Qiagen) and centrifuge at maximum speed for 1 min. Discard flowthrough and place the MinElute column back in the same tube. Centrifuge the column for an additional 1 min at maximum speed. Place the MinElute column in a clean 1.5-mL microcentrifuge tube. Add 100 μL of prewarmed (~50°C) 10 mM Tris–Cl, pH 7.5 and centrifuge for 1 min. Repeat this step with another 100 μL Tris–Cl, pH 7.5.
12. Take a 5-μL aliquot for agarose gel analysis and quantification (**Section 3.4**).
13. Remaining amplified template can be kept at 4°C (for short-term storage) or −20°C (for long-term storage).

3.3. Multiple Displacement Amplification on Sodium Bisulfite DNA Using the Qiagen REPLI-g Mini Kit (Alternative Procedure to 3.2)

1. Preheat incubator or water bath to 30°C.
2. Thaw REPLI-g Mini DNA Polymerase on ice, and all other components at room temperature (*see* **Note 13**).
3. Prepare sufficient Buffer D1 (denaturation buffer) and Buffer N1 (neutralization buffer) for the total number of WGA reactions.
4. Add 5 μL of bisulfite-treated DNA (corresponding to 50 ng original genomic DNA template) to a microcentrifuge tube.
5. Add 5 μL Buffer D1 to the DNA. Mix by vortexing and centrifuge briefly.
6. Incubate the samples at room temperature (15–25°C) for 3 min.
7. Add 10 μL of Buffer N1 to the samples. Mix by vortexing and centrifuge briefly.
8. Make a master mix comprising for each sample to be amplified including 29 μL of REPLI-g Mini Reaction Buffer and

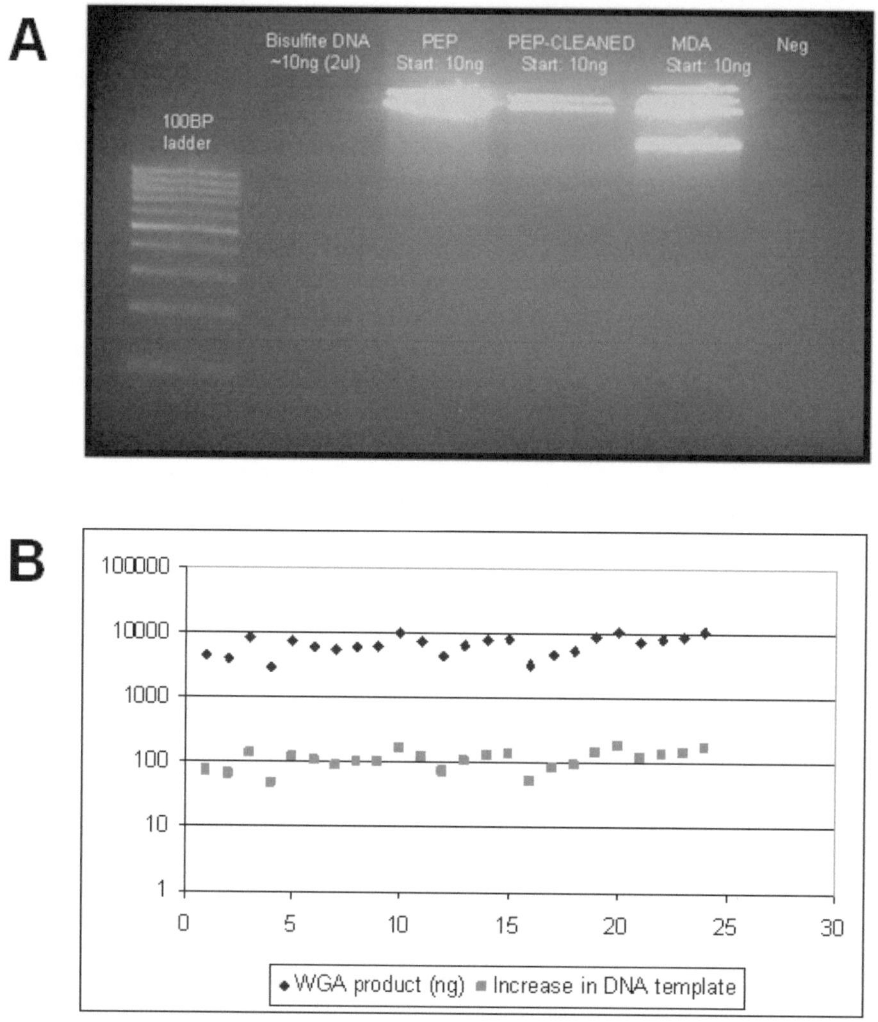

Fig. 27.2. Whole genome amplification (WGA) typically results in a large increase in bisulfite-treated DNA template. Representative WGA products obtained from standard sodium bisulfite-treated DNA starting material. Products in lanes 3 and 4 [primer-extension preamplification (PEP)] and lane 5 [multiple displacement amplification (MDA)] were produced from approximately 10 ng original bisulfite-converted DNA.;10 ng non-WGA bisulfite-treated DNA is not visible (lane 2), but both WGA methods produce clearly visible DNA smears of high molecular weight, demonstrating efficient amplification. Marker (lane 1) is the GeneRuler^{TM} 100-bp DNA ladder (Fermentas, Hanover, MD). (**B**) Typical results from spectrophotometric analysis of 25 representative PEP-amplified products. The average yield from the pooled PEP replicates was 6.8 μg (± 2.2 μg). Assuming no degradation during the sodium bisulfite conversion reaction, this represents a 113-fold (± 37) increase from the initial starting genomic DNA (60 ng).

1 μL of REPLI-g Mini DNA Polymerase, and add 30 μL of this master mix to each DNA sample.

9. Incubate at 30°C for 16 h.
10. Inactivate REPLI-g Mini DNA Polymerase by heating the sample for 3 min at 65°C.
11. Take a 5-μL aliquot for agarose gel analysis and quantification (**Section 3.4**).

12. Store amplified DNA at either 4°C (for short-term storage) or −20°C for long-term storage.

3.4. Quality Control Analysis of WGA Products

1. Before using WGA products for downstream bisulfite-PCR, it is advisable to check that the amplification reaction has worked efficiently and that there is no contamination present in the reaction. Make a 1% agarose gel by adding 1.0 g biotechnology grade agarose to 100 mL of 1 × TBE buffer and microwaving the mixture for ~2 min. Allow the gel to cool for a few minutes and add ethidium bromide to a final concentration of 5 μg/mL. Mix well. When the gel is cool, it can be poured into a small gel tray and allowed to polymerize.
2. Add 1 μL of gel loading buffer to 5 μL of each WGA-enriched sample, and load each into a separate well on the gel. Include the negative control WGA products, and also a suitable DNA size marker in the first well on the gel.
3. Allow products to run for a suitable distance on the gel and then visualize under UV light. You should observe clearly visible DNA smears of high genomic weight, demonstrating efficient amplification. The negative sample should not have such a smear (**Fig. 27.2A**).
4. Spectrophotometric analysis can be performed on the samples to determine the average yield from each reaction. **Fig. 27.2B** illustrates spectrophotometric readings from 25 representative PEP-amplified products generated in our laboratory. The average yield from pooled PEP reactions (performed in triplicate) was 6,767 ng (± 2,206 ng). Assuming no degradation during the sodium bisulfite conversion reaction, this represents a 113-fold (±37) increase from the initial starting genomic DNA (60 ng) for these samples.

4. Notes

1. Free oxygen in water can reduce the efficiency of sodium bisulfite conversion.
2. It is possible that the sodium bisulfite will not entirely dissolve. If any substrate remains, centrifuge solution before use and use the supernatant.
3. The use of the Montage PCR96 96-well filtration plate (Millipore) for high-throughput sample processing requires a suitable vacuum manifold.
4. It is imperative that all reagents used for WGA are clean and sterile – the WGA process will amplify any DNA in solution, even when present at very low concentration.
5. While we do not observe too many problems using standard Taq polymerase, the proofreading Taq polymerase mix improves the error rate inherent in WGA.

6. In our experience, the optimal starting amount of genomic DNA is between 250 ng and 1 μg. Small amounts of DNA (e.g., 50 ng or below) may not provide enough useable template for direct use following sodium bisulfite treatment, but can be used for DNA methylation profiling following WGA of the sodium bisulfite-treated DNA.
7. It can be beneficial to ramp the reaction up to 95°C for 1 min every hour to ensure that the DNA remains single stranded.
8. PEP amplification is highly efficient, and contamination problems can be an issue. For this reason, we suggest including always a negative water control in each set of WGA reactions. Use the products of these as a negative control for all subsequent PCR amplifications. Note well: even contamination from the researcher performing the PEP experiments can be a problem.
9. If this exceeds the amount of available template, it is better to do three replicates on smaller amounts of template than one replicate on a larger amount of template.
10. The high fidelity PCR enzyme mix contains a proofreading polymerase in addition to *Taq* and is thus less prone to sequence errors, but is considerably more expensive. The choice of enzyme depends upon your ultimate downstream application of the WGA material – occasional sequence errors may be less important if the products are to be used only for assay optimization, etc.
11. If possible, these postamplification steps should be performed in a separate room (or fume hood) away from the location of preamplification steps. The release of post-WGA aerosols, even in minute quantities, can lead to contamination problems that are hard to rectify.
12. This step removes excess degenerate primers and deoxynucleotides that can interfere with certain downstream applications if not removed fully. One downside is that some loss of amplified template may occur due to the maximum binding capacity of the Qiagen columns. If this is an issue, then the sample can be split and extra columns utilized.
13. The REPLI-g mini reaction buffer may form a precipitate after thawing. The precipitate can be dissolved by vortexing for 10 s.

References

1. Henikoff, S., Matzke, M. A. (1997) Exploring and explaining epigenetic effects. *Trends Genet* 13, 293–295.
2. El-Maarri, O. (2003) Methods: DNA methylation. *Adv Exp Med Biol* 544, 197–204.
3. Clark, S. J., Harrison, J., Paul, C. L., et al. (1994) High sensitivity mapping of methylated cytosines. *Nucleic Acids Res* 22, 2990–2997.
4. Schumacher, A., Kapranov, P., Kaminsky, Z., et al. (2006) Microarray-based DNA methylation profiling: technology and applications. *Nucleic Acids Res* 34, 528–542.

5. Tost, J., El Abdalaoui, H., Gut, I.G. (2006) Serial pyrosequencing for quantitative DNA methylation analysis. *Biotechniques* 40, 721–726.
6. Grunau, C., Clark, S.J., Rosenthal, A. (2001) Bisulfite genomic sequencing: systematic investigation of critical experimental parameters. *Nucleic Acids Res* 29, e65
7. Olek, A., Oswald, J., Walter, J. (1996) A modified and improved method for bisulphite based cytosine methylation analysis. *Nucleic Acids Res* 24, 5064–5066.
8. Ballantyne, K.N., van Oorschot, R.A., Mitchell, R.J. (2007) Comparison of two whole genome amplification methods for STR genotyping of LCN and degraded DNA samples. *Forensic Sci Int* 166, 35–41.
9. Zhang, L., Cui, X., Schmitt, K., et al. (1992) Whole genome amplification from a single cell: implications for genetic analysis. *Proc Natl Acad Sci USA* 89, 5847–5851.
10. Dean, F.B., Hosono, S., Fang, L., et al. (2002) Comprehensive human genome amplification using multiple displacement amplification. *Proc Natl Acad Sci USA* 99, 5261–5266.
11. Kuivaniemi, H., Yoon, S., Shibamura, H., et al. (2002) Primer-extension preamplified DNA is a reliable template for genotyping. *Clin Chem* 48, 1601–1604.
12. Hosono, S., Faruqi, A.F., Dean, F.B., et al. (2003) Unbiased whole-genome amplification directly from clinical samples. *Genome Res* 13, 954–964.
13. Mill, J., Yazdanpanah, S., Guckel, E., et al. (2006) Whole genome amplification of sodium bisulfite-treated DNA allows the accurate estimate of methylated cytosine density in limited DNA resources. *Biotechniques* 41, 603–607.

INDEX

D

E

N

O

P

S

T

Color Plates

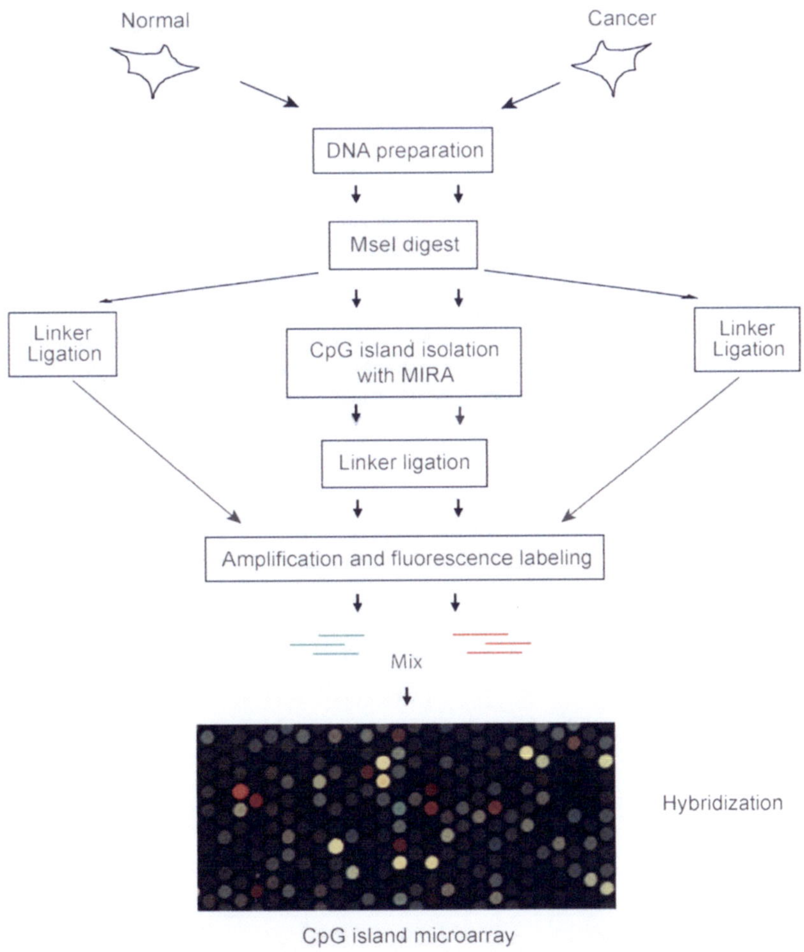

Color Plate 1. Schematic diagram of the MIRA microarray method. Input and MIRA-enriched fractions are prepared, labeled with different dyes, mixed, and hybridized to the microarray slides. In a simplified version, MIRA-enriched DNA from normal and tumor cells can be mixed and hybridized directly. (*see* complete caption and discussion on p. 67)

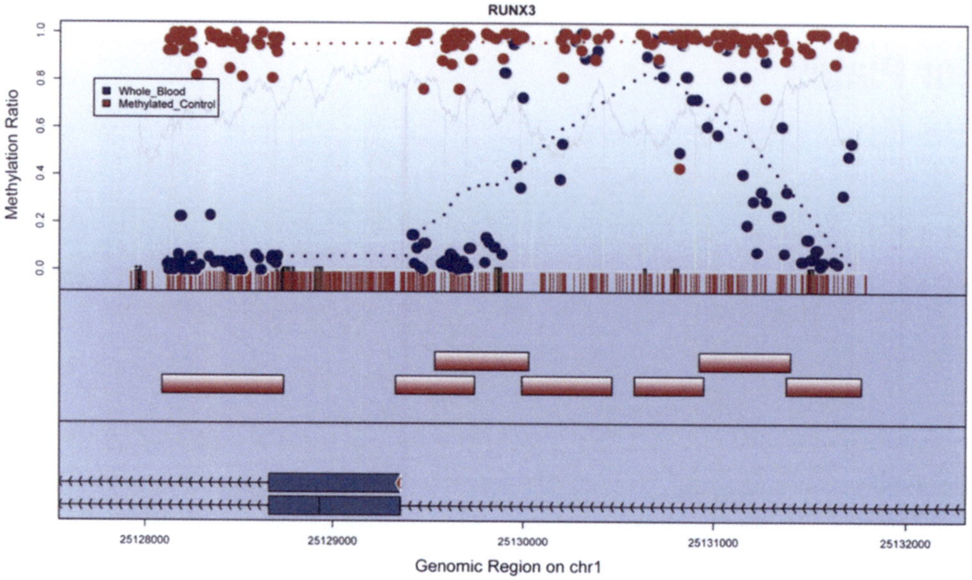

Color Plate 2. Depicted are regional differences that can be found in CpG island promoter regions. A set of methylated DNAs (*red*) and control DNAs derived from whole blood samples (*blue*) were measured across a total seven amplification targets covering approximately 3.5 kb of genomic sequence. The results are shown as circles with respect to their genomic location in the upper panel. The red bars at the bottom of the upper panel indicate the position of the CpG sites in this area and green rectangles indicate the presence of a conserved transcription factor binding site. The second panel indicates the location of the seven amplification targets and the third panel shows the location of the RUNX3 exons (blue rectangles) including the transcription start site (TSS, red triangle). The methylation levels around the transcription start site are separated into low and highly methylated. A good example of regional methylation differences can be seen around 700 bp upstream of the TSS, where methylation levels in whole blood control are increasing and drop to low levels again around 2 kb upstream of the TSS. (*see* complete caption and discussion on p. 209)

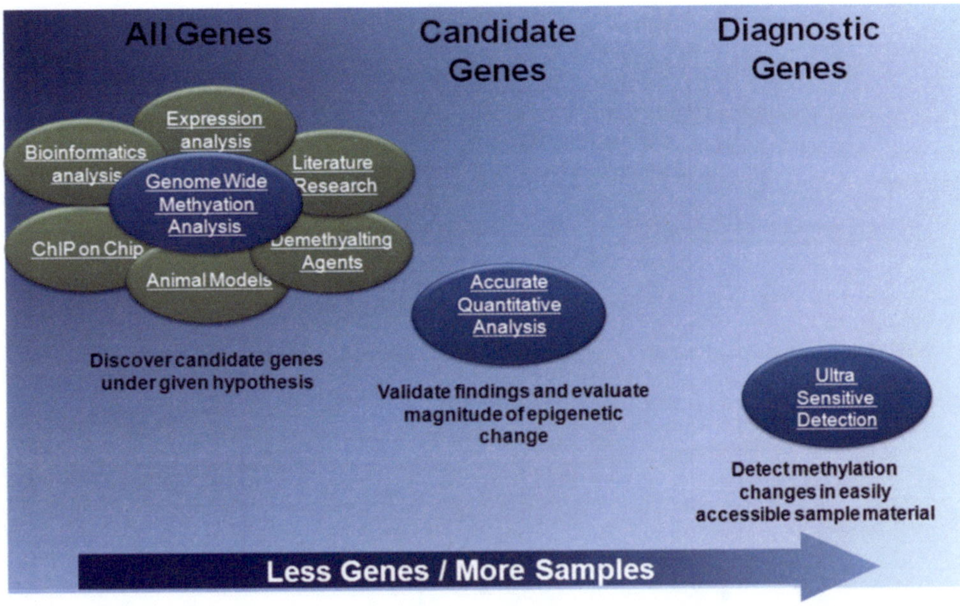

Color Plate 3. Schematic overview of a typical methylation biomarker discovery experiment. Usually a list of candidate genes is selected using either genome-wide interrogating methods or focused literature search or pathway analysis. These candidate genes are followed up in a larger number of samples using a highly accurate cost efficient method, like EpiTYPER. Once the clinically relevant genes are identified, a method that is able to detect a small number of methylated molecules, against the background of unmethylated molecules can be employed for clinical testing (e.g., MGMT methylation testing in brain cancer). (*see* complete caption and discussion on p. 210)

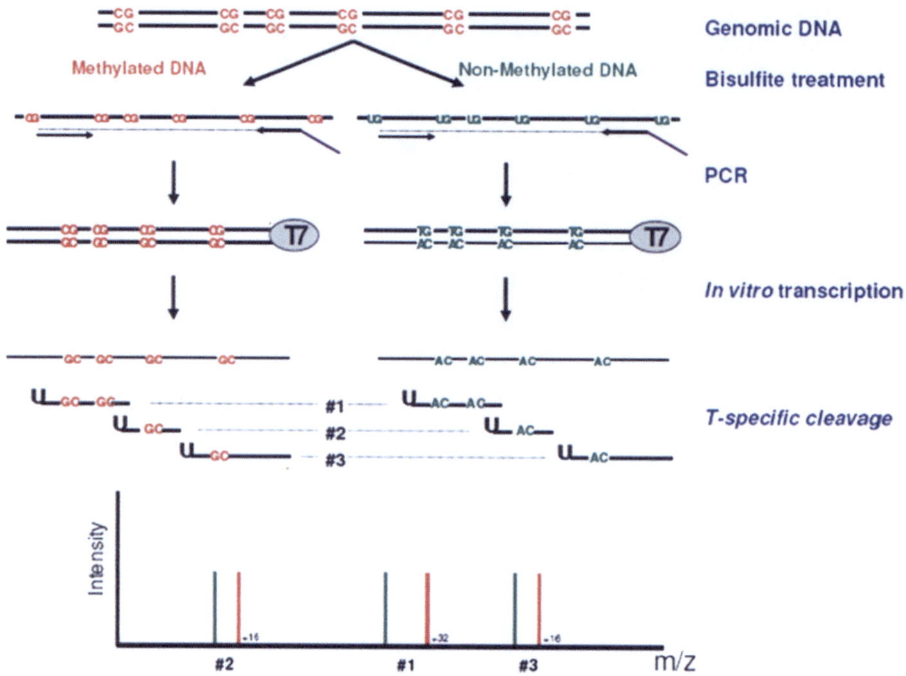

Color Plate 4. Base-specific cleavage process for quantitative analysis for cytosine methylation by MALDI-TOF MS. As a first step genomic DNA is bisulfite treated to introduce methylation-dependent sequence changes in the genomic DNA, which then can be amplified as a stable signal during PCR amplification. To avoid biased amplification, PCR primers should be designed devoid of CG sites. For base-specific cleavage PCR primers are tagged with a T7 promoter tag (for the reverse primer). Amplified regions can span anywhere from 100 bp to 500 bp, depending on the quality of the genomic DNA used for bisulfite treatment. After PCR amplification, the PCR product is transcribed into a single-stranded RNA from the reverse. The C/T changes introduced during the bisulfite treatment should be represented as G/A changes in this RNA transcript. The RNA transcription can utilize substitution of single rNTP by its dNTP analog (e.g., CTP fully substituted by dCTP). The RNA transcript is then cleaved base specifically by RNase A. This enzyme cleaves at every C and U. If dCTP was used during the transcription, the RNase will cleave base specifically at every uracil. The cleavage process is driven to completion and yields a complex mixture of short oligonucleotides (the base-specific cleavage products), which is then analyzed by MALDI-TOF MS analysis. Detected mass signals can be easily interpreted to identify methylation events in the target sequence. In the depicted case three cleavage products are generated, which each carry one or two CpG sites. A difference in methylation will lead to an A/G sequence change within each of the cleavage products. The corresponding mass signals will therefore shift 16 Da (the mass difference between A and G) if a methylation event occurs. If two or more CpG sites are embedded within a cleavage product mass may shift by multiples of 16 Da. For samples containing a mixture of methylated and nonmethylated target sequence (here indicated in red and green) both mass signals will be observed and the peak area ratio of mass signals representing methylated and nonmethylated sequence can be utilized to estimate the degree of methylation for each CpG site. (*see* complete caption and discussion on p. 211)

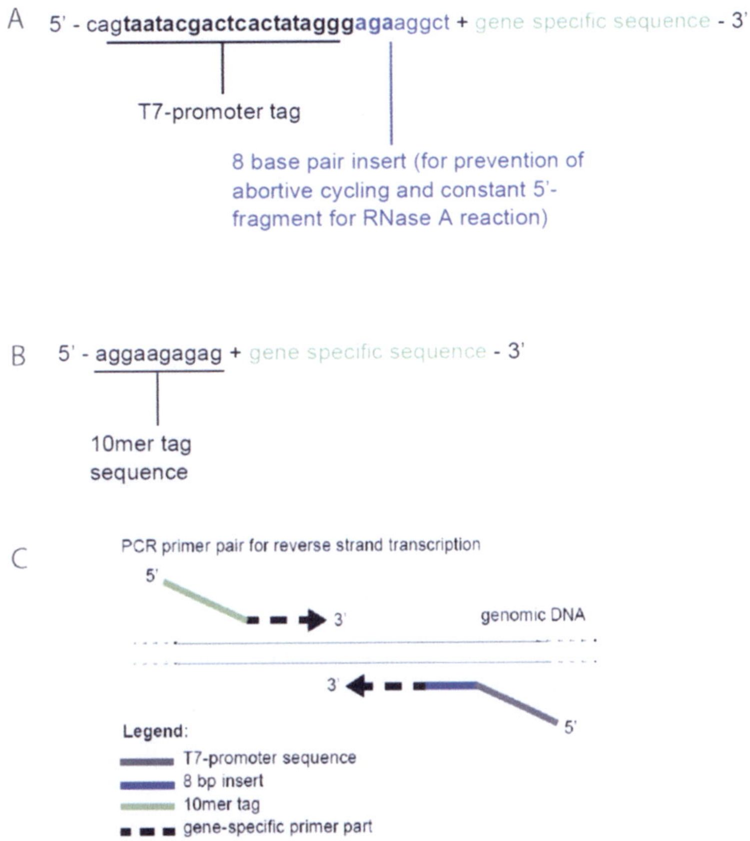

Color Plate 5. Schematic representation of primer design and tag structure for MassCLEAVE-based methylation analysis. (**A**) Depiction of the T7 promoter tag sequence including an 8 bp tag reducing abortive cycling in the transcription process; (**B**) depiction of the 10mer tag sequence; (**C**) depiction of the PCR primer pair composition and orientation for the generation of reverse transcripts enabling analysis of methylation as G/A sequence changes. (*see* complete caption and discussion on p. 215)

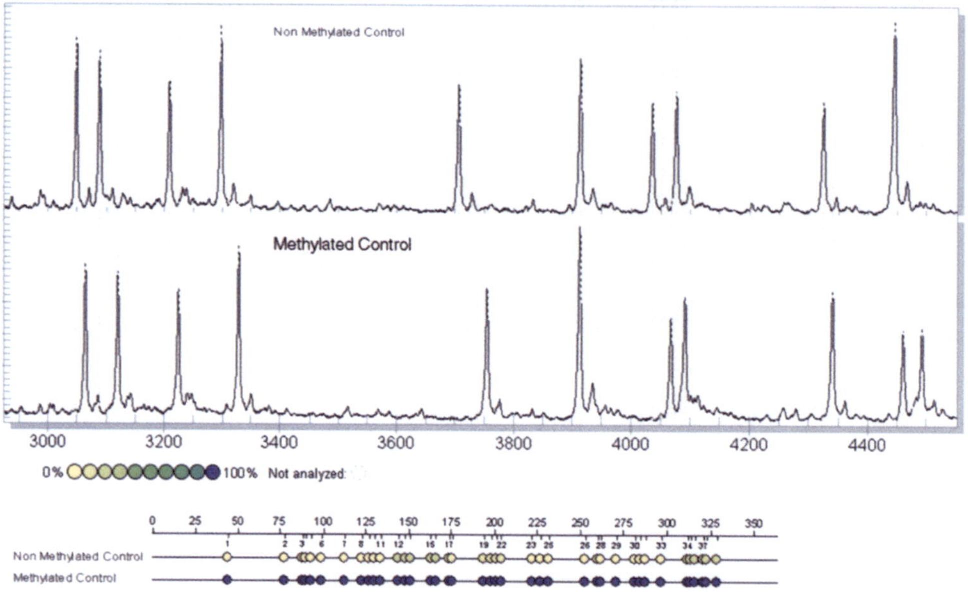

Color Plate 6. The upper panel shows mass spectra derived from nonmethylated DNA (*top*) and fully methylated DNA (*bottom*). While the start tag, which does not contain CpG residues, remains unchanged at 3,913 Da, all other signal are shifted by 16, 32, or 48 Da indicating the presences of one, two, or three methylated CpG sites, respectively. The *lower panel* shows how this data is visualized in an Epigram (see text). (*see* complete caption and discussion on p. 222)

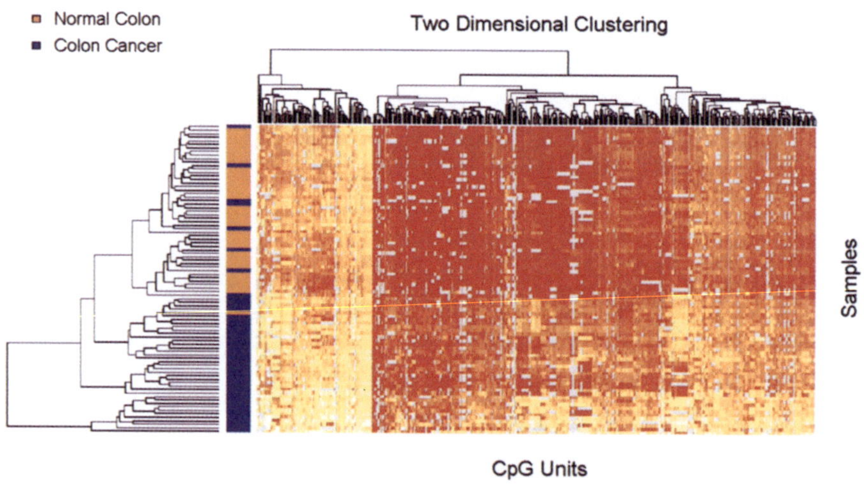

Color Plate 7. We analyzed CpG island promoter methylation in 48 pairs matched colon cancer and adjacent normal tissue samples. The methylation data was filtered to exclude poor quality data (less than 90% of all data available) and then used in a two-dimensional hierarchical clustering algorithm. This figure presents the results in a pseudo color image (heatmap), where red indicates no methylation and highly methylated CpG sites are shown in yellow. The type of tissue samples is marked as a colored sidebar on the left side of the figure (*orange* = normal colon tissue, *blue* = colon cancer tissue). The results indicate generally higher methylation levels in the group of colon cancer tissues, which leads to a separation of the two groups in the unsupervised clustering. (*see* complete caption and discussion on p. 223)

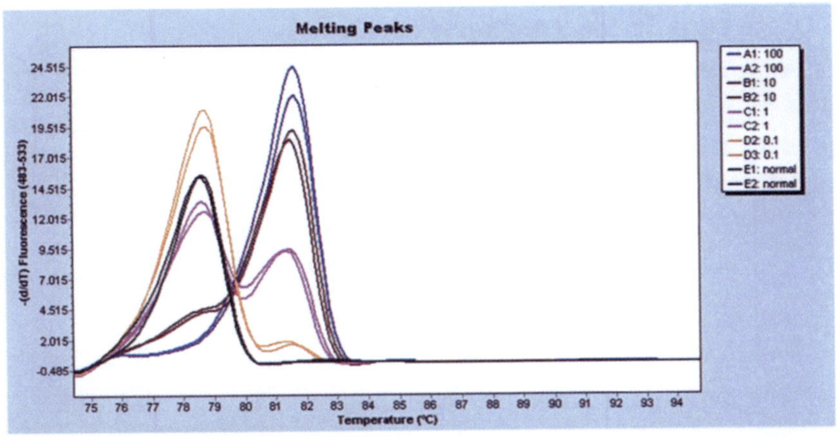

Color Plate 8. Melting peak analyses (first derivative of the melting curves) of the *MGMT* gene. Details of the MS-HRM assay are given in *(6)* and **Note 7**. The melting curves were derived from samples with known methylated to unmethylated template ratios (ranging from 100% to 0.1%) of fully methylated template diluted in unmethylated template. The experiments were performed on the LC480. (*see* complete caption and discussion on p. 237)

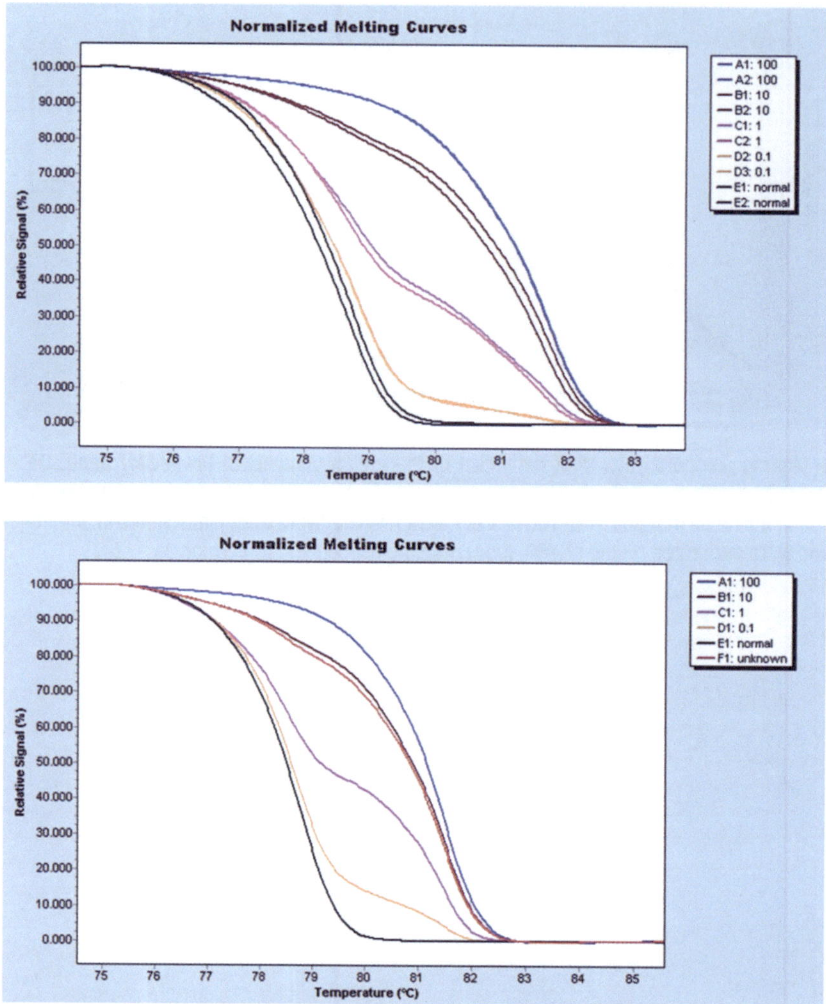

Color Plate 9. (**a**) An example of "Gene Scan" analyses (*see* **Section 3.8.1**) of melting curves of the *MGMT* MS-HRM assay (*see* **Note 7**), in which melting curves derived from mixtures of methylated and unmethylated template were normalized for input fluorescence. The use of normalization allows similarly shaped curves to be grouped together. (**b**) An example of the estimation of the methylation levels of an unknown sample (*red*) on the basis of the similarity of its normalized melting profile to normalized melting profiles of standards of known methylated to unmethylated template ratios. The unknown sample shows methylation level at around 10% as its normalized melting profile is similar to the melting profile of the standard with 10% methylated template. (*see* complete caption and discussion on p. 238)

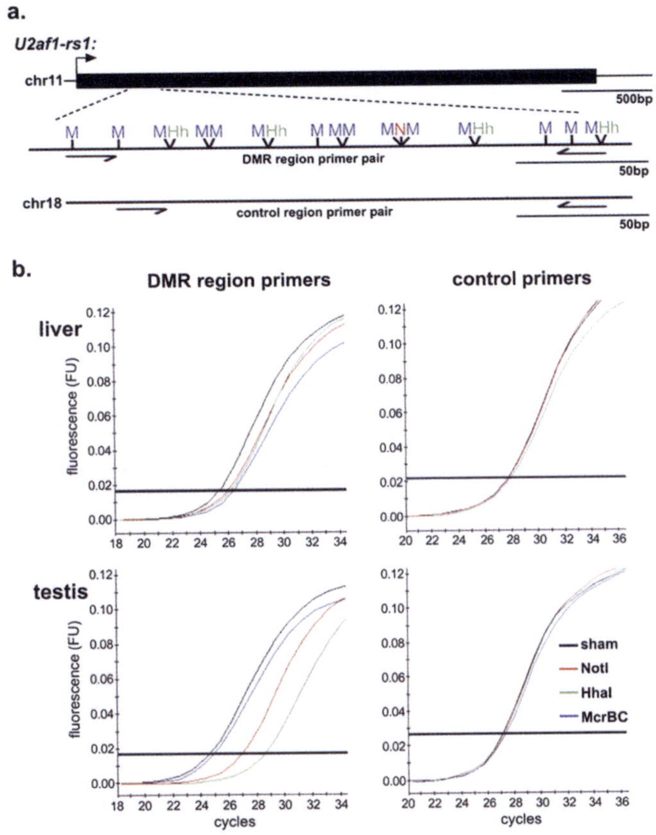

c.

liver

	Ct		ΔCt (sham)	
Template Digest	DMR region	Control	DMR region	Control
sham	25.3	27.8	\	\
NotI	26.1	27.8	0.8	0.0
HhaI	26.0	28.0	0.7	0.2
McrBC	26.3	27.8	1.0	0.0

testis

	Ct		ΔCt (sham)	
Template Digest	DMR region	Control	DMR region	Control
sham	24.8	27.1	\	\
NotI	27.1	27.3	2.3	0.2
HhaI	28.8	27.1	4.0	0.0
McrBC	25.1	27.4	0.3	0.3

Color Plate 10. The evaluation of the differentially methylated region (DMR) of the imprinted gene, *U2af1-rs1*, in testis and liver of mouse. (**a**) Primers are designed to flank *NotI* (N), *HhaI* (Hh), and *McrBC* (M) restriction sites in the DMR region. (**b**) Liver and testis genomic DNA templates are PCR amplified following digestion with no enzyme (sham), *NotI*, *HhaI*, or *McrBC*. Amplification using a second primer pair that has been designed to a sequence devoid of restriction sites demonstrates that the templates are of equal concentration. (**c**) The difference in the Ct value of an enzyme-digested template relative to the sham-digested template determines the levels of DNA methylation in each tissue. For clarity, in (**b**) only single values are shown, whereas values in (**c**) are means of three replicates (reproduced from *(8)*; with permission). (*see* complete caption and discussion on p. 275)

MIX
Papier aus verantwortungsvollen Quellen
Paper from responsible sources
FSC® C105338

If you have any concerns about our products,
you can contact us on
ProductSafety@springernature.com

In case Publisher is established outside the EU,
the EU authorized representative is:
Springer Nature Customer Service Center GmbH
Europaplatz 3, 69115 Heidelberg, Germany

Printed by Libri Plureos GmbH
in Hamburg, Germany